T0206998

Robotic Process Automation using UiPath StudioX

A Citizen Developer's Guide to Hyperautomation

Adeel Javed
Anum Sundrani
Nadia Malik
Sidney Madison Prescott

Apress®

Robotic Process Automation using UiPath StudioX: A Citizen Developer's Guide to Hyperautomation

Adeel Javed
Lake Zurich, IL, USA

Anum Sundrani
Chicago, IL, USA

Nadia Malik
Austin, TX, USA

Sidney Madison Prescott
New York, NY, USA

ISBN-13 (pbk): 978-1-4842-6793-6
ISBN-13 (electronic): 978-1-4842-6794-3
https://doi.org/10.1007/978-1-4842-6794-3

Managing Director, Apress Media LLC: Welmoed Spahr
Acquisitions Editor: Natalie Pao
Development Editor: James Markham
Coordinating Editor: Jessica Vakili

Distributed to the book trade worldwide by Springer Science+Business Media New York, 233 Spring Street, 6th Floor, New York, NY 10013. Phone 1-800-SPRINGER, fax (201) 348-4505, e-mail orders-ny@springer-sbm.com, or visit www.springeronline.com. Apress Media, LLC is a California LLC and the sole member (owner) is Springer Science + Business Media Finance Inc (SSBM Finance Inc). SSBM Finance Inc is a **Delaware** corporation.

For information on translations, please e-mail booktranslations@springernature.com; for reprint, paperback, or audio rights, please e-mail bookpermissions@springernature.com.

Apress titles may be purchased in bulk for academic, corporate, or promotional use. eBook versions and licenses are also available for most titles. For more information, reference our Print and eBook Bulk Sales web page at http://www.apress.com/bulk-sales.

Any source code or other supplementary material referenced by the author in this book is available to readers on GitHub via the book's product page, located at www.apress.com/978-1-4842-6793-6. For more detailed information, please visit http://www.apress.com/source-code.

Printed on acid-free paper

To my daughter Alaia, the light of my life.

—Adeel

To my father Ahmed and my family, for the invaluable support and inspiration.

—Anum

To my father, for always loving and mentoring me.

—Nadia

To my siblings, I'm forever blessed to be your big sister. And to that little girl with the big glasses, keep dreaming and achieving.

—Sidney

Table of Contents

About the Authors

Adeel Javed is an intelligent automation architect, an author, and a speaker. He helps organizations automate work using low-code, business process management (BPM), robotic process automation (RPA), analytics, integrations, and ML. He loves exploring new technologies and writing about them. He published his first book, *Building Arduino Projects for the Internet of Things*, with Apress back in 2015. He shares his thoughts on various technology trends on his personal blog (adeeljaved.com).

Anum Sundrani is a business systems analyst and technology enthusiast who specializes in business process management and robotic process automation. Anum is a Certified Appian Analyst, Tableau Author, Six Sigma Green Belt, and Scrum Master, alongside her several trainings in the areas of RPA development and the automation delivery life cycle. She has an inquisitive eye for simplifying complex business processes and has focused on implementing automation solutions for business users since 2017.

Nadia Malik is a presales engineer with a background in software development. She started her journey as a software engineer at IBM developing cloud storage applications and then joined the UiPath rocketship in June of 2018 helping in designing, implementing, and providing training to customers in robotic process automation. Today, she continues to evangelize RPA and mentor young women in STEM.

Sidney Madison Prescott is a senior technology leader, keynote speaker, and robotics evangelist specializing in the creation of Robotic Process Automation Centers of Excellence for Fortune 500 companies. Sidney currently heads up the Global Intelligent Automation initiative at the music streaming powerhouse Spotify. In addition to her enterprise technology expertise, Sidney is an executive board member for three global nonprofit organizations, where she contributes valuable automation insights to enhance overall program objectives. To round out her career accolades, Sidney was also named a global recipient of the 2020 Top 50 Technology Visionaries award.

About the Technical Reviewer

Rayudu Addagarla has 20 years of experience in web/mobile application development, the cloud, and solution architecture. He has been programming with Microsoft and LAMP stack since 1998. He has always been a full-stack technologist. His passion is toward digital transformation and business process automation, and he is a certified specialist in UiPath RPA. Along with Level 3 Advanced Certification in UiPath, he holds certifications in Pega RPA, WorkFusion, Tricentis TOSCA, Appian, and Pega BPM.

He holds a master of science degree in computer science from the University of Louisville, Kentucky, USA, and a bachelor of technology degree from JNTUACEA, Ananthapuramu, India. He has worked in the roles of Software Engineer, Business Process Consultant, Senior Manager, Industry Principal, Delivery Manager, and Senior Solution Architect. He has expertise in Healthcare, Manufacturing, Banking and Financial Services, Retail, Ecommerce, and Telecommunications domains.

He has proven experience in building successful Automation COEs. He currently works as an *Intelligent Automation Consultant* for EPAM Systems, Inc., a global consulting firm in Toronto, Canada.

Rayudu teaches Scratch and Python for kids in his spare time and shares his knowledge on LinkedIn. He believes strongly in servant leadership and lifelong learning.

He can be reached at `http://bit.ly/raylnkd`.

Acknowledgments

Thank you to Andrew Hall, Brandon Nott, Corneliu Niculite, Cosmin Voicu, Ovidiu Ponoran, Robert Love, Teodora Baciu, and Tom Merkle from the UiPath Team for providing valuable feedback.

Thank you to Rayudu Addagarla for agreeing to become a technical reviewer for our book and executing all the exercises to ensure accuracy.

Thank you to Natalie Pao and Jessica Vakili, our editors at Apress, for guiding us through the entire publishing process.

PART I

Overview

CHAPTER 1

Robotic Process Automation: Overview

To remain competitive in today's hyper-automated world, digital transformation initiatives have become a primary focus across various industries. Traditionally, C-suite executives are increasingly interested in lowering operational expenditures, particularly costs associated with the human workforce. In addition, business leaders are simultaneously focused on driving increased efficiencies and employee satisfaction across the enterprise. As a result, companies are undergoing a higher level of scrutiny surrounding existing business processes to seek out opportunities for automation at a global scale. One area of process optimization is that which exists for desk-level procedures typically executed by business stakeholders. With such a high percentage of automation opportunities executed by the business stakeholders, a bottom-up approach is commonly seen in which the workforce chooses tasks to automate based on their individual needs. Not only can they help in identifying opportunities to automate but also create automations themselves so that they may focus on high-value tasks. With the advent of intelligent automation, specifically Robotic Process Automation (RPA), companies now have a proven way to automate business processes at the keystroke level.

A. Javed et al., *Robotic Process Automation using UiPath StudioX*,
https://doi.org/10.1007/978-1-4842-6794-3_1

UiPath provides a technology that enables the automation of business processes traditionally performed by business users, using configurable software referred to as "robots." UiPath's development platform, StudioX, is extremely flexible and user-friendly as it is a low-code/no-code solution. The software allows users the ability to interact with systems via a robot which leverages the users' own credentials or can be configured with distinct credentials and specific permissions. Robotic Process Automation tools interact with an application in the same way as end users do, through interactions with the user interface (UI), as well as through the back end of a system.

From a compliance and risk perspective, robots can only execute tasks that are specifically designed with the virtual robot worker in mind. The robot's access to both internal and external systems is limited to the design of the robot workflow, which demonstrates the rules-based nature of the software. The robot can also be designed to prompt the user for input or incorporate artificial intelligence (AI) to handle more cognitive tasks. As a result, Robotic Process Automation enables automation of the manual, repetitive tasks that are typically a fundamental component of a business user's daily job responsibilities.

A fundamental premise of Robotic Process Automation software is the belief that robotic software is designed to complement the human workforce, by empowering organizations with the ability to upskill employees to build simple automations or route more complex automations to a set of developers reporting into a core automation team. Adding Robotic Process Automation functionality into a business department can maximize the efficiency of employee outputs, minimize the risk of human error, and mitigate the number of tedious, manual processes employees are expected to execute, thus increasing the potential for a higher level of employee satisfaction.

Return on Investment (ROI)

In addition to enabling the automation of repetitive tasks, Robotic Process Automation software can provide substantial return on investment to both business process owners and the enterprise at large. Robotic Process Automation software allows firms to automate manual processes in a cost-efficient manner, due to the fact the price point of the RPA software is typically lower than that of traditional business applications. Robots are beneficial in minimizing the costs typically incurred in automation projects, as Robotic Process Automation tools can leverage existing infrastructure architecture without impacting live systems.

The infrastructure necessary to support robots is considerably minimal when compared to other tools, as robots can either run on an end user's desktop (attended automation) or a virtual machine (unattended automation). One of the many benefits of Robotic Process Automation is the ability users to dictate whether a human or a robot will be responsible for executing a particular step of the process within a given workflow. In addition, workflows can be customized to indicate when robots encounter changes in each system including routine software upgrades whereby elements of the user interface might deviate from previous versions. A significant benefit of Robotic Process Automation is the ability users to create workflows to support a dynamically changing environment with minimal impact to underlying infrastructure capabilities.

Automation Types

Moving forward, we will dive into the nuances of unattended robots and attended robots, to understand how the distinction between the two types of automation is driving a new approach to enabling business process automation through citizen development (business users with the ability to build automations). Robotic Process Automation can be leveraged to automate a wide variety of processes including but not limited to payroll processing, customer service, advertising operations, report aggregation, and vendor onboarding. Robotic Process Automation also offers a wide variety of automation deployment models which can be used interchangeably to automate processes across the business including

- Attended robots that reside on the end user's computer or virtual machine for the purpose of automating simple manual processes that can be triggered by the actions of the user.
- Unattended robots that can be provisioned to reside on machines based on-premises (physical server based) or off-premises (virtual machines/cloud based) for the purpose of automating more complex back-office functions commonly scheduled to run based on a time or queue. Typically, unattended automation lends itself to more data-intensive tasks and processes with higher transaction volumes such as batch jobs.
- Hybrid robots that reside on a combination of end user and on-premises/off-premises solutions to enable a combination of attended and unattended style processing to enable the end-to-end automation of processes that require both human support and back-end functionality.

Each automation deployment model allows the end user the ability to determine the best way to interact with a robot based on the task at hand, alongside careful consideration of the existing variables in each environment. The various automation deployment models can be leveraged interchangeably as a part of a holistic enterprise-level automation platform and digital transformation strategy. As we move forward, we'll focus on features and hands-on exercises specific to the Robotic Process Automation industry leader, UiPath, to discuss the unique value proposition the company offers citizen developers through the use of StudioX.

UiPath StudioX

UiPath is a global Robotic Process Automation software company based out of Romania. The company was founded in 2005 by Daniel Dines. The company originally offered automation libraries and software as an outsourced service, but quickly positioned itself to become an industry leader through a customer-centric model designed to democratize access to Robotic Process Automation capabilities. Through a robust product road map and unique approach to empower business users with the ability to automate simple business processes via StudioX, UiPath's enterprise platform demonstrates the seamless fusion that exists between business processes and automation capabilities.

StudioX is one product of UiPath's Robotic Process Automation platform designed to enable business users to build automation without the need for a traditional development background. The StudioX functionality includes a no-code interface with out-of-the-box drag-and-drop functionality to facilitate ease of use. In addition, StudioX contains predesigned templates and native integrations with common business applications such as the Microsoft Office suite to facilitate faster development of automation workflows. Business users can deploy a robot directly to a local machine, such as a desktop which removes the need for traditional IT deployment support. In addition, governance functionality is also built into the StudioX framework to allow auditing capabilities to ensure that existing company compliance protocols remain intact. Regarding the scheduling and sharing of automations, users can complete both tasks through the UiPath Assistant and Orchestrator components of UiPath.

One of the key elements that demonstrates the flexibility of StudioX is the fact the tool allows business users a user-friendly way to learn how to build automations that are beneficial to their job functions while simultaneously learning a new technical skill. In a world where technical prowess has become increasingly important, providing employees an opportunity to leverage Robotic Process Automation tools can help individuals to feel empowered and more satisfied, potentially leading to less attrition. The citizen developer model is the methodology by which business users are trained on the skills required to build automations while also being provisioned access to RPA tools to begin the development of robot workflows. As Robotic Process Automation continues to expand across a wide variety of industries, it will be important to continue to expand the knowledge of business users with tools such as StudioX to provide a wealth of benefits at an organizational level.

In the rest of the book, we will explore hands-on exercises with detailed reference guides for various activities and sample files to help you as you work to build your first RPA robots in StudioX. The goal of each chapter is to provide real-world business process scenarios for readers to reference as Robotic Process Automation learning tools. As you work through the exercises, make a note of any challenges you encounter to allow time to reflect on possible ways to solve any roadblocks you may have. This book is intended for both the business user looking to learn how to leverage StudioX for the first time and the experienced RPA developer looking to build upon existing knowledge to automate manual, repetitive tasks across your organization. As you step into the future of working with robots, remember that you have taken an important step in the journey to democratize automation and heighten your technical skill set. So, let us get started.

CHAPTER 2

UiPath StudioX

UiPath's StudioX tool offers a no-code approach to automation, providing citizen developers with the tools necessary to configure and run their automations.

Learning Objectives

At the end of this chapter, you will learn how to

- Download and install UiPath StudioX
- Access and use common interface components in UiPath StudioX

System Requirements

This section provides hardware and software requirements for UiPath StudioX.

Note This book uses **UiPath Studio Community v2020.10.2**. At the time of installation, you might find a slightly different version, and that will not affect the exercises.

The Community version is free and is not limited to a trial period. It is a great starting point for anyone just starting with UiPath or for organizations that are looking to evaluate it for enterprise use.

A. Javed et al., *Robotic Process Automation using UiPath StudioX*,
https://doi.org/10.1007/978-1-4842-6794-3_2

Hardware Requirements

Table 2-1 lists the minimum and recommended hardware requirements.

Table 2-1. *Hardware requirements*

	Minimum	Recommended
CPU	2 x 1.8GHz 32-bit (x86)	4 x 2.4GHz 64-bit (x64)
RAM	4 GB	8 GB

Software Requirements

Table 2-2 lists software versions supported by the current version of UiPath StudioX.

Note UiPath does not support resolutions below 1024 x 768.

Table 2-2. *Software requirements*

	Supported Versions
Operating System	Windows (7, 7 N, 7 SP1, 8.1, 8.1 N, 10, 10 N)
	Windows Server (2012 R2, 2016, 2019)
.NET Framework	Version 4.6.1 or higher
Web Browsers	Internet Explorer v8.0 or greater
	Google Chrome version 64 or greater
	Mozilla Firefox version 52.0 or greater
	Microsoft Edge on Windows 10 version 1803 or greater

Installation and Setup

This section will provide you with step-by-step instructions for downloading and installing UiPath StudioX.

Register

Before you can download and install UiPath StudioX, you will need to register with UiPath.

Note If your company has purchased licenses for UiPath, then you should request this from your IT department.

---

1. Open the UiPath website (`www.uipath.com`) and click the `Try UiPath Free` button, as shown in Figure 2-1.

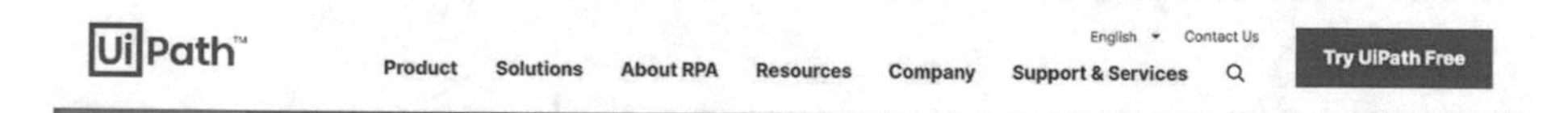

Figure 2-1. *UiPath trial option*

2. If this is the first time you are accessing the UiPath site, you will need to `Sign Up` using one of the provided options. Once you have registered, click the `Log In` link to enter your credentials.

Download

Next, we are going to download the UiPath StudioX installer:

1. Once you have logged in to the UiPath Automation Cloud portal, `https://cloud.uipath.com/`, you will see a `Download Studio/StudioX` button on the right under the `Home` tab, as shown in Figure 2-2. The location can change with updates to the UiPath Automation Cloud portal.

Figure 2-2. *Download the link for UiPath StudioX*

2. Click the `Download Studio/StudioX` button; this will download the installer on your computer. Figure 2-3 shows the name of the installer once it has been downloaded.

Figure 2-3. *Installation package for UiPath StudioX*

Install

Once the UiPath StudioX download is complete, you can start the installation process:

1. To start the installation, double-click the installer, that is, `UiPathStudioSetup.exe`, that you downloaded in the previous section. This will launch the UiPath StudioX installation window, as shown in Figure 2-4.

Figure 2-4. *Start the UiPath StudioX installation process*

2. The next screen allows you to select a license option. This book uses the community edition; thus, you can select the `Community License` option as shown in Figure 2-5. This will activate UiPath StudioX.

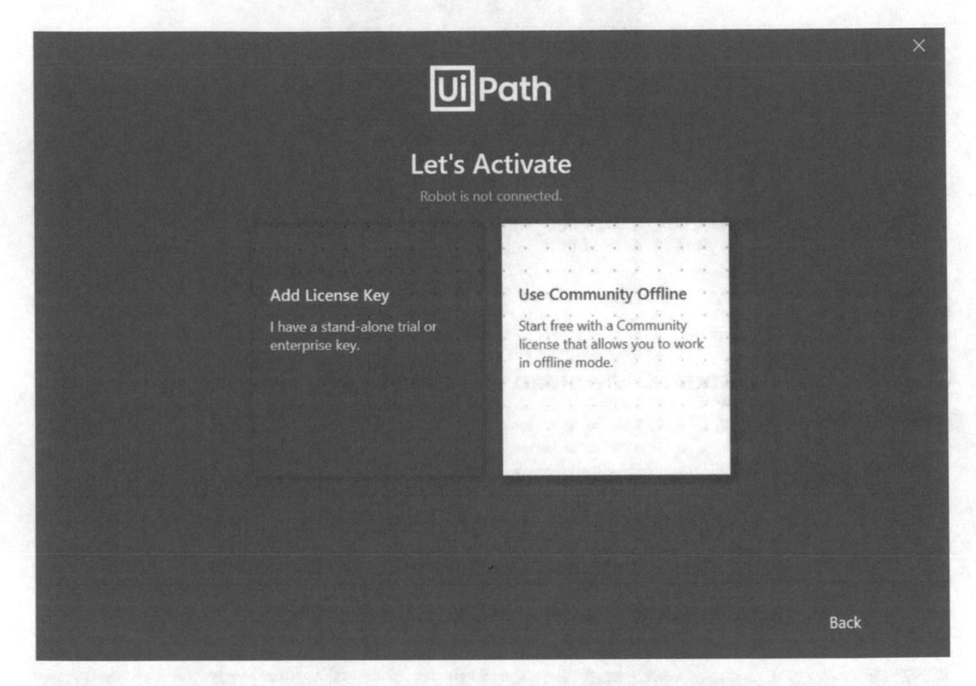

***Figure 2-5.** Activate UiPath StudioX with the Community License*

3. UiPath provides three profiles of their Studio. UiPath Studio is for RPA developers looking to build medium to complex automations. UiPath StudioX focuses on citizen developers and business users who are looking to create automations for simple business tasks. Since the primary audience of this book is the citizen developer and business user community, select the `UiPath StudioX` profile, as displayed in Figure 2-6.

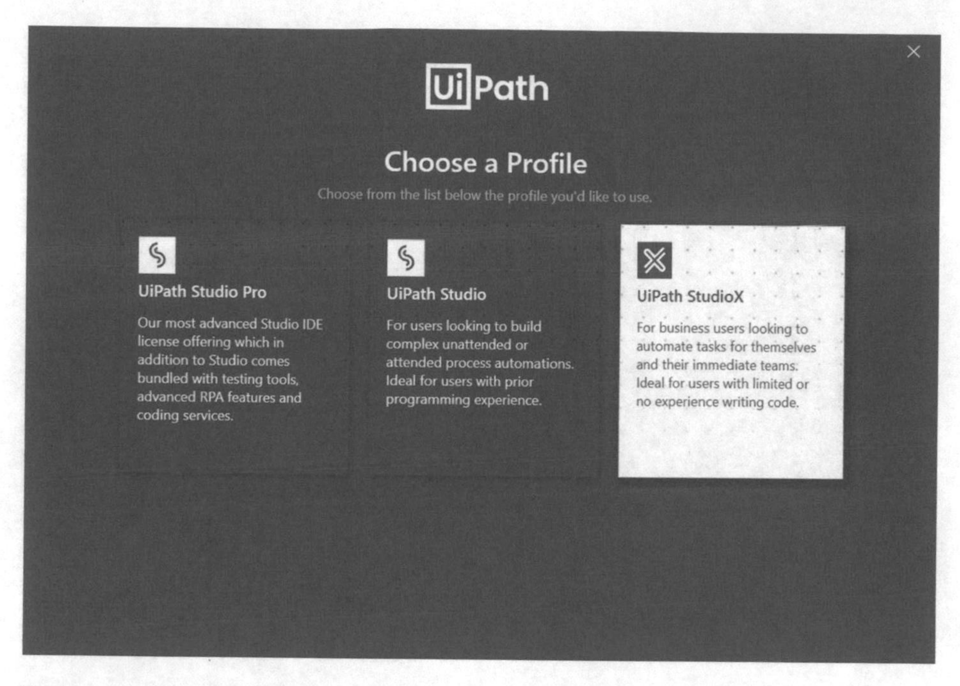

Figure 2-6. *Select UiPath StudioX profile for citizen developers and business users*

4. UiPath rolls out updates regularly, hence the two options on the `Update Channel` screen. For this book, as shown in Figure 2-7, we assume that you are on `Stable` version, as that will introduce changes less frequently compared to the `Preview` version.

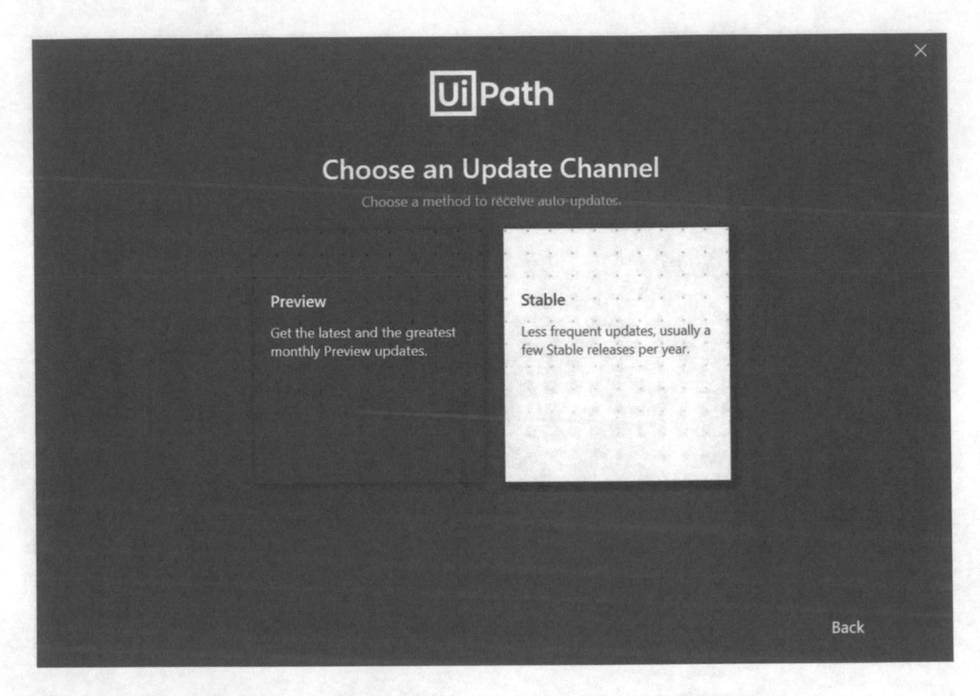

Figure 2-7. *Select less frequently updated Stable version*

5. At this point, you will have configured UiPath StudioX. This wraps up the "Installation and Setup" section, and you will see the Home screen, as shown in Figure 2-8.

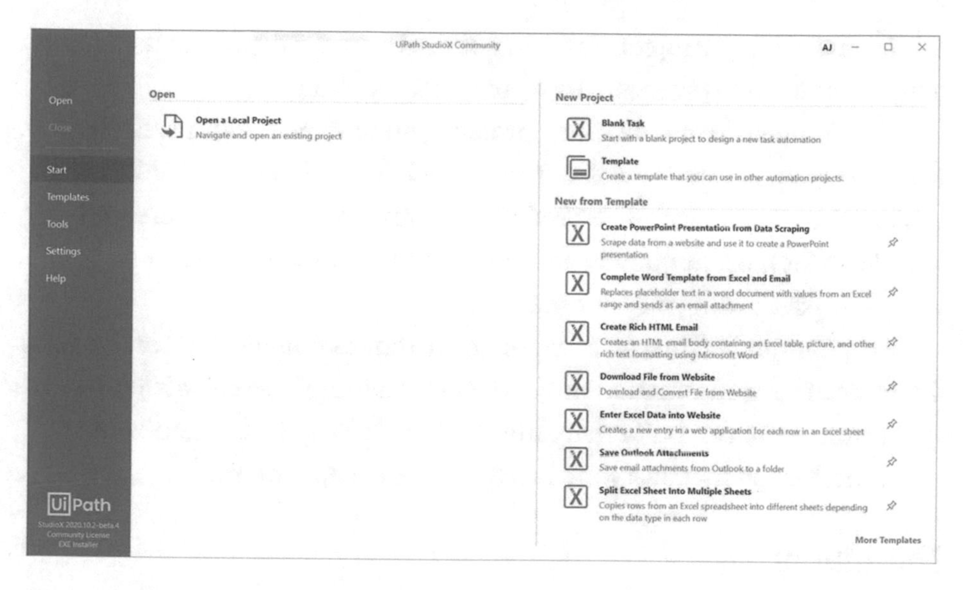

Figure 2-8. *Home screen of UiPath StudioX*

Interface Overview

The UiPath StudioX interface contains different panels, each of which provides specific functionality. This section will give you a brief overview of the UiPath StudioX interface.

Home

`Home` screen, as shown in Figure 2-8, allows you to open and start projects, create or use existing templates, update interface settings, install extensions, and explore documentation.

Start

The Start tab, shown in Figure 2-8, allows you to open an existing automation project or create a new one.

Open Existing Project: You can use the `Open a Local Project` option to open an existing project. This would be a .xaml file.

New Project: To create a new project, you have two options. Either you can create new automation from scratch by clicking the `Blank Task` option, or you can use a predefined template. UiPath has created templates for most frequently automated tasks; these templates can give you a jumpstart with your automation.

New Template: You can also create custom templates that can become the basis of your automation projects. Custom templates can also help with governance. You can build templates that include all your organization's best practices, so that everyone has the same starting point.

Templates

The **Templates** tab, as shown in Figure 2-9, lists all the predefined templates and all the custom templates that you have created.

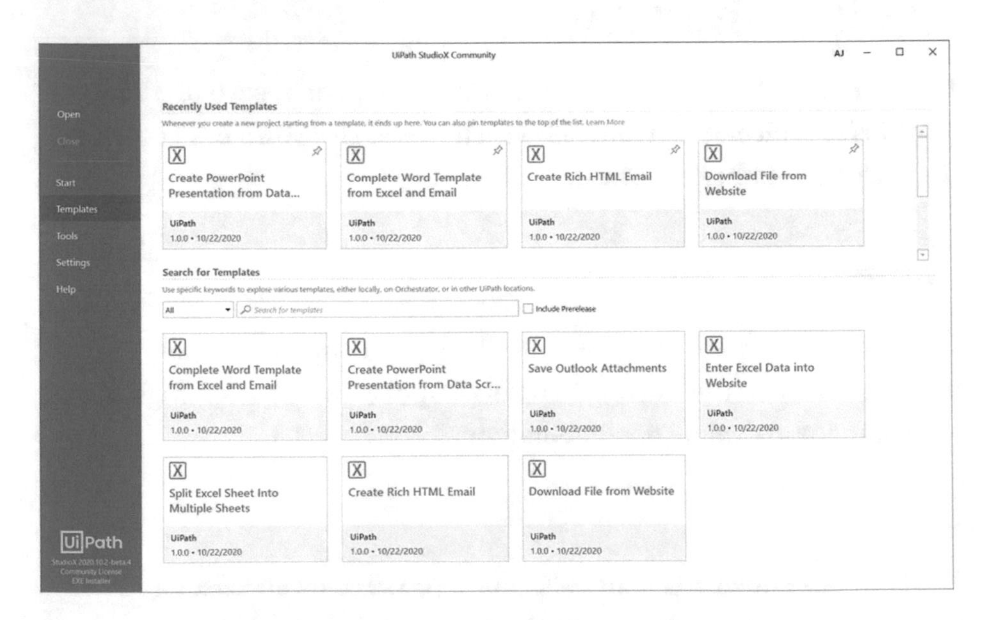

***Figure 2-9.** Select custom and predefined templates from Templates tab*

Tools

The **Tools** tab, as shown in Figure 2-10, provides options for installing extensions such as Chrome, Firefox, Edge, Excel, and a few others. You can also use this tab for launching Project Dependencies Mass Update Tool.

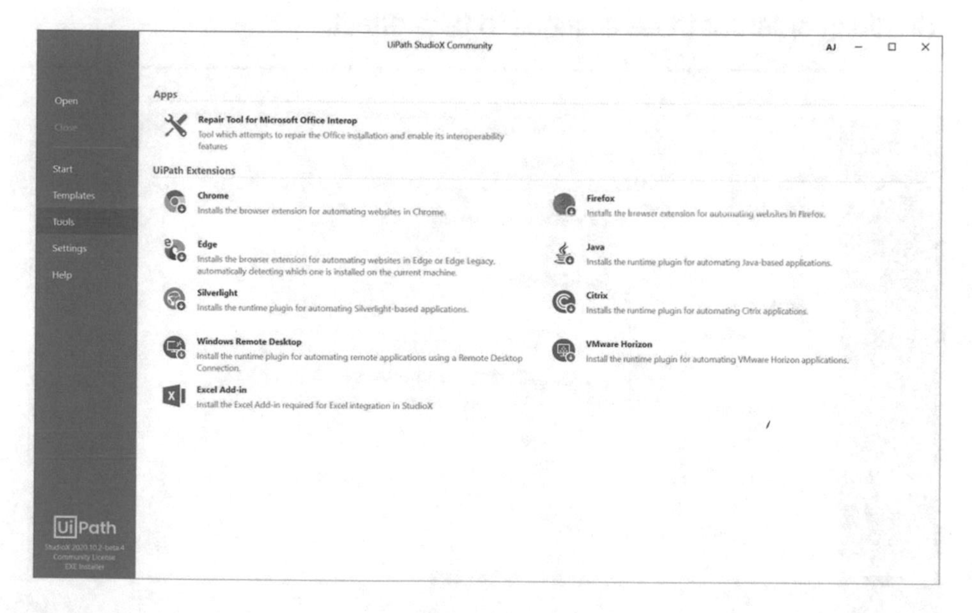

Figure 2-10. *Extensions available from the Tools tab*

Settings

The **Settings** tab, as shown in Figure 2-11, provides options for changing the interface language, theme, design preferences, default locations, and managing your license and profile.

Note To switch profiles between different versions of Studio, click the `License and Profile` option, and then select `View or Change Profile` option. This action will prompt you with available profiles. If you select a new profile, then you will need to restart the Studio in order for those changes to take effect.

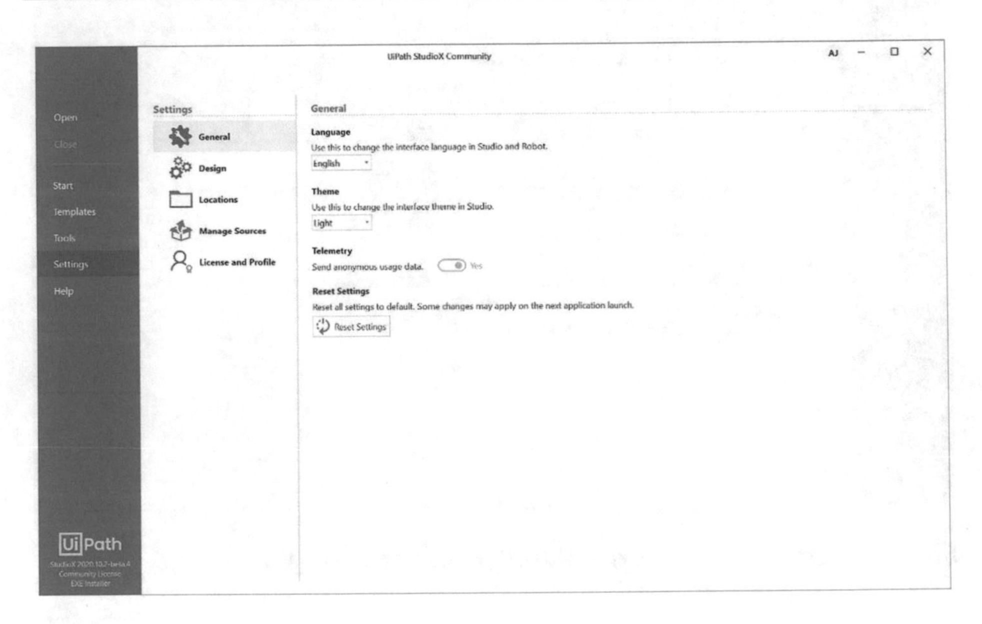

Figure 2-11. *Settings tab*

Help

As shown in Figure 2-12, the **Help** tab provides you with links to the product documentation, community forum, academy, release notes, and other online resources. Additionally, you can access your product version and license information, device ID, and update channel (switch between Stable and Preview versions) from this tab.

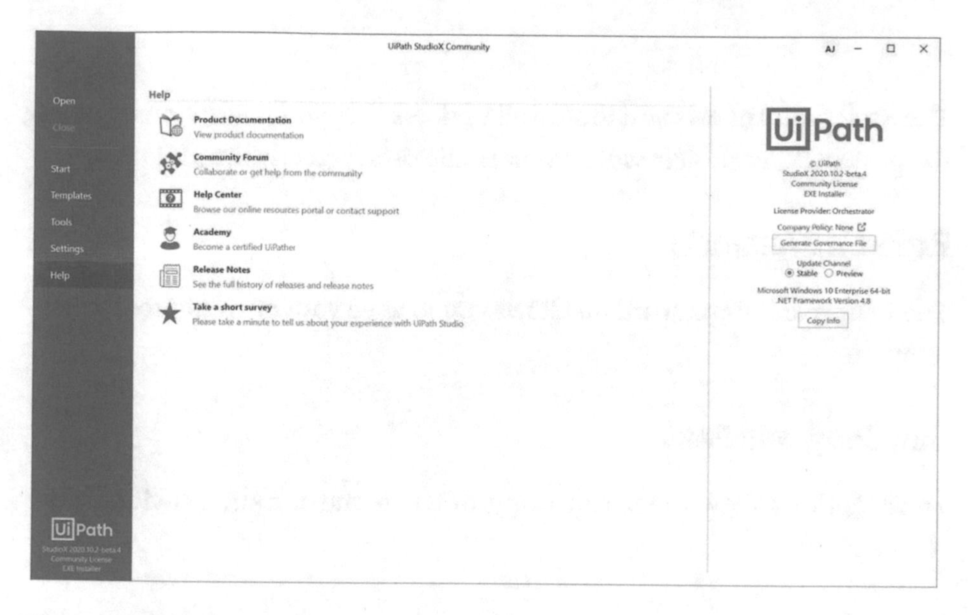

Figure 2-12. *Help tab*

Design View

The **Design** view is where you design, run, and publish your automations. This view also allows you to access your project's Notebook and manage data and packages.

The Ribbon

The ribbon at the top, as shown in Figure 2-13, is only available when you have a project open.

The following sections provide a brief overview of each menu item.

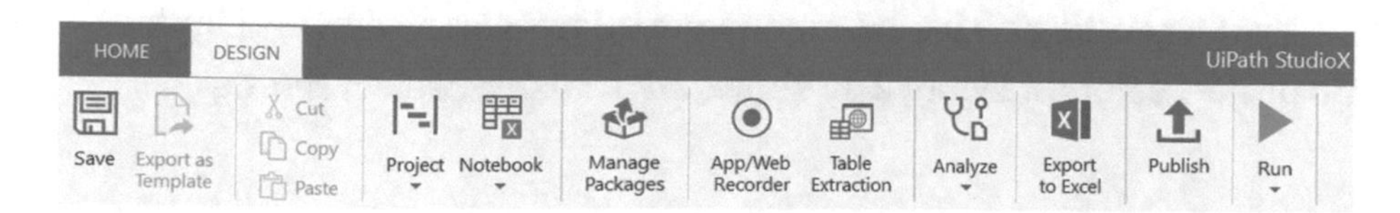

Figure 2-13. *Design tab*

Save

The Save menu allows you to save all updates that you made since opening the project. Your project will automatically be saved any time it is run.

Export as Template

The Export as Template menu allows you to save your current project as a template.

Cut, Copy, and Paste

These options allow you to Cut, Copy, or Paste one or more activities in a project.

Project

The **Project** menu item, shown in Figure 2-14, has two submenu items.

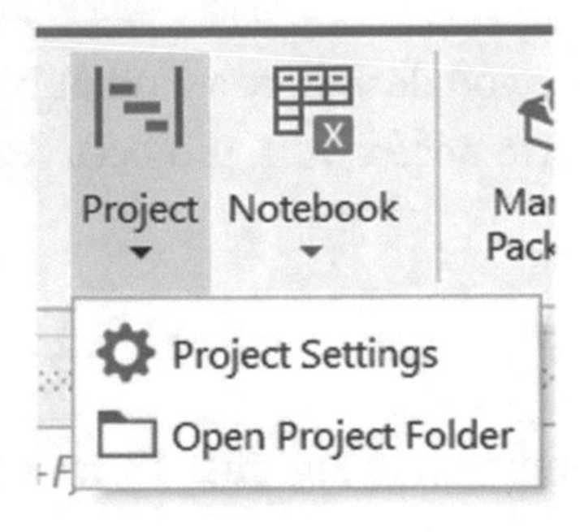

Figure 2-14. *Project submenu items*

Project Settings: This submenu item allows you to view and update the project settings. Figure 2-15 shows all available project settings.

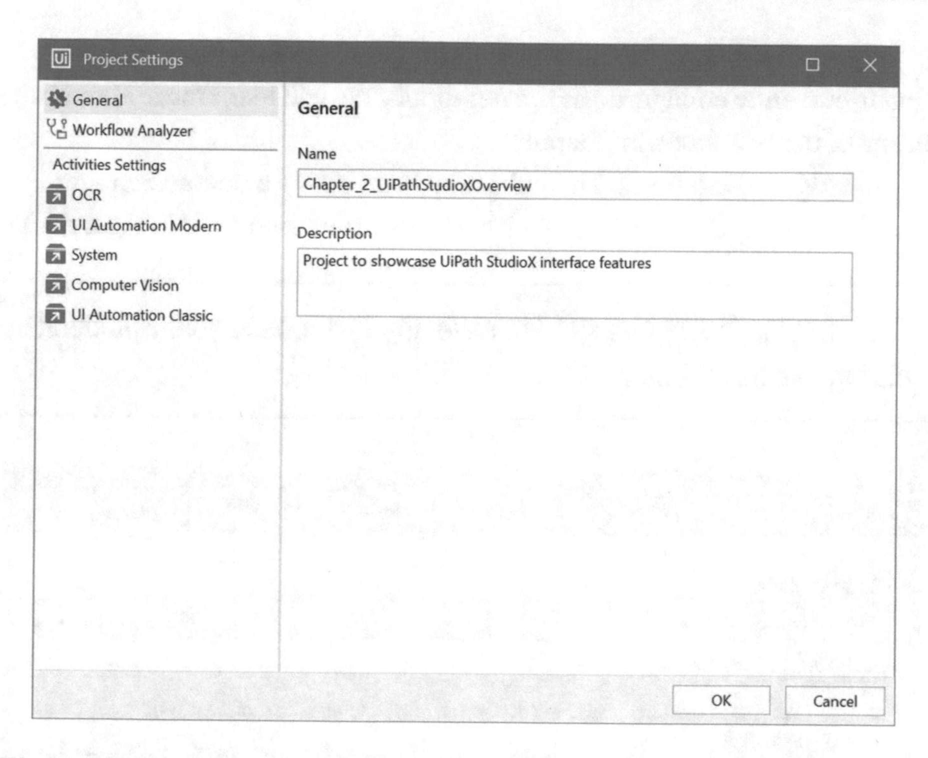

***Figure 2-15.** Project Settings*

Open Project Folder: This submenu item, as the name suggests, opens the folder on your system where StudioX has stored all project files.

Notebook

The **Notebook** menu item, shown in Figure 2-16, has two submenu items.

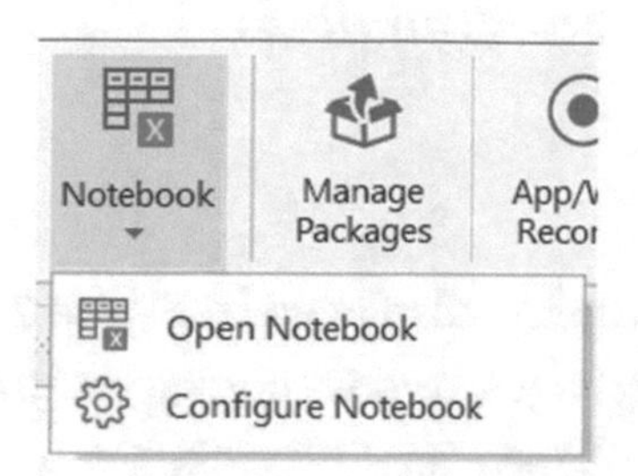

***Figure 2-16.** Notebook submenu items*

Open Notebook: This submenu item opens the Excel spreadsheet that is currently configured as the notebook. We will learn more about the usage of the Notebook in Chapter 3.

Configure Notebook: This submenu item opens a new screen that allows you to update the Notebook configurations displayed in Figure 2-17.

Tip Enable "Save changes" to save any changes to your Notebook during execution time.

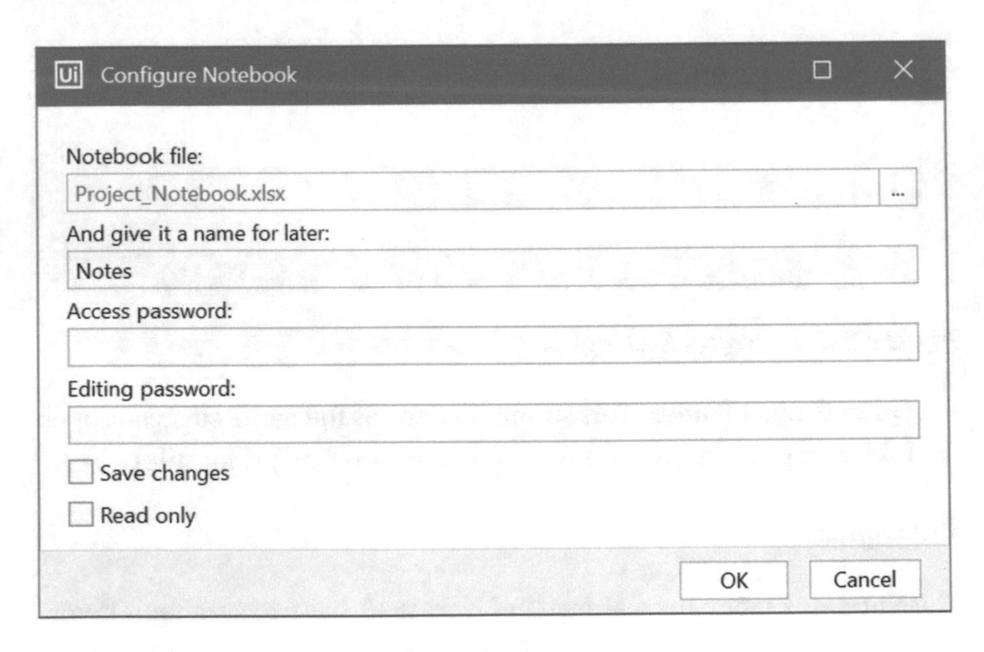

***Figure 2-17.** Notebook configurations*

Manage Packages

The **Manage Packages** menu, displayed in Figure 2-18, allows you to update your existing packages, update package sources, view project dependencies, and install new activity packages.

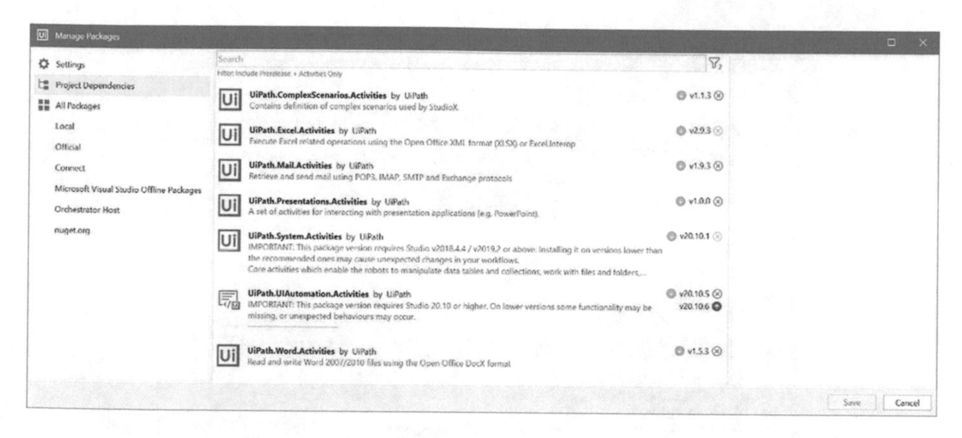

Figure 2-18. *Manage Packages screen*

App/Web Recorder

The concept of the **App/Web Recorder** is that instead of creating your automation from scratch or a template, you allow UiPath to record the actions that you perform. As a result, the recorded actions automatically populate the activities for the automation project. Although the generated activities still require some level of manual updates, the recorder provides a good starting point for your automation. An example to showcase the features of the App/Web Recorder can be found in Chapter 4.

Table Extraction

The Table Extraction menu item allows you to extract data from a table and store it for use later. You can use this functionality to extract tabular data from both desktop and web applications. You will learn more about Table Extraction in Chapter 4. Figure 2-19 shows the configuration of Table Extraction.

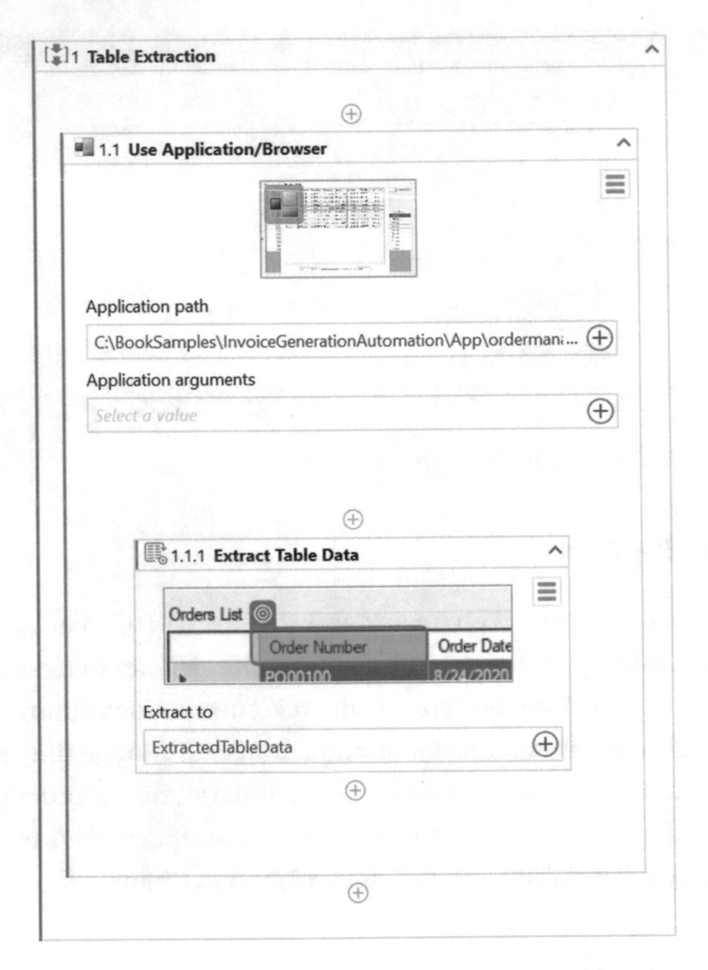

Figure 2-19. *Configuration of Table Extraction*

Analyze

The **Analyze** menu item, shown in Figure 2-20, allows you to check if your project follows automation best practices and defined restrictions. This menu item has three submenu items.

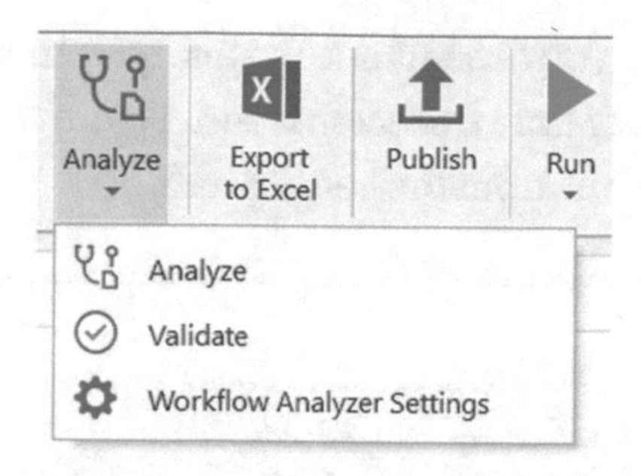

Figure 2-20. *Analyze submenu items*

Workflow Analyzer Settings: This submenu item, shown in Figure 2-20, allows you to define rules that your automation project must adhere to. You can check/uncheck what rules to analyze against. For each rule, you can define what type of action StudioX should take. For example, you can define that an error should be displayed if the automation uses a prohibited URL. You can use these settings to enforce certain best practices across the organization, as shown in Figure 2-21.

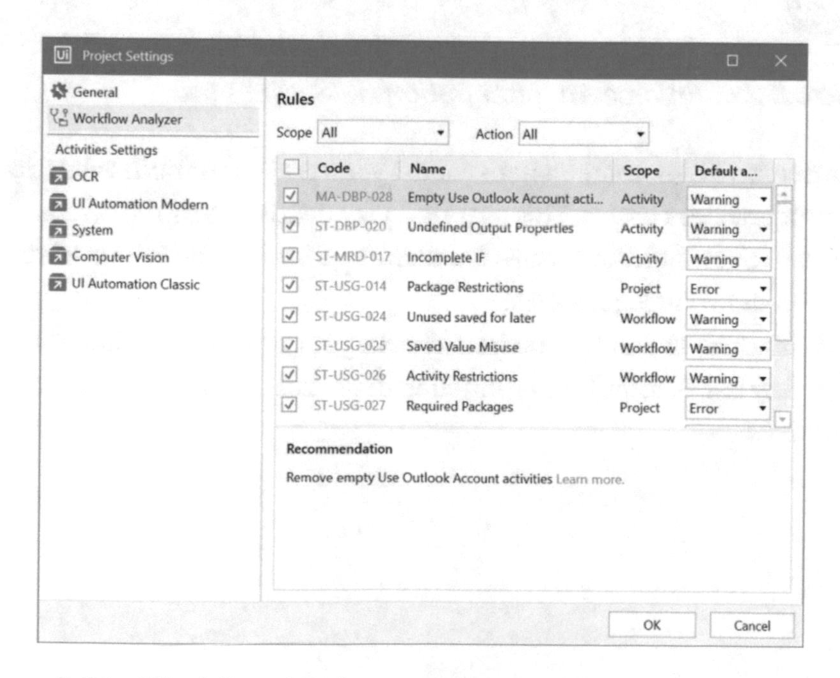

Figure 2-21. *Workflow Analyzer settings*

Through policies, you can enforce the analyzer to be run before publish or before every time a process is run. Figure 2-22 shows how to enforce analyzer options from the Settings tab.

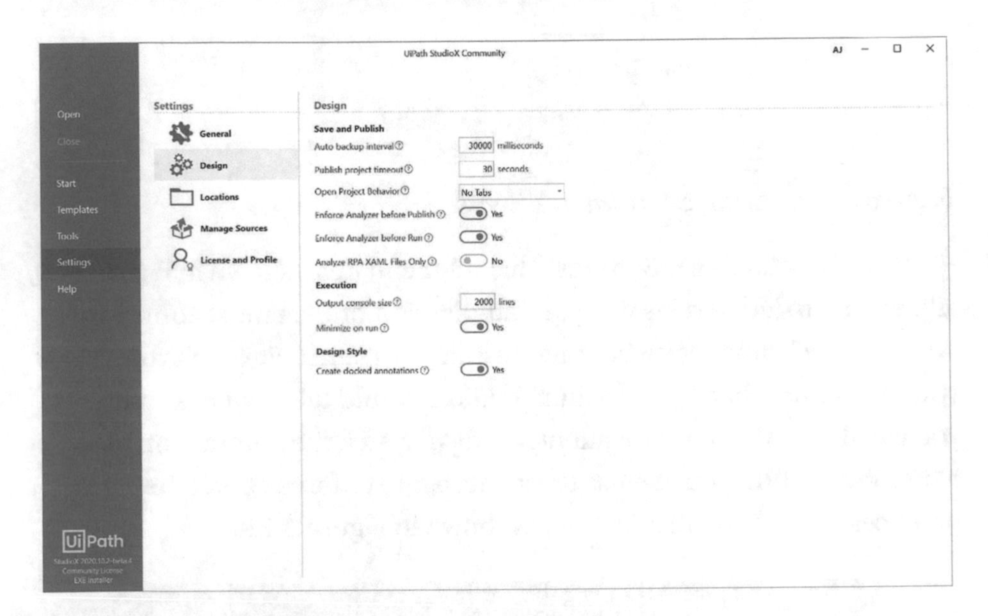

Figure 2-22. *Enforce Analyzer options in Settings tab*

Analyze: This submenu item checks your project for both validation and best practices issues. This uses the rules defined in the `Workflow Analyzer Settings`. Any issues found are reported in the `Error List` panel, as shown in Figure 2-23.

Validation: This submenu item checks your project for validation errors. Any issues found are reported in the `Error List` panel, as shown in Figure 2-23.

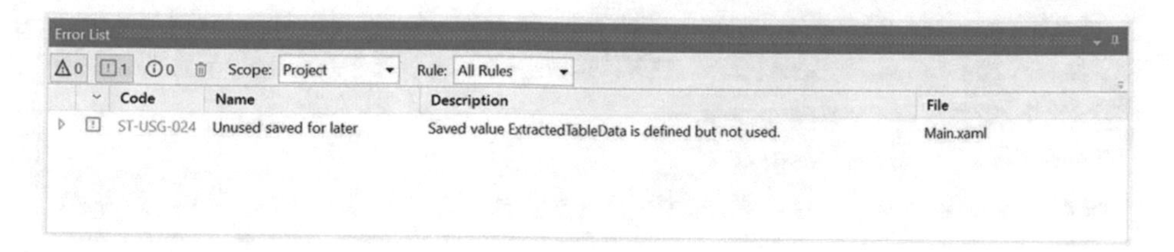

Figure 2-23. *Validation and Analyzer errors*

Export to Excel

The Export to Excel menu allows you to export your automation definition in an Excel format. Figure 2-24 displays a sample export.

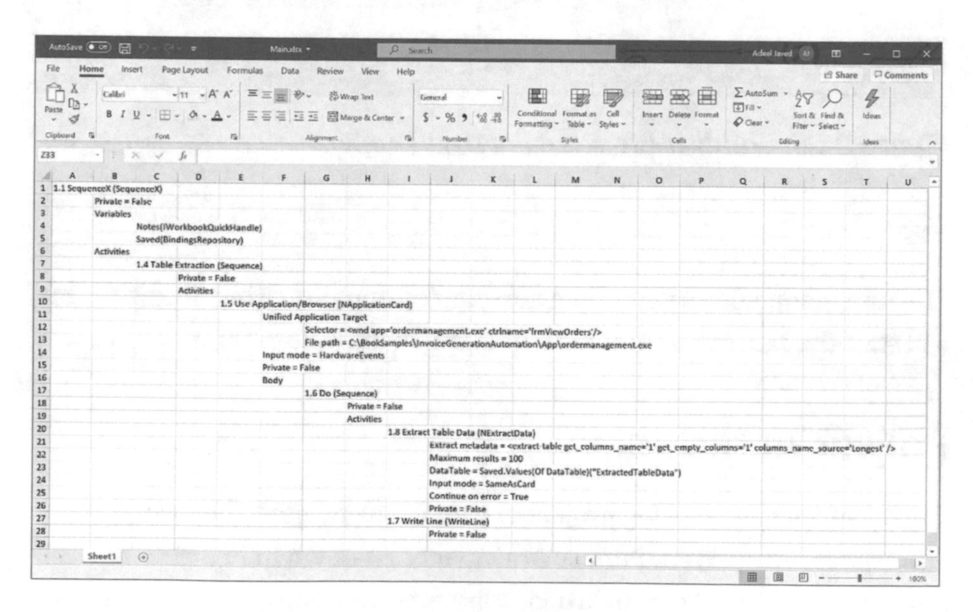

Figure 2-24. *Excel export of an automation*

Publish

The Publish menu allows you to create a new release and publish your automation to Orchestrator (including your Personal Workspace) or a custom folder. Figure 2-25 shows the publish process dialog.

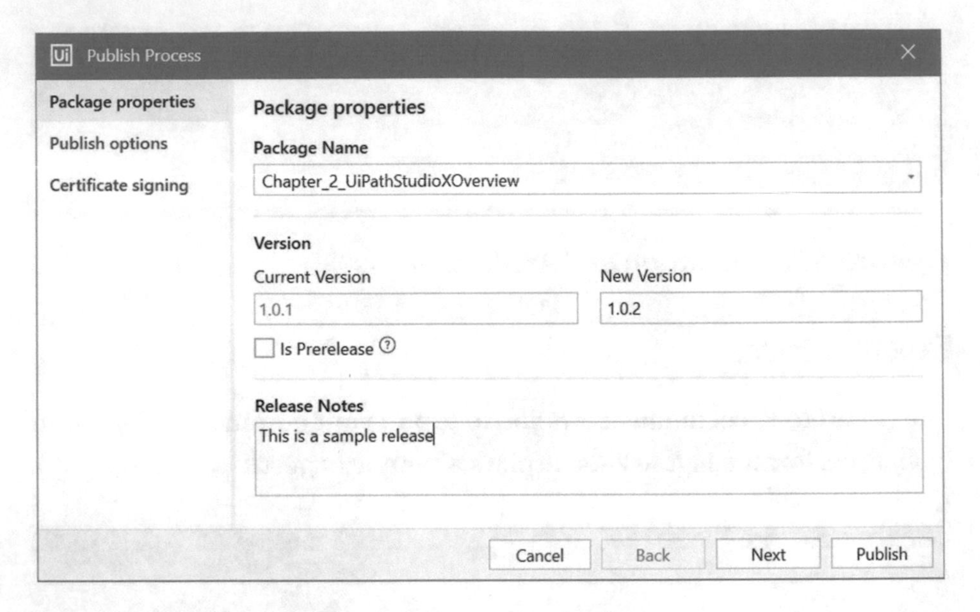

Figure 2-25. *Publish Process – Package properties*

Run

The Run menu allows you to execute your automation from StudioX. This will start the execution from the beginning of the automation.

Project Workspace

Once you open an existing project or start a new project, the Project Workspace is displayed. Project Workspace, as shown in Figure 2-26, provides you with all the tools necessary for developing your automations.

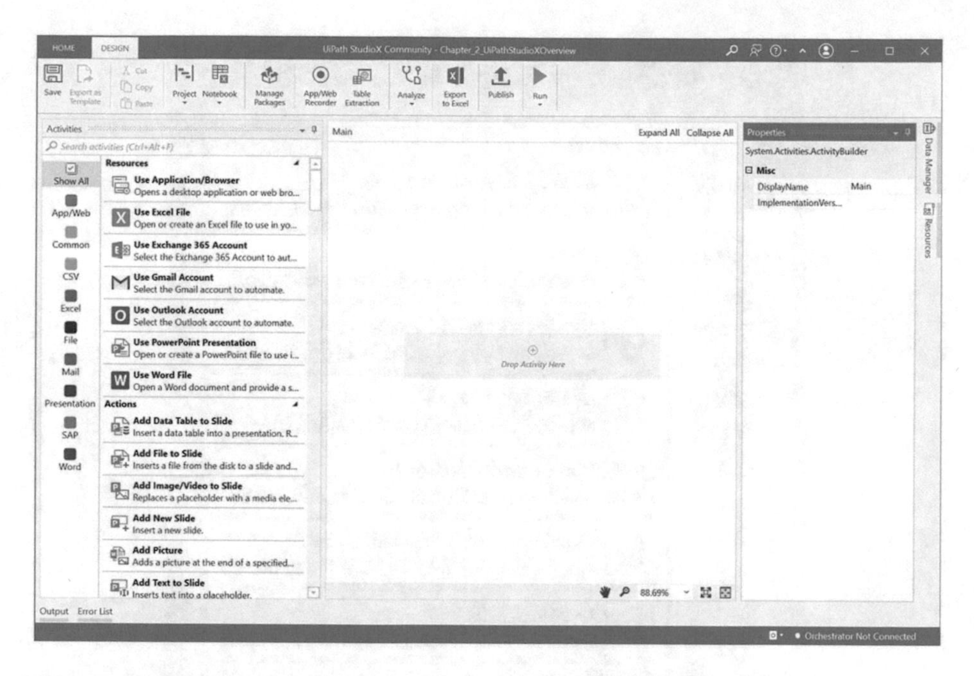

***Figure 2-26.** Project Workspace in UiPath StudioX*

Activities Panel

A great way to understand activities is to think about them in terms of LEGO blocks. You can assemble individual blocks in different ways to build the resulting LEGO model. Similarly, in UiPath StudioX, activities can be considered as individual building blocks that work together to assemble the defined automation.

UiPath StudioX provides approximately 100 activities that enable interaction with various types of apps such as desktop apps, web apps, Outlook, Word, Excel, CSV, and others.

In UiPath StudioX, activities are available in the Activities panel shown in Figure 2-27. You can drag and drop activities from this panel to the designer panel (discussed next). You can also use the search bar to find activities quickly. We will learn more about configuring each of these Activities in Chapters 3 through 9.

Figure 2-27. *Activities panel*

Designer Panel

The Designer panel, as shown in Figure 2-28, is where you will define your automation by organizing and configuring activities.

The Designer panel is where you will define the order of the workflow for your automation activities and is where you will edit and configure each activity. Additionally, the Designer panel will serve as the visual representation of your automation.

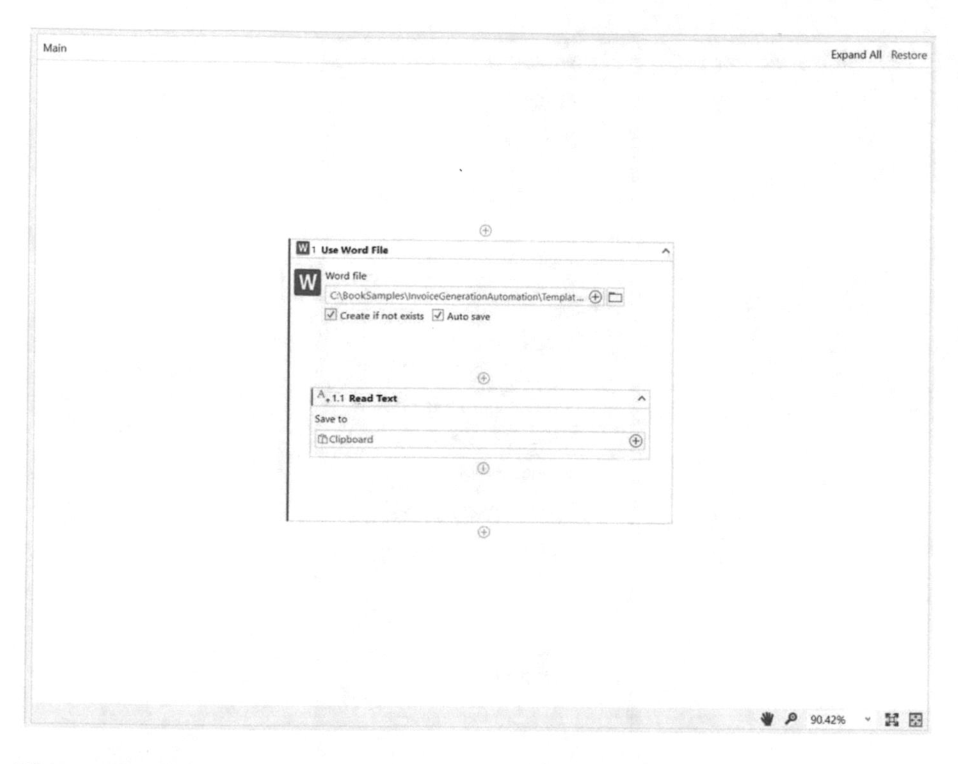

Figure 2-28. *Designer panel*

Context Menu

You can access the Context menu by right-clicking in the Designer panel. Depending on where you perform the right-click, the items on menu enable or disable. Figure 2-29 shows the Context menu.

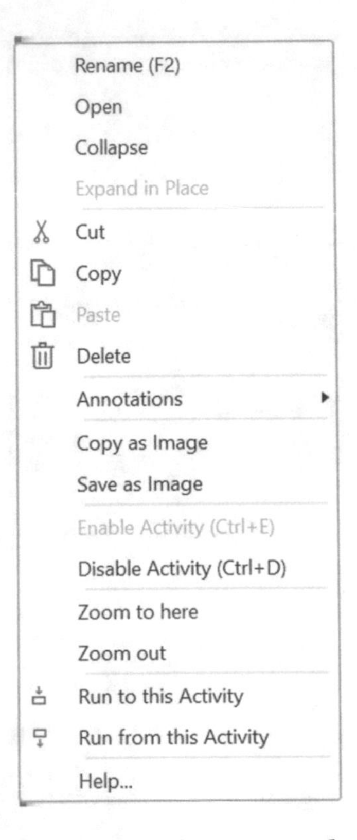

Figure 2-29. *Context menu in Designer panel*

Table 2-3 provides a brief description of each menu item.

Table 2-3. *Context menu item description*

Menu Item	Description
Rename (F2)	Allows you to rename the selected activity.
Open	Allows you to open the activity in its own designer panel, i.e., drill into the activity.
Collapse	Allows you to collapse the currently selected activity.
Expand in Place	Allows you to expand the activity in the same designer panel.

(*continued*)

Table 2-3. (*continued*)

Menu Item	Description
Cut	Allows you to cut the selected activity (and any nested activities).
Copy	Allows you to copy the selected activity (and any nested activities).
Paste	Allows you to paste the activity that was previously cut or copy (and any nested activities).
Delete	Allows you to delete the selected activity (and any nested activities).
Annotations	Allows you to annotate the selected activity; this is really helpful if multiple people are working or reviewing an automation.
Copy as Image	Allows you to copy the selected activity as an image in memory.
Save as Image	Allows you to save the selected activity as an image on file system.
Enable Activity	Allows you to enable the selected activity.
Disable Activity	Allows you to disable the selected activity so that it does not run, commenting it out.
Zoom to here	Allows you to drill into the selected activity; this opens the activity in its own Designer panel.
Zoom out	Allows you to zoom out the selected activity, i.e., close the individual Designer panel and show the activity in the parent Designer panel.
Run to this Activity	Allows you to run the automation till this activity, i.e., this activity and all prior activities are run.
Run from this Activity	Allows you to run the automation from this activity, i.e., this activity and all subsequent activity are run.
Help...	Allows you to open the documentation in a browser.

Properties Panel

The Properties panel, as shown in Figure 2-30, allows you to view and change properties of an activity that is currently selected in the Designer panel.

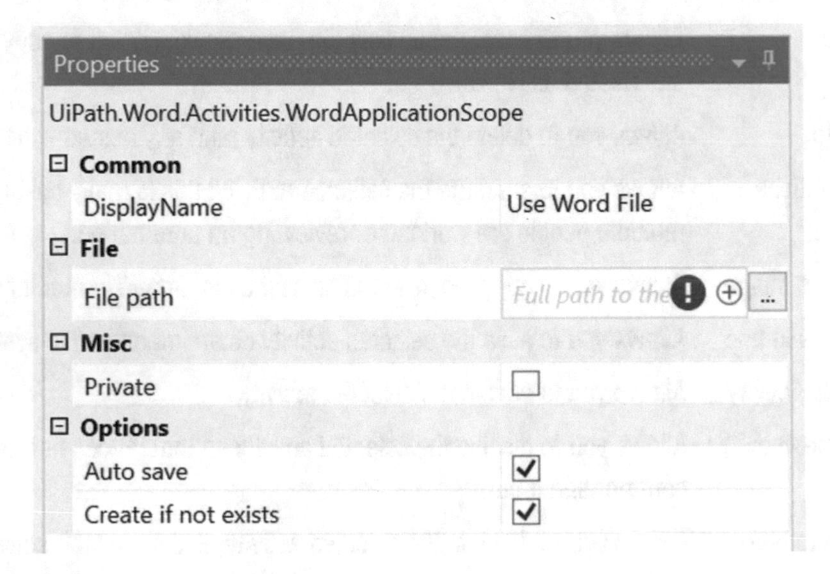

Figure 2-30. *Properties panel*

Data Manager Panel

As shown in Figure 2-31, the Data Manager panel allows you to access the Project Notebook, Resources, and all Saved Values. You can also rename saved values from this panel. We will learn more about Saved Values in Chapter 3.

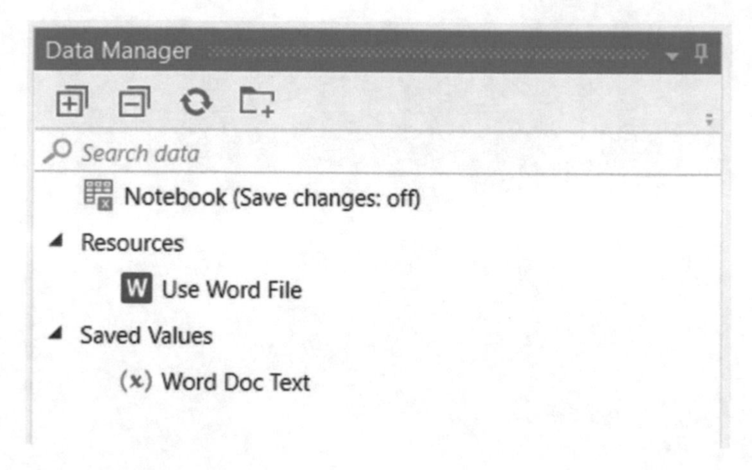

Figure 2-31. *Data Manager panel*

Output Panel

The Output panel, as shown in Figure 2-32, displays all log messages, the output from `Write Line` activity, and all errors generated by the automation. You can select the log level (e.g., error, info) of messages that you want to view in the Output panel. By default, StudioX has audit logs for some activities.

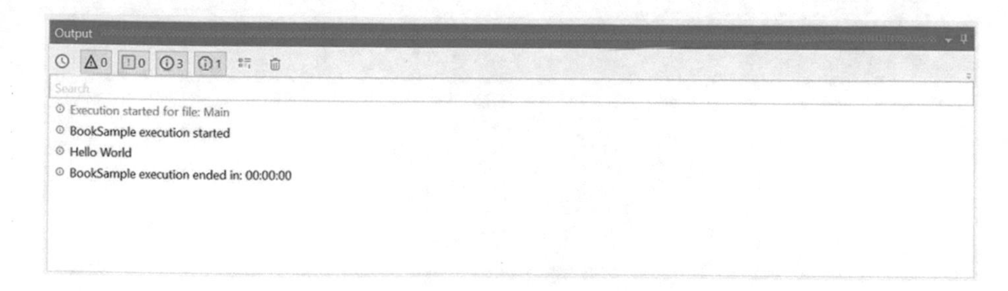

Figure 2-32. *Output panel*

PART II

Building Blocks

CHAPTER 3

Common Concepts

This chapter is going to introduce you to the concepts, properties, and activities commonly used for building automation in UiPath StudioX.

Learning Objectives

At the end of this chapter, you will learn how to

- Use project Notebook for using formulas and creating custom formulas
- Pass data to activities
- Store data returned from activities for use later
- Configure common activities in automation

Notebook

The Notebook is a great concept that makes UiPath StudioX indeed no code.

Most business users and citizen developers are familiar with Excel. By default, each UiPath StudioX automation has access to an Excel file, known as Notebook. You can use the default Notebook in your project or configure a custom Excel file as your project Notebook.

A. Javed et al., *Robotic Process Automation using UiPath StudioX*,
https://doi.org/10.1007/978-1-4842-6794-3_3

Default Notebook

The default Notebook has some pre-built formulas for manipulating Date, Text, Number, and File values. Figure 3-1 shows the Text worksheet of the Notebook that contains some pre-built formulas to manipulate text values.

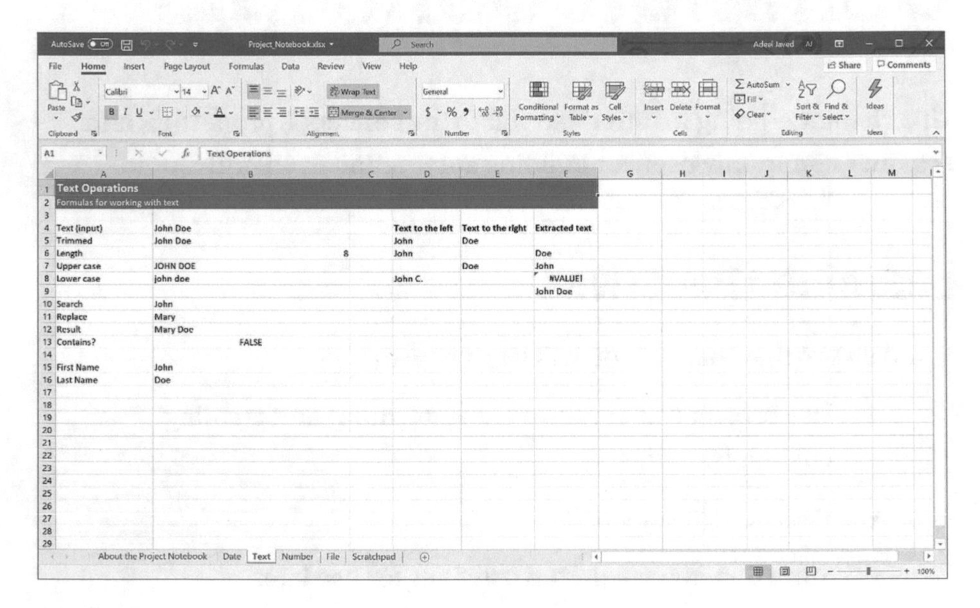

Figure 3-1. Text worksheet of Project Notebook

For additional custom fields and formulas, you can use the Scratchpad worksheet or add more worksheets to the Notebook and define your formulas.

To pass data to your custom formulas in Notebook, you can use the `Use Excel` and `Write Cell` activities to write data to a specific cell (see Chapter 7).

Any activity that requires input data can reference the Notebook to retrieve the data inputs.

Figure 3-2 shows an example of passing data to a custom formula in Notebook and then reading the result from Notebook. In this case, we are writing a name, John Doe, to a cell in the Text worksheet. Multiple formulas use this cell. In our case, we are interested in splitting the name into first name and last name and then printing it in the `Output` panel, so we output the cell that contains the First Name.

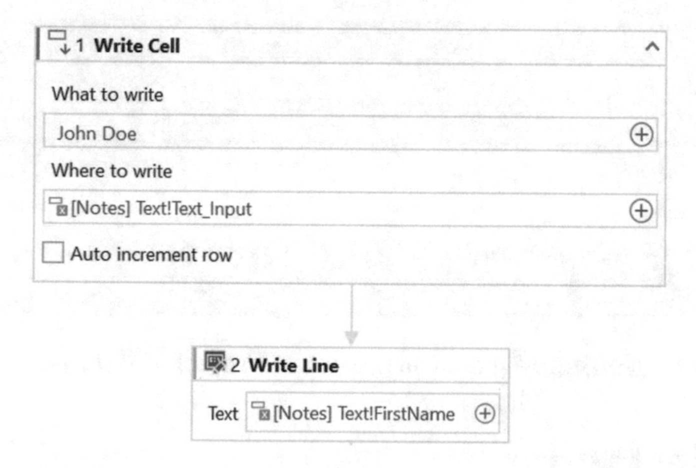

***Figure 3-2.** Passing data to and reading data from Project Notebook*

Custom Notebook

You have the option to configure a custom Excel file as your project Notebook.

To configure a custom Excel as your project Notebook, click the `Configure Notebook` option from the Notebook menu item in the ribbon, and then select the custom Excel file in `Notebook file` field, shown in Figure 3-3.

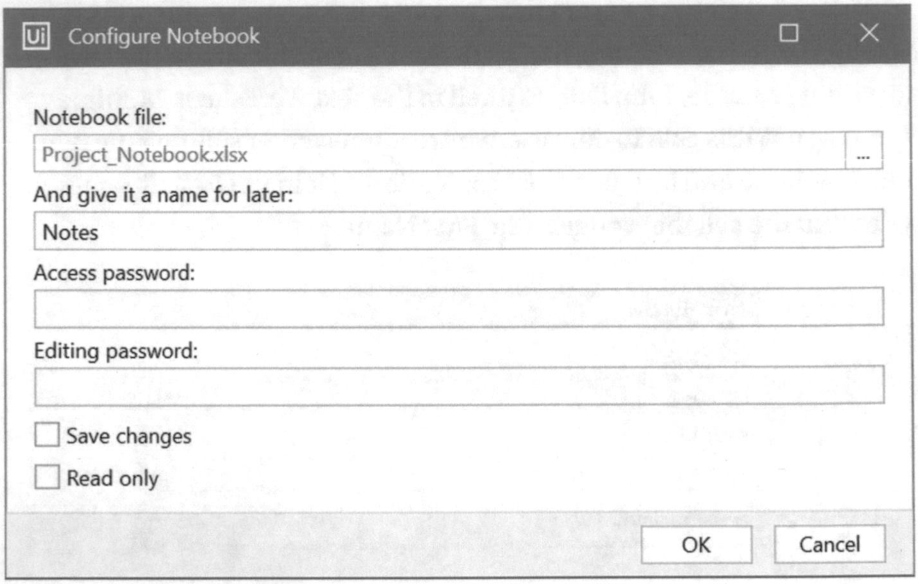

Figure 3-3. *Configure a custom Excel as Project Notebook*

Activity Inputs

Some of the activities require data inputs for processing. For example, you can provide text as an input to a `Write Line` activity. It will print the value in the Output panel, or you can provide text as an input for the `Type Into` activity, and it will enter the value in a form field. Each activity that requires input data gives you options to specify the input data. Figure 3-4 shows a sample menu for data input options. The data input options menu becomes available when you click the `Plus` icon in the field.

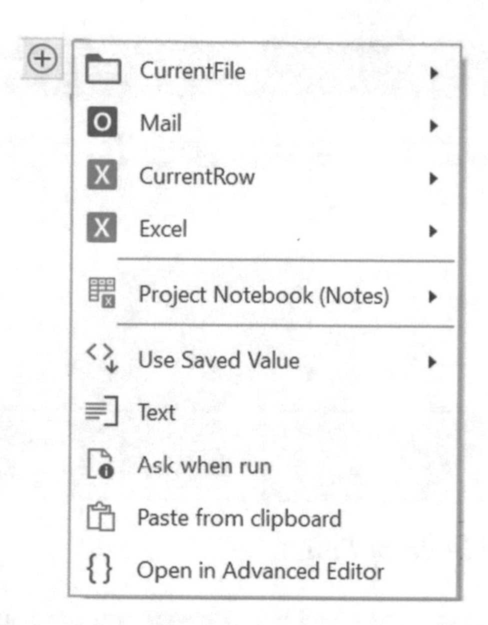

Figure 3-4. *Sample menu of data input options*

Excel: Allows you to specify a cell address, range, table, custom input, or even a sheet from the Notebook or any other Excel as an input to an activity. You can also select values from an individual Excel row as data inputs (see Chapter 7 for details).

Mail: Allows you to specify data from an email as an input to an activity (see Chapter 5 for details).

File: Allows you to specify attributes of a file as an input to an activity (see Chapter 9 for details), for example, the last modified date of a file or the full path of the file.

Use Saved Value: Allows you to specify data returned from a previous activity as an input to another activity.

Text: Allows you to specify a static or dynamic text string as an input to an activity. This option opens the `Text Builder` dialog. Using the `Text Builder`, you can combine multiple inputs into a single text string. For example, Figure 3-5 shows that two data inputs have been combined with some static text to return a single text input.

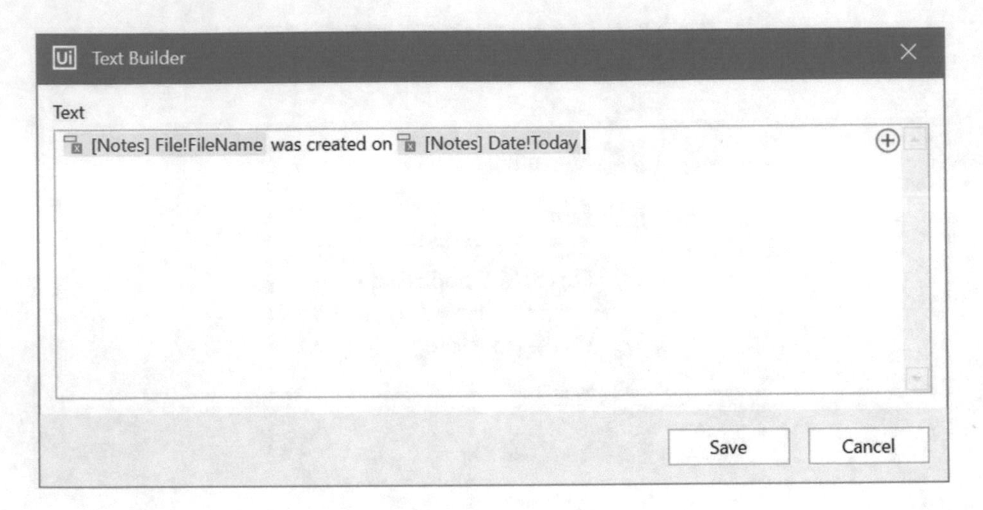

Figure 3-5. *Text input option*

Number: Allows you to specify a numeric calculation that can be used to input a number, basic calculations, or complex formulas from Excel to an activity.

Choose Date/Time: Allows you to specify a date or time as an input to an activity. Figure 3-6 shows the option to specify the date, while Figure 3-7 shows the option to specify the time.

***Figure 3-6.** Date input option*

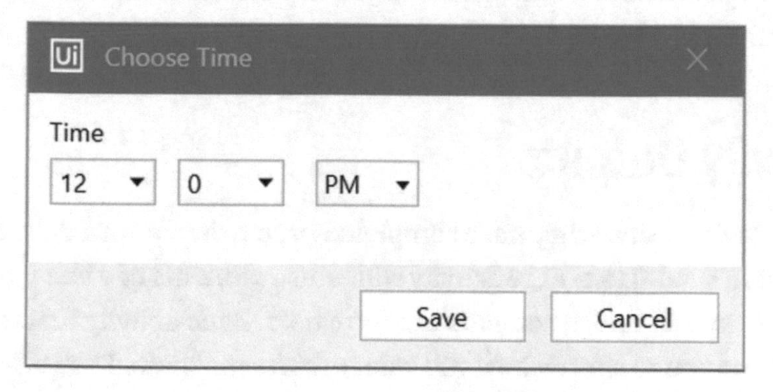

***Figure 3-7.** Time input option*

Choose Duration: Allows you to specify a duration as an input to an activity. Figure 3-8 displays the option to specify the duration.

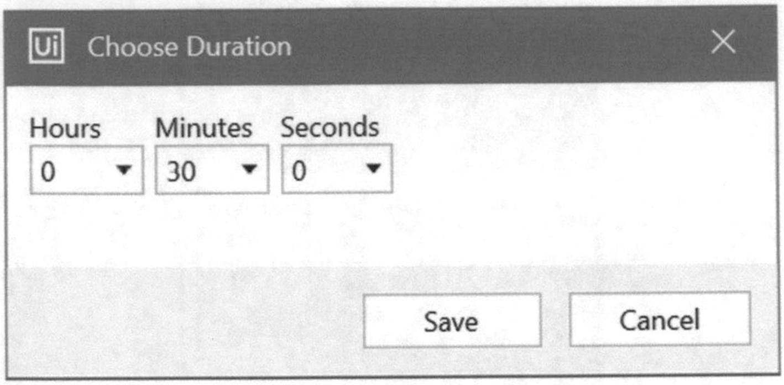

Figure 3-8. Duration input option

Ask when run: Allows you to specify a value as an input to activity at runtime. This option works well for attended robots.

Paste from clipboard: Allows you to paste clipboard data as an input to an activity.

Open in advanced editor: Allows you to enter a VB expression to specify the input value.

Activity Outputs

Some activities, once they have completed processing, return data. For example, a `Read Text File` activity will return contents of a text file, or `Get Text` activity will return contents of a form field. Each activity that returns data allows you to specify how you want to store that data. Figure 3-9 shows a sample menu of data output options. This data output options menu becomes available when you click the `plus` icon.

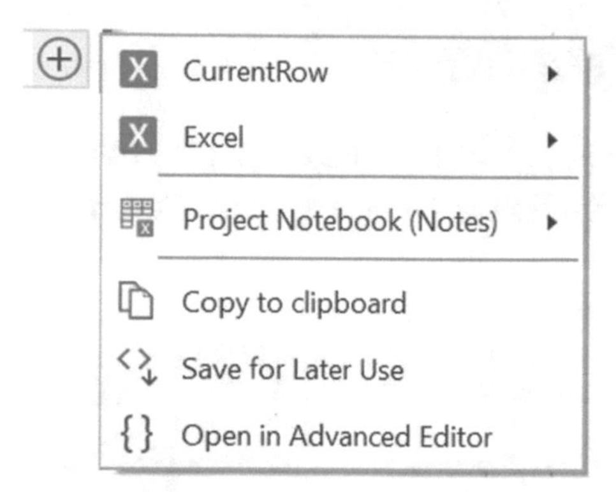

Figure 3-9. *Sample menu of data output options*

Excel: Allows you to specify a cell address, range, table, or sheet in the Project Notebook or Excel file from the parent `Use Excel File` activity where you want to store the data returned by an activity. Cells in individual Excel rows can also be specified (see Chapter 7 for details).

Save for Later Use: Allows you to specify a variable that will store data returned by an activity. Variables are accessible from the Data Manager panel. You can name or rename a variable to make it easy to understand. Figure 3-10 shows the option to store data in a variable.

Save value for later
Give your value a name:
FirstNameText
OK
Cancel

Figure 3-10. *Save for Later Use option*

Copy to clipboard: Allows you to copy data returned by an activity to your clipboard. This data can be accessed later using the `Paste from clipboard` option.

Common Properties

Common properties, as shown in Figure 3-11, can be accessed from the `Properties` panel under the `Common` section. Common properties can vary between activities.

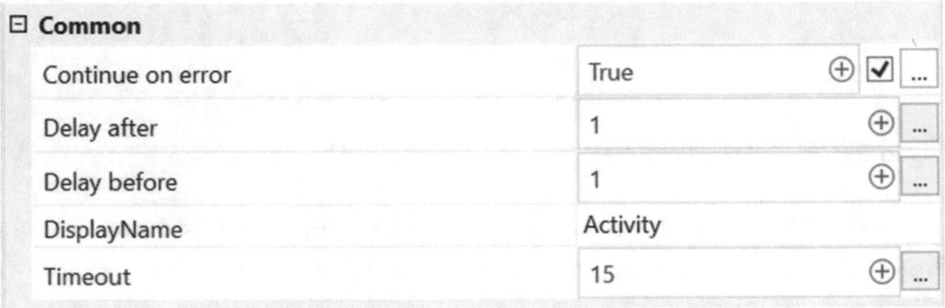

***Figure 3-11.** Common properties for activities*

Table 3-1 provides a brief description of each property.

***Table 3-1.** Common properties*

Property	Description
Continue on error	This property allows you to specify (true or false) if the automation should continue running, i.e., proceed to the next activity even if the selected activity encounters an error during execution.
Delay after	This property allows you to specify, in seconds, if the automation should wait after completing the selected activity and before starting the next activity. By default, the value is 0.3 seconds.
Delay before	This property allows you to specify, in seconds, if the automation should wait before starting the selected activity. By default, the value is 0.2 seconds.

(*continued*)

Table 2-2. (*continued*)

Property	Description
DisplayName	This property allows you to update the display name of the selected activity in the Designer panel.
Timeout	This property allows you to specify, in seconds, how long should the automation wait for the selected activity to complete before timing out. By default, the value is 30 seconds.

Common Activities

Activities that are commonly used across all automation types can be found under the **Common** tile, as shown in Figure 3-12. The following sections will provide instructions on how to configure and use each of these activities.

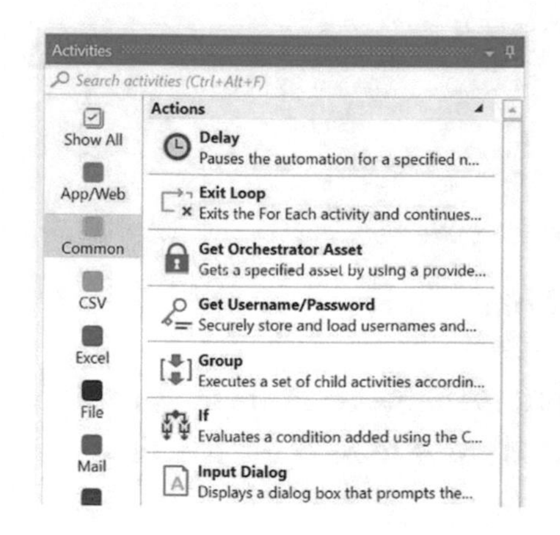

Figure 3-12. *Common activities for automation*

Write Line

The **Write Line** activity allows you to print a specified message to the Output panel.

Tip The Write Line activity is beneficial during and after the development phase of automation. Adding this activity can help you follow the flow of automation and debug any issues that might arise during development. Once you have developed the automation, this activity can help you create an audit log of your automation, which is essential in regulated industries.

Configuration

This section provides instructions on how to configure a **Write Line** activity, shown in Figure 3-13.

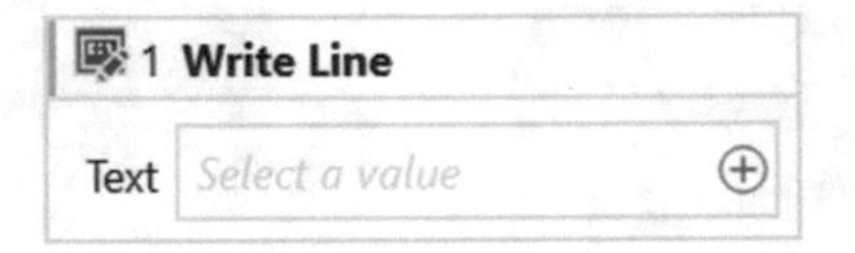

Figure 3-13. *Activity card for Write Line*

Text: This is an optional configuration available on the activity card. The text you specify in this configuration is going to be printed in the `Output` panel.

EXERCISE

Goal: Use the `Write Line` activity to print "Automation has started processing data for <Current Date>." message in the Output panel. The <Current Date> should be replaced with the actual date when the automation runs.

Source Code: Chapter_3-WriteLineExercise

Setup: Here are step-by-step implementation instructions:

1. In `StudioX`, add the `Write Line` activity to a blank process.
2. Next, click the `Plus` icon and select the `Text` option. In the `Text Builder`, enter `Automation has started processing data for`.
3. Next, from within the `Text Builder`, click the `Plus` icon and select `Project Notebook` ➤ `Date` ➤ `Today`. Your final text should be like `Automation has started processing data for [Notes] Date!Today`.

Once you have completed the exercise, the final configuration of the **Write Line** activity should resemble Figure 3-14. Once you run this automation, the automation will print the message in the Output panel, shown in Figure 3-15.

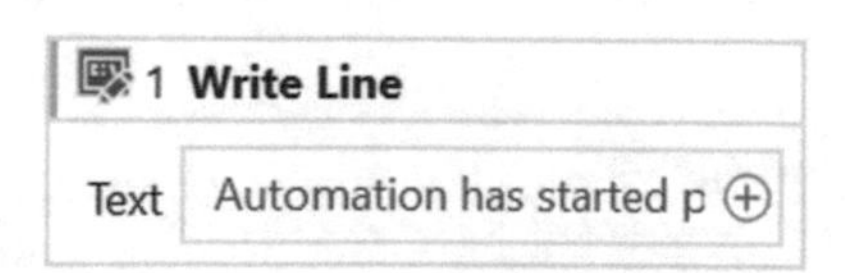

Figure 3-14. Configuration of Write Line activity exercise

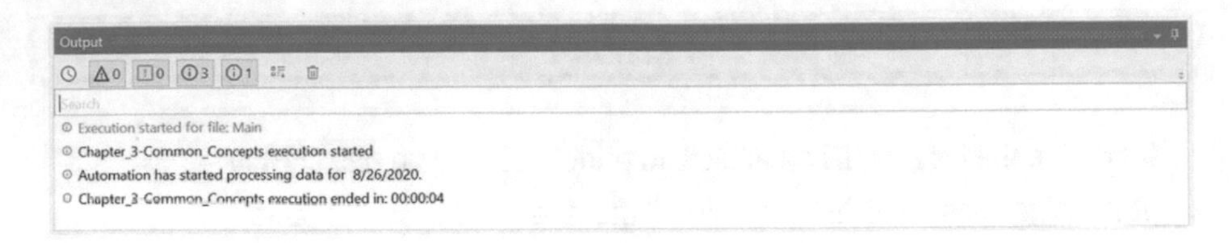

Figure 3-15. *The output of the Write Line activity exercise*

Message Box

The **Message Box** activity allows you to pause the automation and display text with specified buttons in a message box.

Tip The Message Box activity is particularly useful during the development phase of automation. Adding this activity can help you follow the flow of automation and debug any issues that might arise during development.

Configuration

This section provides instructions on how to configure a **Message Box** activity, shown in Figure 3-16.

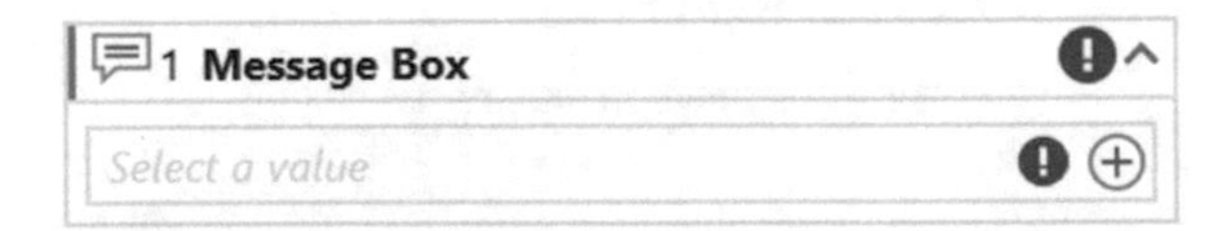

Figure 3-16. *Activity card for Message Box*

Text: This is a required configuration available on the activity card. The text you specify here is going to be displayed in the message box.

Buttons: This is an optional configuration available on the Properties panel. This configuration allows you to specify buttons to display on the message box. Figure 3-17 shows all available options for buttons. By default, the selection is set to the Ok button.

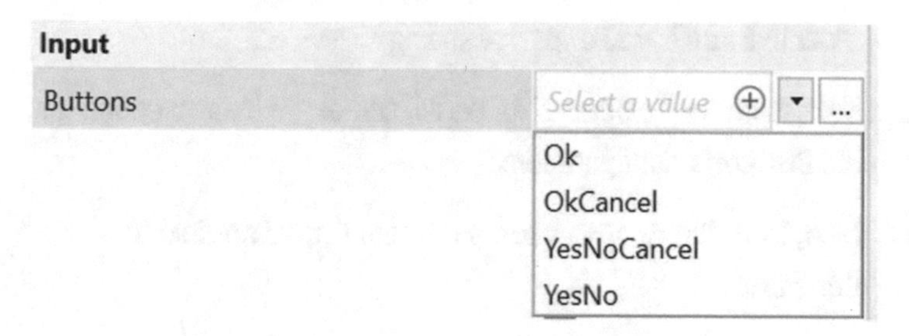

***Figure 3-17.** Button options for the Message Box activity*

Caption: This is an optional configuration available on the Properties panel. This configuration allows you to specify a title/label for the message box. The default caption is Message Box.

ChosenButton: This is an optional configuration available on the Properties panel. This configuration allows you to save the value of the button clicked (from the Buttons configuration). You can later use this value to make decisions for the automation workflow path.

TopMost: This is an optional configuration available on the Properties panel. This configuration allows you to specify if the message box should always be brought to the foreground.

EXERCISE

Goal: Use the Message Box activity to display a message box that asks the user to confirm this message "Word file processing is complete; Do you want to start Excel file processing?".

Source Code: Chapter_3-MessageBoxExercise

Setup: Here are step-by-step implementation instructions:

1. In StudioX, add the Message Box activity to a blank process.
2. Next, click the Plus icon, select Text option, and enter Word file processing is complete; Do you want to start Excel file processing?.
3. Next, from the Properties panel, select the YesNo option for the Buttons configuration.
4. Next, from the Properties panel, set Caption field to Confirmation.
5. Next, from the Properties panel, click the Plus icon in the ChosenButton field and select the Save for Later Use option. Name the variable as ChosenButton.
6. Next, add a Write Line activity, after the Message Box activity. Click the Plus icon and select Use Saved Value ➤ ChosenButton.

Once you have completed the exercise, the final configuration of the **Message Box** activity should resemble Figure 3-18.

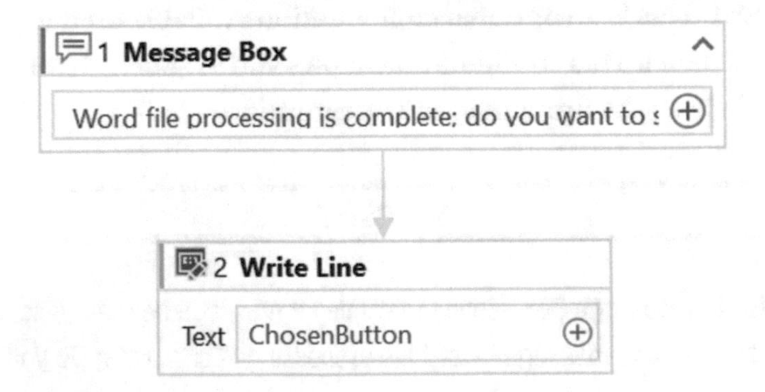

***Figure 3-18.** Configuration of Message Box activity exercise*

Once you run the automation, it will display a message box with Yes/No confirmation buttons, as shown in Figure 3-19. Whichever button you click, the automation will print its value in the Output panel, as shown in Figure 3-20.

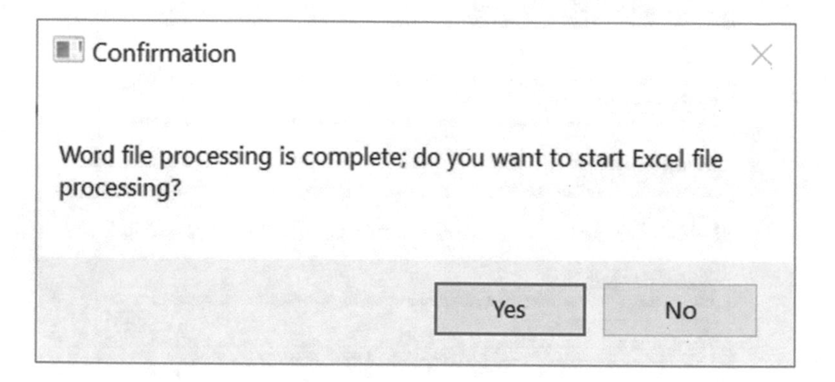

Figure 3-19. *Confirmation message box*

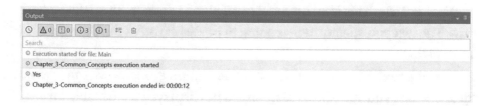

Figure 3-20. *The output of the automation*

Input Dialog

The **Input Dialog** activity allows you to prompt the user to provide input. This activity is particularly useful in attended automation for processes that require user input.

Configuration

This section provides instructions on how to configure an **Input Dialog** activity, shown in Figure 3-21.

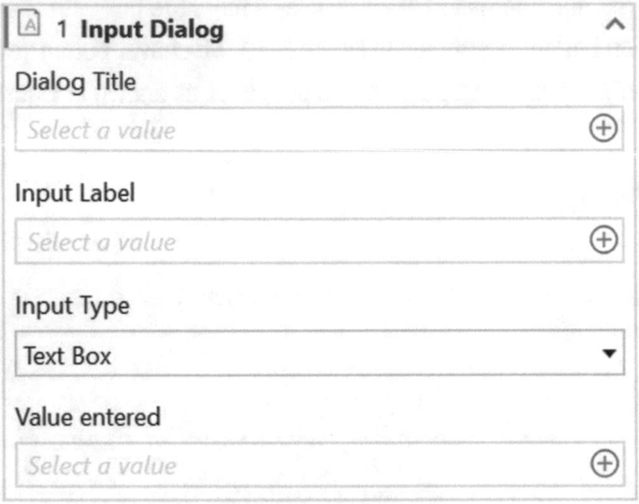

1 **Input Dialog**

Dialog Title

Select a value

Input Label

Select a value

Input Type

Text Box

Value entered

Select a value

Figure 3-21. *Activity card for Input Dialog*

Dialog Title: This is an optional configuration available on the activity card. The text you specify here is going to be displayed in the title of the input dialog.

Input Label: This is an optional configuration available on the activity card. The text you specify here is going to be displayed as the label for input field.

Input Type: This is a required configuration available on the activity card. There are two options, `Text` and `Multiple Choice`. By default, the value is set to `Text`, which prompts the user with a freeform text field to provide input. The `Multiple Choice` option allows you to specify a set of choices, and the user is required to select from one of the options.

Input options (separate with ;): This is an optional configuration available on the activity card. This configuration is only available when you select `Multiple Choice` option in the `Input Type` field. This configuration allows you to specify a list of choices for the user to choose from. Multiple choice options are separated by ;.

Value entered: This is an optional configuration available on the activity card. This configuration allows you to save the provided input value in a variable. You can later use this value to make decisions for the automation workflow path.

EXERCISE

Goal: Use the `Input Dialog` activity to prompt the user to select the next action for an automation.

Source Code: Chapter_3-InputDialogExercise

Setup: Here are step-by-step implementation instructions:

1. In `StudioX`, add an `Input Dialog` activity to a blank process.
2. In the `Dialog Title` field, click the `Plus` icon, select `Text` option, and enter `Next Action`.
3. In the `Input Label` field, click the `Plus` icon, select `Text` option, and enter `Please select how you want to proceed with the automation`.
4. From the `Input Type` field, select `Multiple Choice`.
5. In the `Input options` field, click the `Plus` icon, select `Text` option, and enter `Process now;Process later;Exit`.
6. In the `Value entered` field, click the `Plus` icon, and select the `Save for Later Use` option. Name the variable as `InputDialogResult`.
7. Next, add a `Write Line` activity after the `Input Dialog` activity. Click the `Plus` icon, and hover over `Use Saved Value` to select `InputDialogResult`.

Once you have completed the exercise, the final configuration of the **Input Dialog** activity should resemble Figure 3-22.

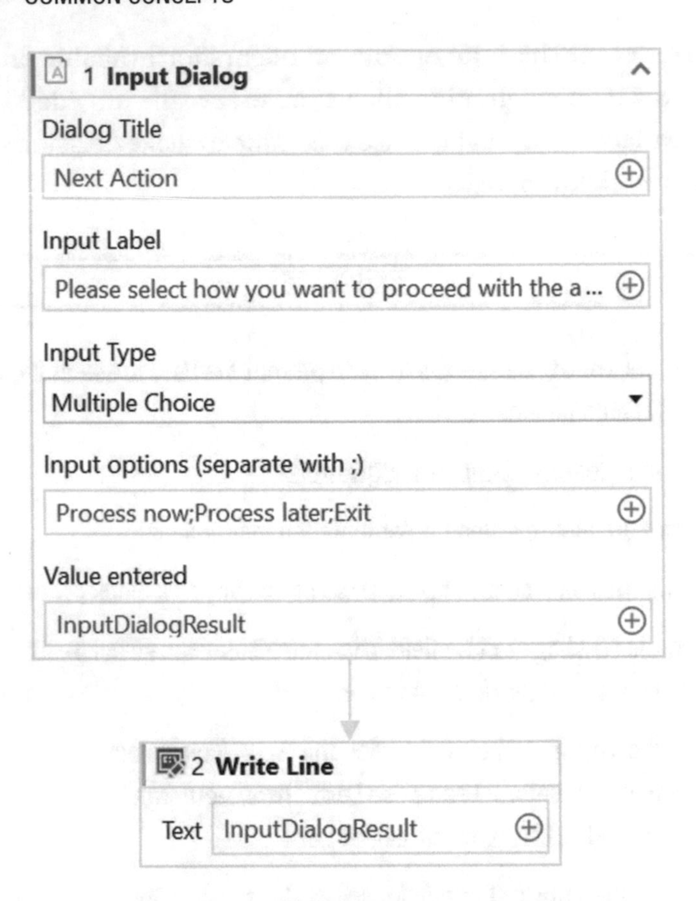

Figure 3-22. *Configuration of Input Dialog activity exercise*

Once you run the automation, it will display an input dialog box with three choices, as shown in Figure 3-23. Whichever option you choose, the automation will print its value in the Output panel.

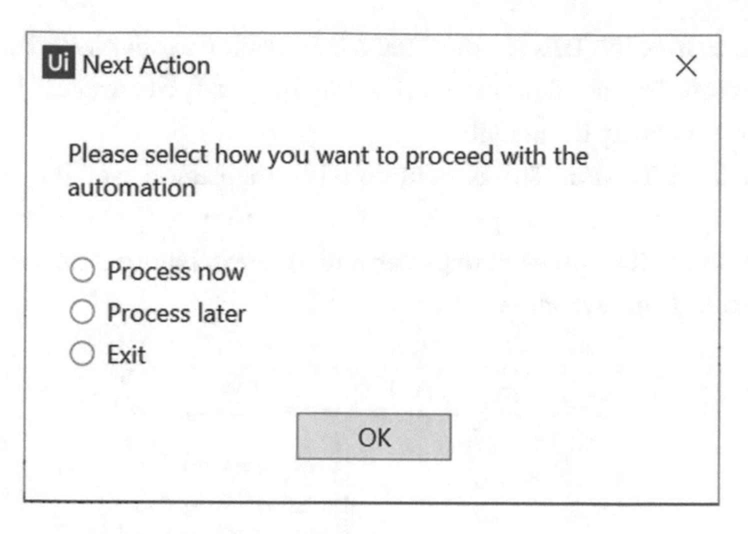

Figure 3-23. *Input dialog*

Modify Text

The **Modify Text** activity allows you to modify the input text value by performing different operations.

Configuration

This section provides instructions on how to configure a **Modify Text** activity, shown in Figure 3-24.

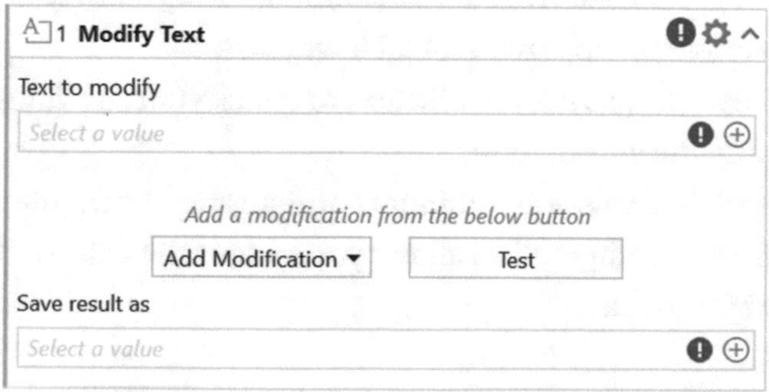

1 Modify Text
Text to modify
Select a value
Add a modification from the below button
Add Modification
Test
Save result as
Select a value

Figure 3-24. *Activity card for Modify Text*

Text to modify: This is a required configuration available on the activity card. This configuration allows you to specify the original text that will be modified by the activity.

Add Modification: This is an optional configuration available on the activity card. This configuration allows you to specify one or more modifications that you want to perform on the text. Figure 3-25 shows all the modifications available.

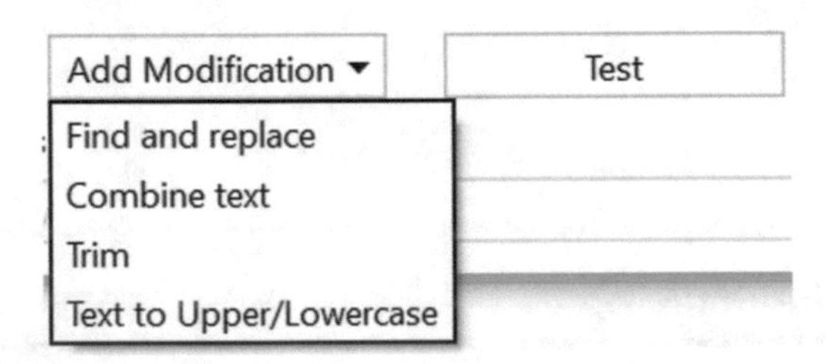

Figure 3-25. *Button options for Message Box activity*

`Find and replace` modification allows you to specify text to search for in the original text and text to replace it with. You can also specify if it should be an exact match.

`Combine text` modification allows you to add new text at the beginning or end of text.

`Trim` modification allows you to trim white spaces from left and right of text.

`Text to Upper/Lowercase` modification allows you to change the case of the text to either all uppercase or all lowercase.

You can also choose the `Test` button to check your modification logic while designing the automation.

Save results as: This is an optional configuration available on the activity card. This configuration allows you to save the value of the modified text for use later.

EXERCISE

Goal: Use the `Modify Text` to change replace % with _ in the input value, trim the text, convert it to uppercase, and then print the result.

Source Code: Chapter_3-TextManipulationExercise

Setup: Here are step-by-step implementation instructions:

1. In `StudioX`, add the `Modify Text` activity to a blank process.
2. In the `Text to modify` field, click the `Plus` icon, and select `Ask when run` option.
3. Next, from the `Add Modification` dropdown, select `Find and replace` option. This will add a child activity.
4. In the `Search for` field of `Find and replace text` child activity, click the `Plus` icon, select `Text` option, and type %. This specifies what text to look for.
5. In the `Replace with` field of `Find and replace text` child activity, click the `Plus` icon, select `Text` option, and type _. This specifies, that if found, what to replace % with.
6. Next, from the `Add Modification` dropdown, select `Trim` option. This will add a child activity. Leave the configurations as is.
7. Next, from the `Add Modification` dropdown, select `Text to Upper/Lowercase` option. This will add a child activity. Leave the default configurations of `Uppercase` as is.
8. Next, in the `Save result as` field, click the `Plus` icon and select the `Save for Later Use` option. Name the variable as `OutputText`.

9. Next, add a `Write Line` activity after the `Modify Text` activity. Click the `Plus` icon and select `Use Saved Value` ➤ `OutputText`.

Once you have completed the exercise, the final configuration of the **Modify Text** activity should resemble Figure 3-26.

***Figure 3-26.** Configuration of Modify Text activity exercise*

Once you run the automation, it will display an input dialog. Type `New%Hires%2020.xlsx` as the value, as shown in Figure 3-27, and click OK. Once all the text modifications have been performed, the result will be displayed in the Output panel, as shown in Figure 3-28.

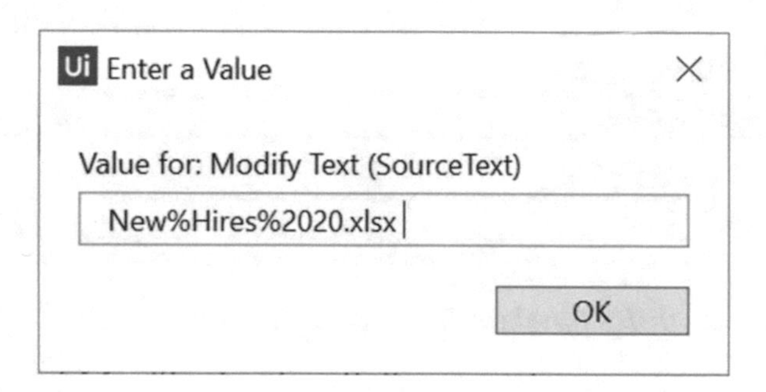

Figure 3-27. *Input value*

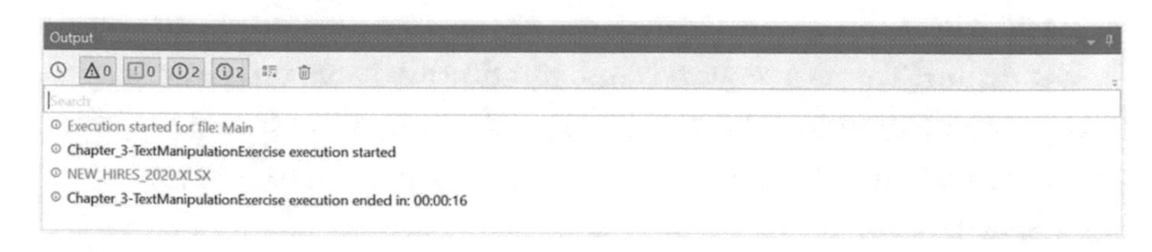

Figure 3-28. *The output of the automation*

Text to Left/Right

The **Text to Left/Right** activity allows you to split a text value into two text values using a separator.

Configuration

This section provides instructions on how to configure a **Text to Left/Right** activity, shown in Figure 3-29.

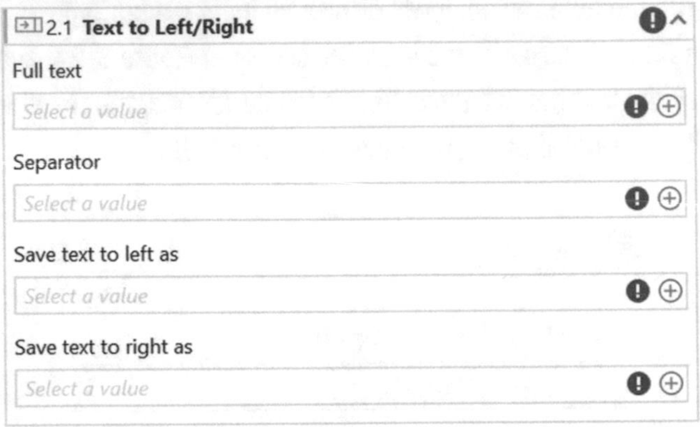

2.1 Text to Left/Right
Full text
Select a value
Separator
Select a value
Save text to left as
Select a value
Save text to right as
Select a value

Figure 3-29. *Activity card for Text to Left/Right*

Full text: This is a required configuration available on the activity card. This configuration allows you to specify the text that will be split into two by the activity.

Separator: This is a required configuration available on the activity card. This configuration allows you to specify the text to use as a separator. The activity will use the first occurrence of the separator to split the original text.

Save text to left as: This is a required configuration available on the activity card. This configuration allows you to save the text to the left of the separator for later use.

Save text to right as: This is a required configuration available on the activity card. This configuration allows you to save the text to the right of the separator for later use.

EXERCISE

Goal: Use the `Text to Left/Right` to split the input text value using % as a separator and print the two new text values.

Source Code: Chapter_3-TextManipulationExercise

Setup: Here are step-by-step implementation instructions:

1. In `StudioX`, add the `Text to Left/Right` activity to a blank process.
2. In the `Full text` field, click the `Plus` icon, and select `Ask when run` option.
3. In the `Separator` field, click the `Plus` icon, select `Text` option, and type %.
4. In the `Save to left as` field, click the `Plus` icon, and select the `Save for Later Use` option. Rename the value as `TextToLeft`.
5. In the `Save to right as` field, click the `Plus` icon, and select the `Save for Later Use` option. Rename the value as `TextToRight`.
6. Next, add a `Write Line` activity after the `Text to Left/Right` activity. Click the `Plus` icon and select `Use Saved Value` ➤ `TextToLeft`.
7. Next, add a `Write Line` activity after the `Write Line` activity. Click the `Plus` icon and select `Use Saved Value` ➤ `TextToRight`.

Once you have completed the exercise, the final configuration of the **Text to Left/Right** activity should resemble Figure 3-30.

Figure 3-30. *Configuration of the Text to Left/Right activity exercise*

Once you run the automation, it will display an input dialog. Type `New%Hires%2020` as the value, as shown in Figure 3-31, and click OK. Once the activity has completed execution, the output will be displayed in the Output panel, as shown in Figure 3-32. As mentioned in the configuration, only the first occurrence is used as a separator; that is why you still see the second % sign in the right text value.

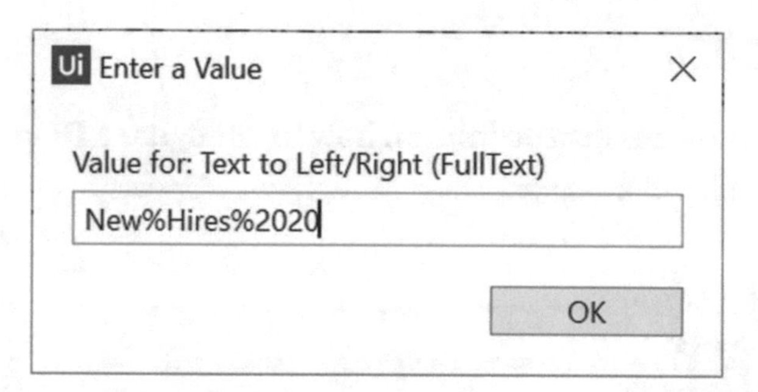

Figure 3-31. *Confirmation message box*

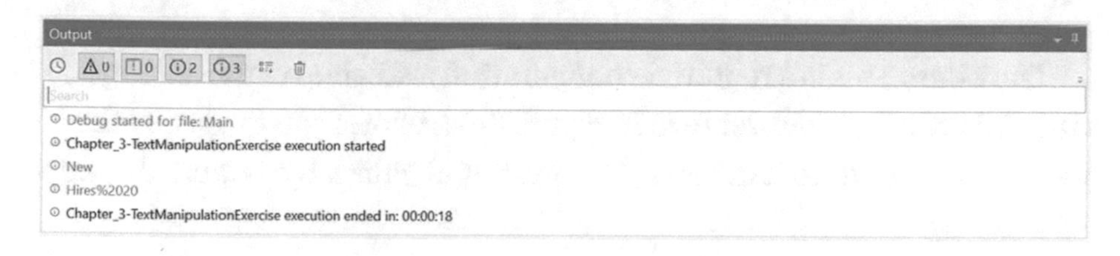

Figure 3-32. *The output of the automation*

Delay

The **Delay** activity allows you to pause your automation for a specified number of seconds.

Tip At times, issues such as slow system performance and latency in the network are unavoidable. The Delay activity can be useful in ensuring that such issues do not cause automation failures. This activity will help make your automation more robust and less prone to failure.

Configuration

This section provides instructions on how to configure a **Delay** activity, shown in Figure 3-33.

Figure 3-33. *Activity card for Delay*

Duration: This is a required configuration available on the activity card. This configuration, shown in Figure 3-34, allows you to specify the number of seconds that the automation should pause. By default, the value is 5 seconds.

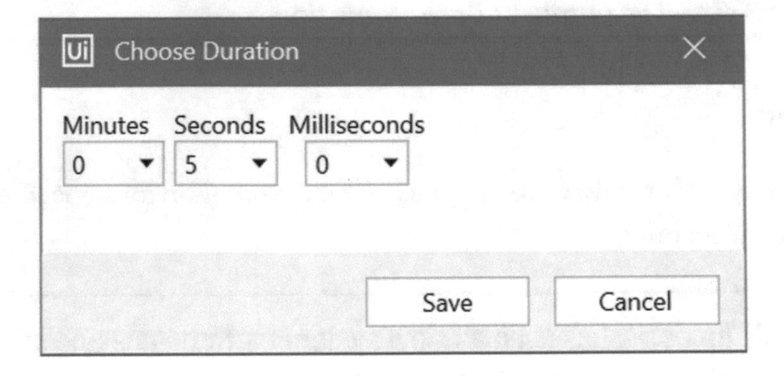

Figure 3-34. *Duration configuration for the Delay activity*

If

The **If** activity allows you to check for a condition and, based on the result, chooses which path the automation should follow.

The automation can either follow the **Then** path or the **Else** path. The automation will follow the **Then** path in cases where the condition is true. Otherwise, the automation will follow the **Else** path.

Note The If activity is also known as a decision activity, condition activity, or if-then-else activity.

Configuration

This section provides instructions on how to configure an **If** activity, shown in Figure 3-35.

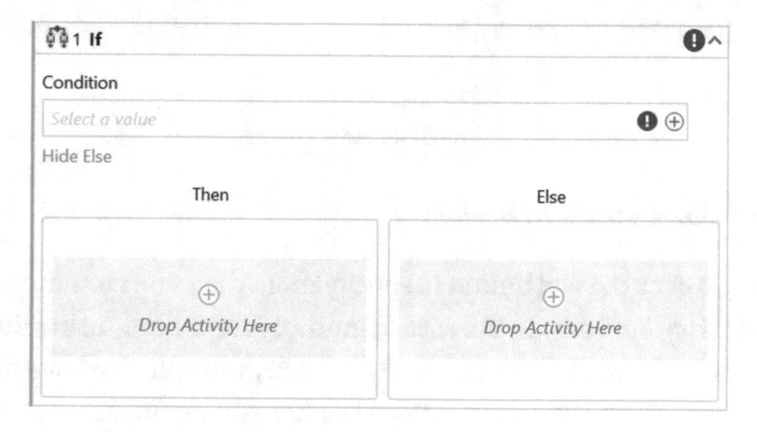

Figure 3-35. *Activity card for If activity*

Condition: This is a required configuration available on the activity card. You can use the `Condition Builder` shown in Figure 3-36 to build a condition that the automation will evaluate.

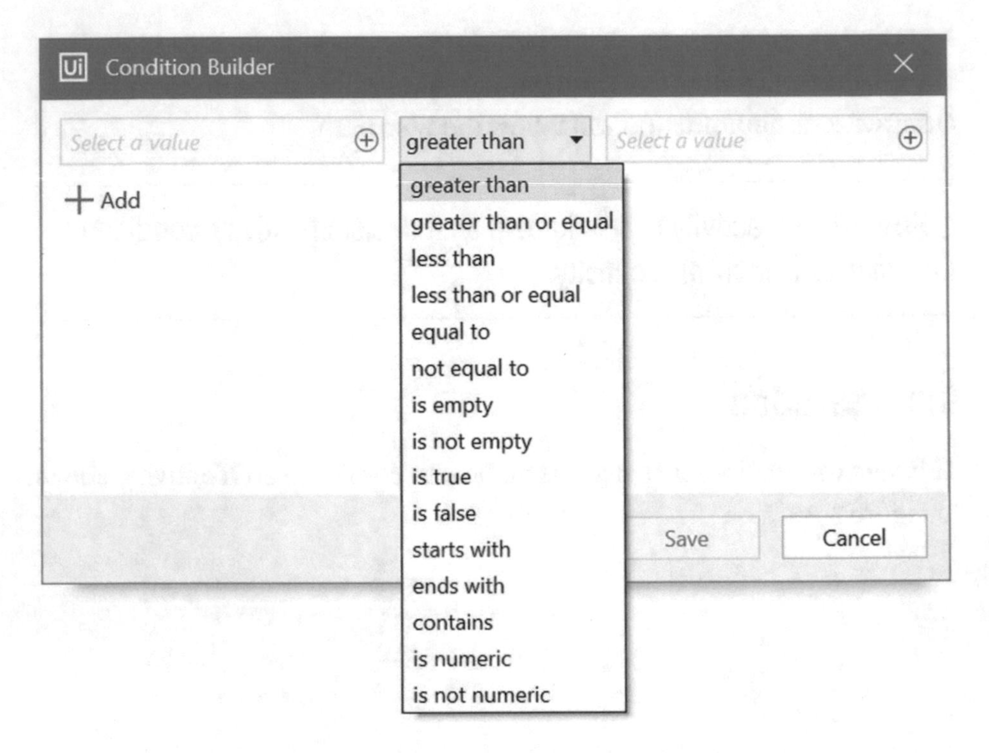

Figure 3-36. *Condition Builder*

You can click the `Add` button to specify more than one condition in the `Condition Builder`. In the case of multiple conditions, using the dropdown on the top left, shown in Figure 3-37, you will also have the option to specify how the automation should evaluate them.

The AND option means that all conditions must be true for the overall condition to be considered true.

The OR option means that any one of the conditions can be true for the overall condition to be considered true.

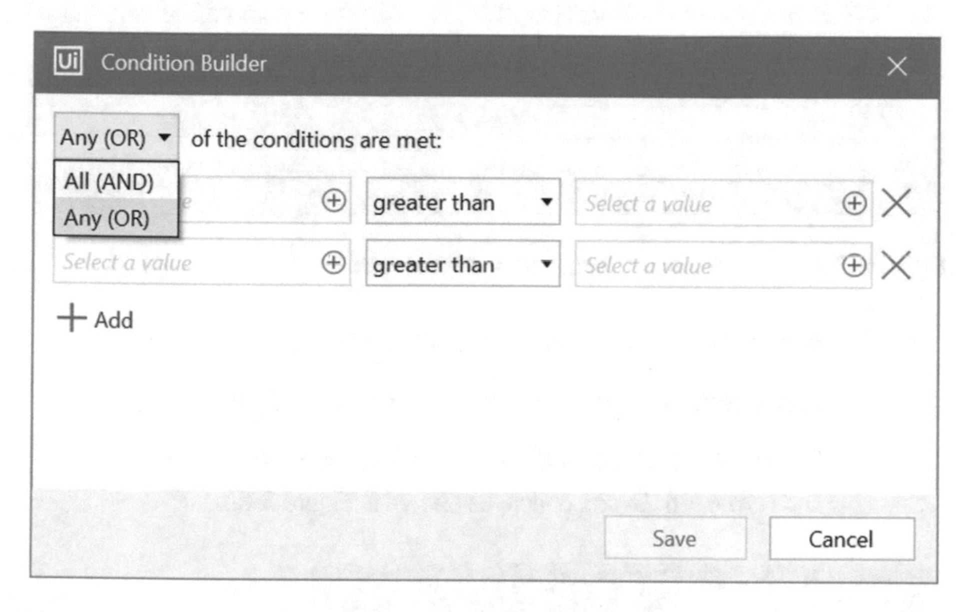

Figure 3-37. *Configuration for multiple conditions*

EXERCISE

Goal: Use the If activity to check if the address in project Notebook is a street address or a building address.

Source Code: Chapter_3-IfConditionExercise

Setup: Here are step-by-step implementation instructions:

1. In StudioX, click Notebook ➤ Open Notebook and open the Scratchpad worksheet.
2. Next, in cell A1, enter Address; this is just a label.

3. Next, in cell `B1`, enter a test address, `123 Main Street, Suite # 100, City, State`. Figure 3-38 shows the `Notebook`. Click `Save` and close the `Notebook`.

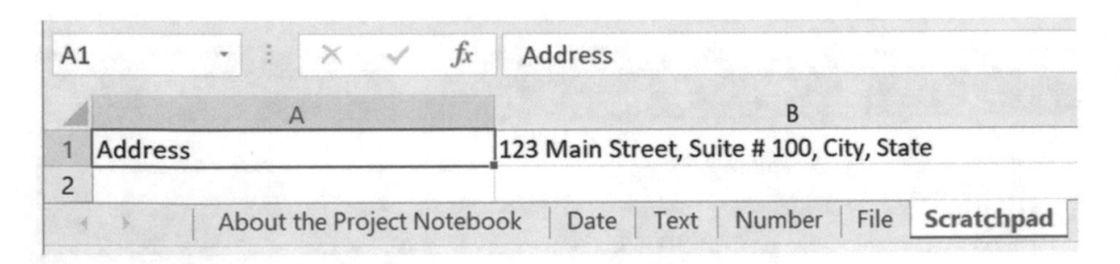

***Figure 3-38.** Notebook containing input data*

4. Next, add an `If` activity to the `Designer` panel.

5. Open `Condition Builder` by clicking the `Plus` icon.

6. On the left side of the condition, select `Notebook` ➤ `Indicate in Excel` option, as shown in Figure 3-39.

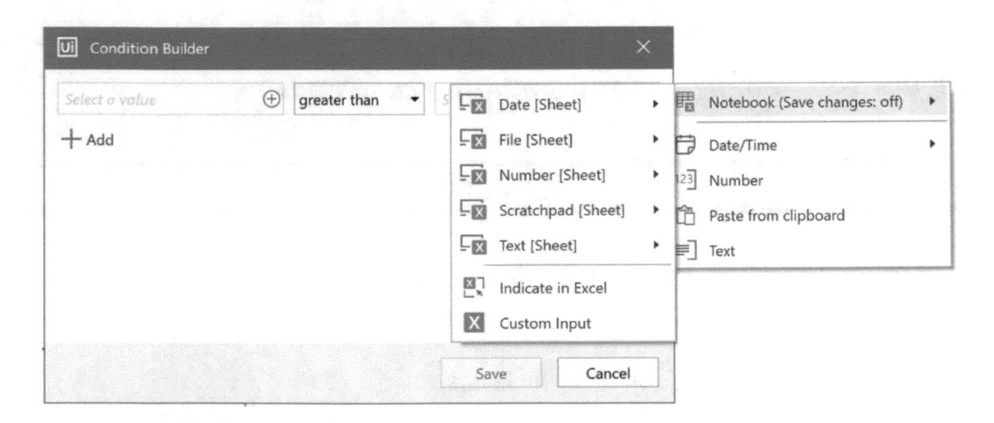

***Figure 3-39.** Select input data from the Notebook*

7. Select cell `B1` and click `Confirm` from the ribbon in Excel.

8. From the condition dropdown, select `contains`.

9. On the right side of the condition, click the `Text` option, and enter `Apt #`. This condition will check if the provided address contains Apt #.

10. Click `+ Add` to add another condition.

11. On the left side of the condition, select `Project Notebook` ➤ `Indicate in Excel` option. Select cell `B1` and click `Confirm` from the ribbon in Excel.

12. From the condition dropdown, select `contains`.

13. On the right side of the condition, click the `Text` option, and enter `Suite #`. This condition will check if the provided address contains Suite #.

14. Select `Any (OR)` from the top-left dropdown. At this point, the `Condition Builder` should look like Figure 3-40. Click `Save`.

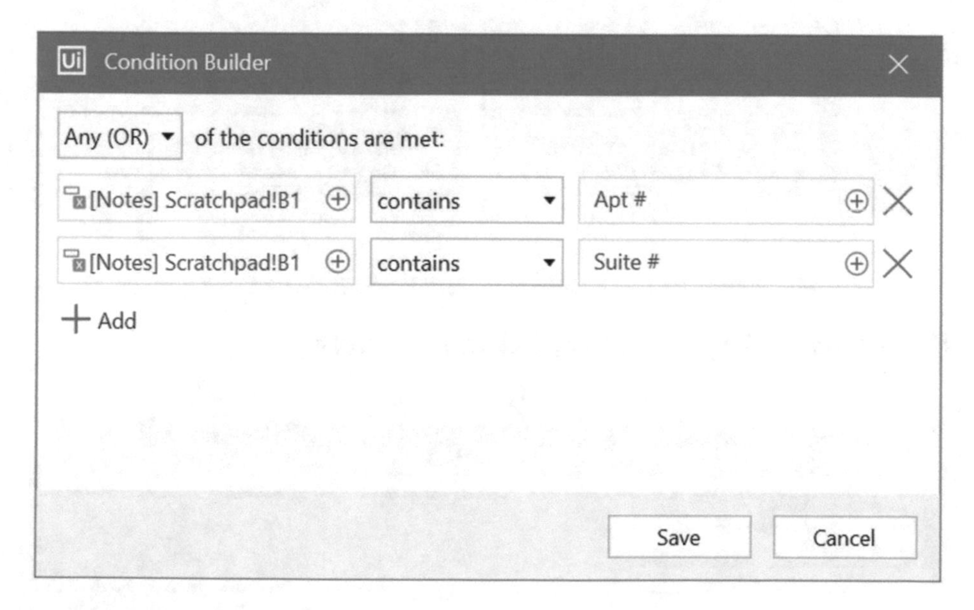

Figure 3-40. *Condition configuration*

15. Next, add a `Write Line` activity in the `Then` body of `If` activity.

16. Click the `Plus` icon, select the `Text` option, and enter `This is a building address`.

17. Next, add a `Write Line` activity in the `Else` body of `If` activity.

18. Click the `Plus` icon, select the `Text` option, and enter `This is a street address`.

Once you have completed the exercise, the final configuration of the **If** activity should resemble Figure 3-41. Figure 3-42 shows the output of the run.

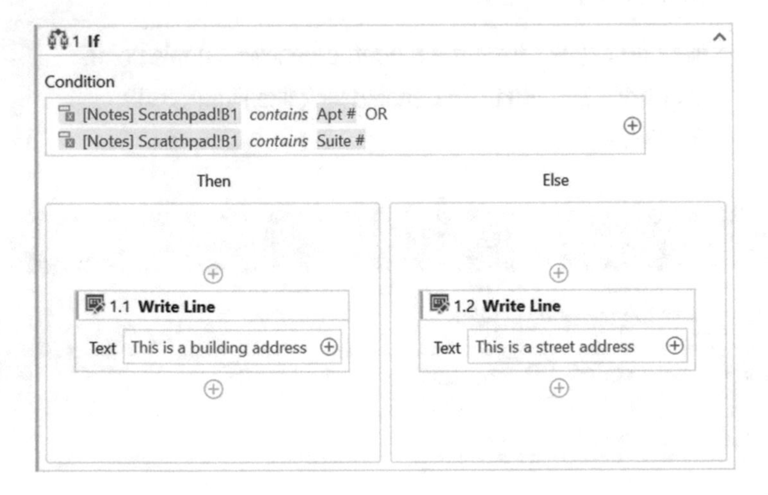

***Figure 3-41.** Configuration of If activity exercise*

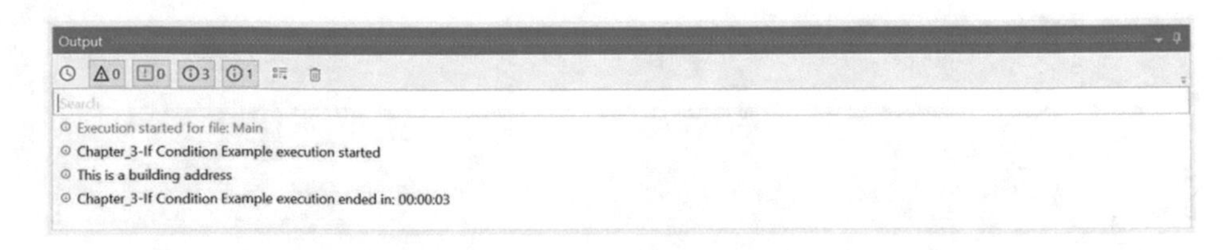

***Figure 3-42.** The output of the automation run*

Switch

The **Switch** activity allows you to define multiple cases, and then based on an expression, it executes a single case.

The If activity and Switch activity are conditional activities, that is, based on the expression results, they follow one of the defined paths. You can use an If activity when there are two choices, while a Switch activity is better suited for multiple choices.

Configuration

This section provides instructions on how to configure a **Switch** activity, shown in Figure 3-43.

Figure 3-43. *Activity card for Switch*

TypeArgument: This is an optional configuration available on the `Properties` panel. From the dropdown shown in Figure 3-44, you need to select the type of value that your expression will return.

If your expression result is true or false, then you will select `Boolean`. If your expression result is a number, that is, 1, 2, 3, and so on, then you will select `Int32`. If your expression result is text, that is, January, February, March, and so on, then you will select `String`.

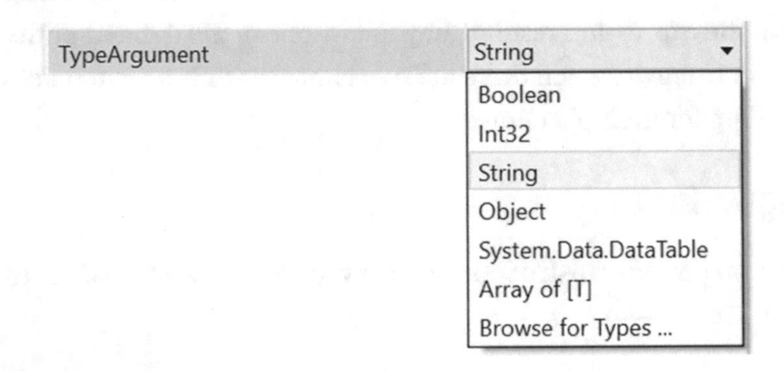

Figure 3-44. *Argument type options*

Expression: This is a required configuration available on the activity card. This expression will return one of the values defined as a case. For example, this could point to a column in Excel that has status values, and based on the status value, you want to perform specific actions.

Add new case: This is a required configuration available on the activity card, that is, a Default case must be provided. This configuration allows you to add a new Case value to the Switch activity, as shown in Figure 3-45.

Figure 3-45. *Add new case in Switch activity*

Once you have added a new `Case` value, you can add activities to the body of this `Case` value, as shown in Figure 3-46. During execution, if the value returned by the expression matches this case, then the activities in the body of this case will be executed.

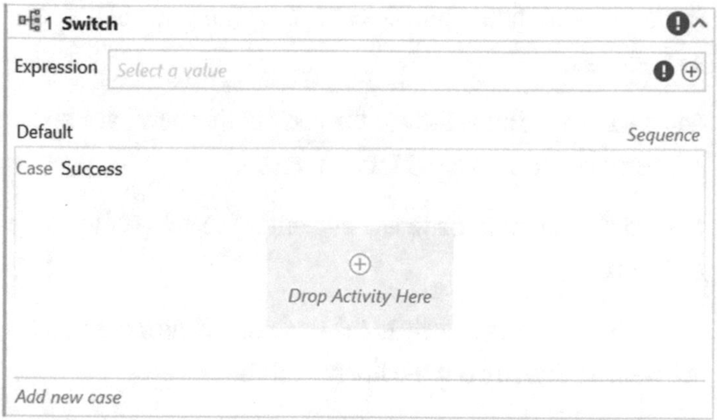

***Figure 3-46.** Switch activity with a case*

EXERCISE

Goal: Use the `Switch` activity to execute a different set of activities depending on the status of data.

Source Code: Chapter_3_SwitchRepeatActivitiesExercise

Setup: Here are step-by-step implementation instructions:

1. In `StudioX`, add the `Switch` activity to a blank process.
2. From the `Properties` panel of `Switch` activity, under `Misc` section, update `TypeArgument` to `String`.
3. Select `Ask when run` from `Expression`.

4. Next, click `Add new case` link and enter the value of the case as `In Progress`.
5. Add a `Write Line` activity to the body of this new case and set `Text` to `Executing In Progress case`.
6. Next, click `Add new case` link and enter the value of the case as `Success`.
7. Add a `Write Line` activity to the body of this new case and set `Text` to `Executing Success case`.
8. Next, click `Add new case` link and enter the value of the case as `Failure`.
9. Add a `Write Line` activity to the body of this new case and set `Text` to `Executing Failure case`.
10. Next, add a `Write Line` activity to the body of `Default` case and set `Text` to `Executing Default case`.

Once you have completed the exercise, the final configuration of the **Switch** activity should resemble Figure 3-47. Once you execute the activities shown in Figure 3-48, the automation will prompt you to either select or enter a case value. Based on your selection, the automation will print the value to the Output panel, as shown in Figure 3-49.

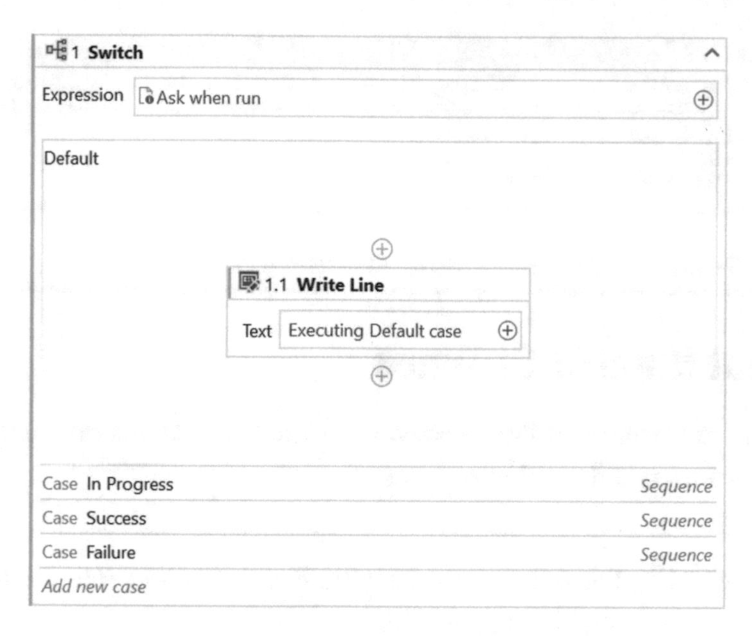

***Figure 3-47.** Configuration of Switch activity exercise*

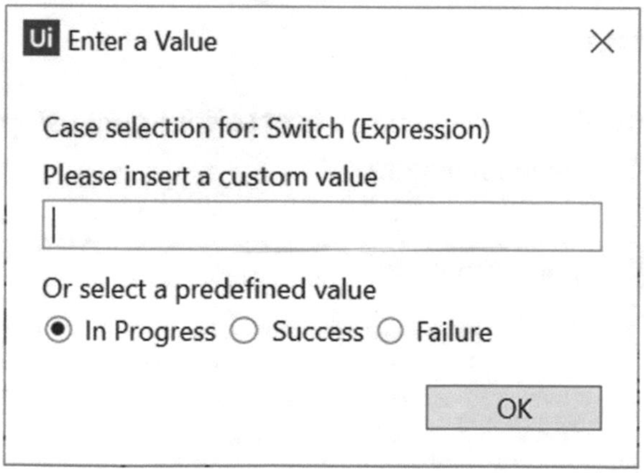

Ui Enter a Value

Case selection for: Switch (Expression)

Please insert a custom value

Or select a predefined value

In Progress Success Failure

OK

***Figure 3-48.** Runtime prompt for the value*

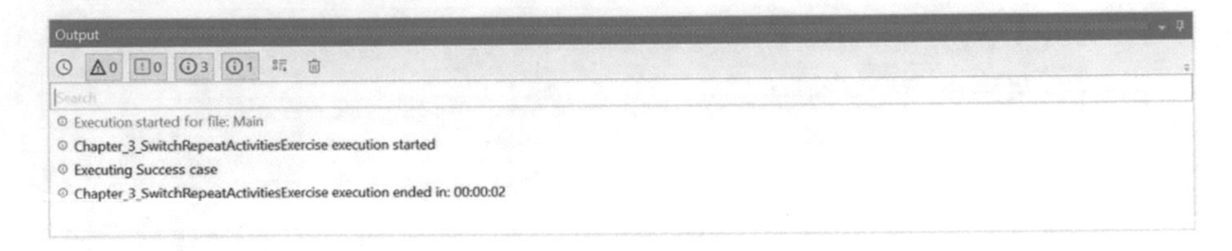

Figure 3-49. *The output of the automation run*

Repeat Number Of Times

The **Repeat Number Of Times** activity allows you to repeat one or more activities a specified number of times.

Tip Typically, this activity will be used in conjunction with the If, Skip Current, and Exit Loop activities.

Configuration

This section provides instructions on how to configure a **Repeat Number Of Times** activity, shown in Figure 3-50.

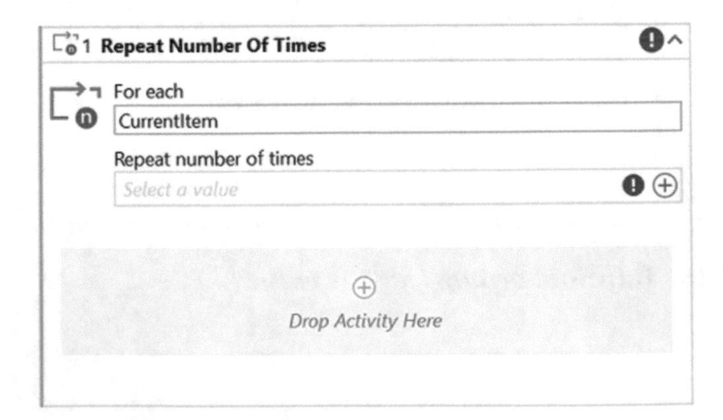

Figure 3-50. *Activity card for Repeat Number Of Times*

Repeat number of times: This is a required configuration available on the activity card. This configuration allows you to specify the number of times this activity should repeat. Commonly, this configuration is specified through the Number builder option.

For each: This is a required configuration available on the activity card. As the activity repeats, the index will keep increasing; for example, when the activity is repeating the second time, the index will be 2. This configuration allows you to name the index so that you can reference it in the body of the activity. You can use this index to decide if you need to skip a specific iteration or simply stop the iterations before reaching the `Repeat Number Of Times` value. By default, the text is set to `CurrentItem` and can be updated to suit your automation; for example, you can rename `CurrentItem` to Vendor or Employee.

EXERCISE

Goal: Use the `Repeat Number Of Times` activity to repeat the Switch activity created in the previous exercise three times. This exercise builds upon the previous exercise for `Switch` activity.

Source Code: Chapter_3_SwitchRepeatActivitiesExercise

Setup: Here are step-by-step implementation instructions:

1. In `StudioX`, add the `Repeat Number Of Times` activity before the `Switch` activity in the `Designer` panel.
2. In the `Repeat number of times` field, click the `Plus` icon, select the `Number` option, and type `3`.
3. Move the `Switch` activity (from the previous exercise) to the body of `Repeat Number Of Times` activity.

Once you have completed the exercise, the final configuration of the **Repeat Number Of Times** activity should resemble Figure 3-51.

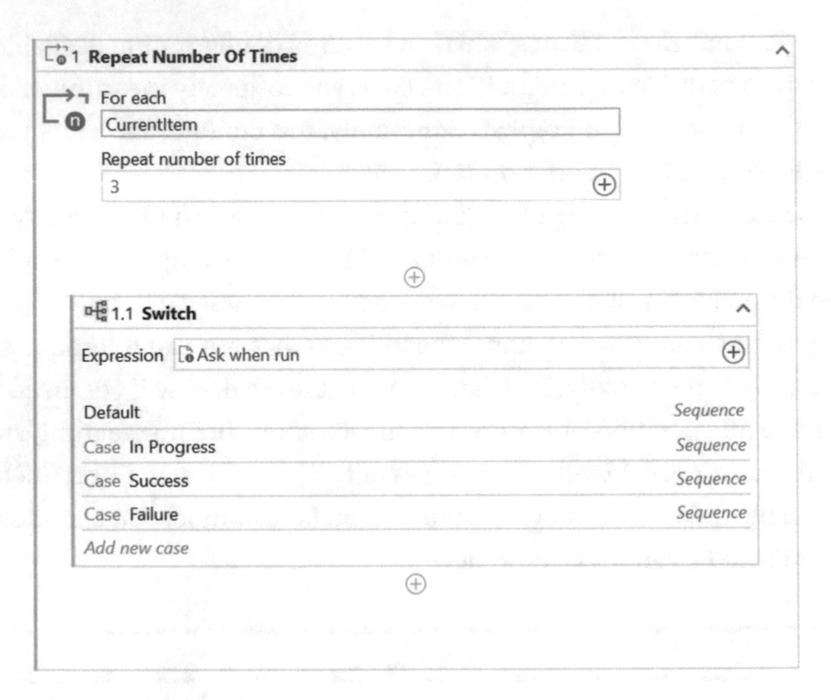

Figure 3-51. *Configuration of Repeat Number Of Times activity exercise*

Once you execute the activities shown in Figure 3-51, the automation will prompt you three times to either select or enter a case value, shown in Figure 3-48. Based on your selection, the automation will print the value to the Output panel, as shown in Figure 3-52.

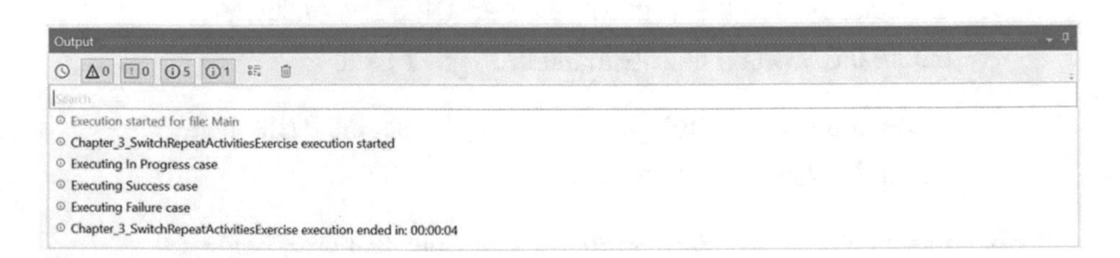

Figure 3-52. *The output from the automation run*

Skip Current

The **Skip Current** activity allows you to skip the current iteration in a repeating activity.

This activity is useful when you are iterating through a list of items using a repeating activity. Based on certain conditions, you may only want to process some items while ignoring the rest; then, you can use the `Skip Current` activity to skip the current iteration.

Configuration

The **Skip Current** activity, shown in Figure 3-53, does not have any configurations. The only requirement for this activity is that it must be used inside a repeating activity such as the **Repeat Number Of Times** activity or the **For Each** activity.

1.1 **Skip Current**

Figure 3-53. *Activity card for Skip Current*

EXERCISE

Goal: Use the `Skip Current` activity to skip the current iteration if the status is In Progress. This exercise builds upon the previous exercises for `Switch` and `Repeat Number Of Times` activities.

Source Code: Chapter_3_SwitchRepeatActivitiesExercise

Setup: Here are step-by-step implementation instructions:

1. In `StudioX`, add the `Skip Current` activity to the body `In Progress` case before the `Write Line` activity. This `Write Line` activity will not execute.

Once you have completed the exercise, the final configuration of the **Skip Current** activity should resemble Figure 3-54.

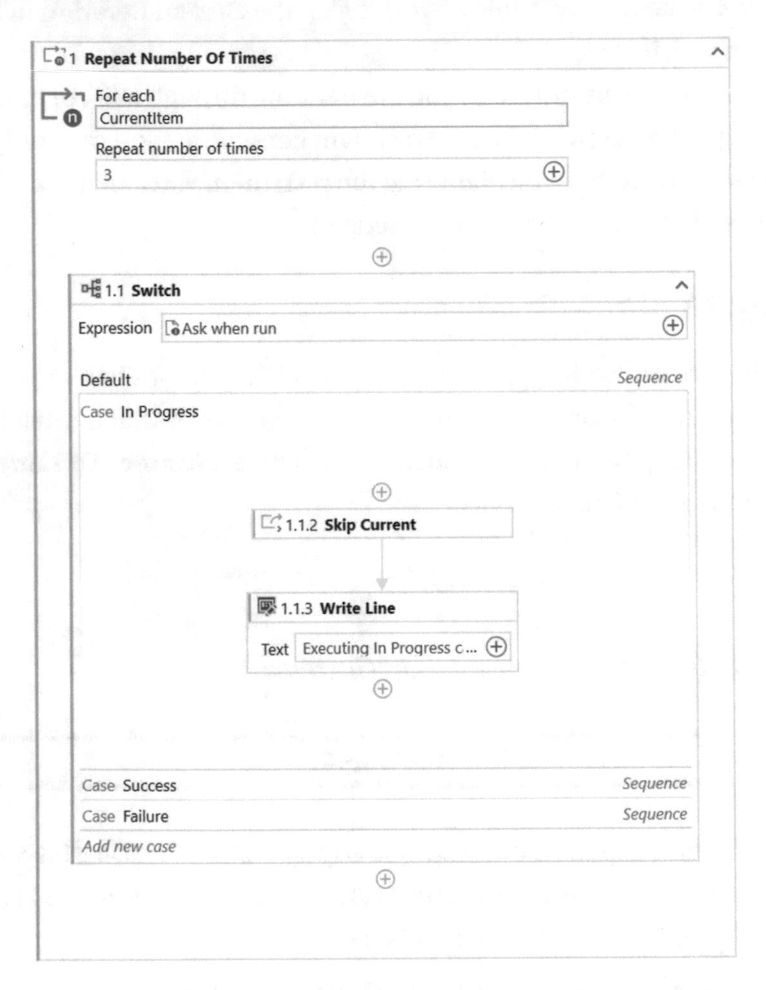

***Figure 3-54.** Configuration of Skip Current activity exercise*

Once you execute the activities shown in Figure 3-54, the automation will prompt you three times to either select or enter a case value, shown in Figure 3-48. Based on your selection, automation will print the values to the Output panel, as shown in Figure 3-55. If you select `Success`, then `In Progress`, and then `Failure`, you will see that `In Progress` was skipped and hence was not printed in the Output panel.

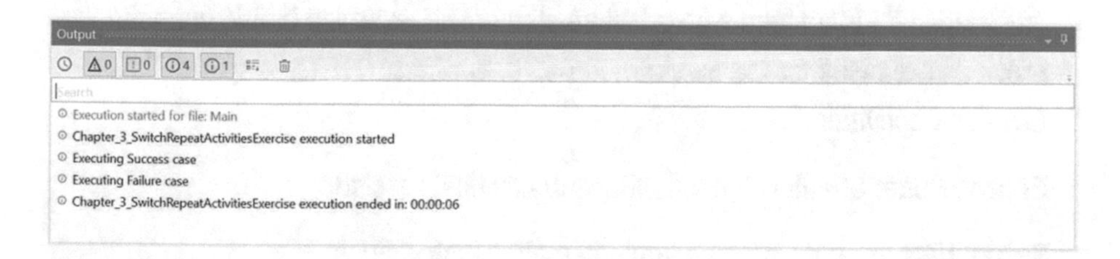

Figure 3-55. *The output of the automation run*

Exit Loop

The **Exit Loop** activity allows you to terminate the repeating activity.

This activity is useful when you are iterating through a list of items using a repeating activity. Based on certain conditions, if you want to terminate the repeating activity, then you can use `Exit Loop` activity.

Configuration

The **Exit Loop** activity, shown in Figure 3-56, does not have any configurations. The only requirement for this activity is that it must be used inside a repeating activity such as the **Repeat Number Of Times** activity or the **For Each** activity.

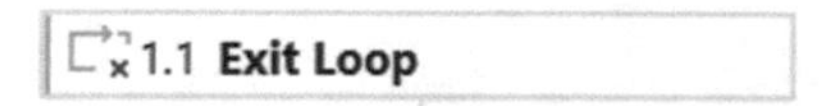

Figure 3-56. *Activity card for Exit Loop*

EXERCISE

Goal: Use the `Exit Loop` activity to terminate the loop as soon as the automation finds an item with a Failure status. This exercise builds upon the previous exercises for the `Switch`, `Repeat Number Of Times`, and `Skip Current` activities.

Source Code: Chapter_3_SwitchRepeatActivitiesExercise

Setup: Here are step-by-step implementation instructions:

1. In `StudioX`, add the `Exit Loop` activity to the body `Failure` case before the `Write Line` activity. This `Write Line` activity will not execute.

Once you have completed the exercise, the final configuration of the **Exit Loop** activity should resemble Figure 3-57.

Once you execute the activities shown in Figure 3-57, the automation will prompt you three times to either select or enter a case value, shown in Figure 3-48. Based on your selection, automation will print the values to the `Output` panel, as shown in Figure 3-58. If you select `Success` and then `Failure`, you will notice that the automation only printed one line to the `Output` panel, and you were never prompted for the third run.

Figure 3-57. Configuration of Exit Loop activity exercise

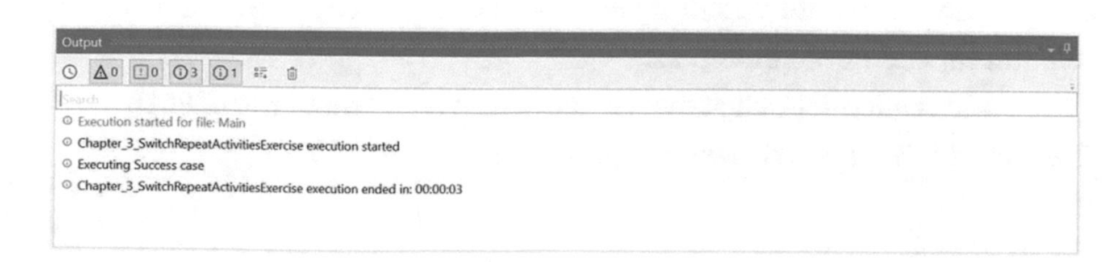

Figure 3-58. The output of the automation run

Get Username/Password

The **Get Username/Password** activity allows you to retrieve credentials for a specific application from the Windows Credentials Manager.

Note To log in to an application, utilize the Get Username/Password activity in conjunction with the Type Into activity to type in the Username and Password saved from this activity execution.

Configuration

This section provides instructions on how to configure a **Get Username/Password** activity, shown in Figure 3-59.

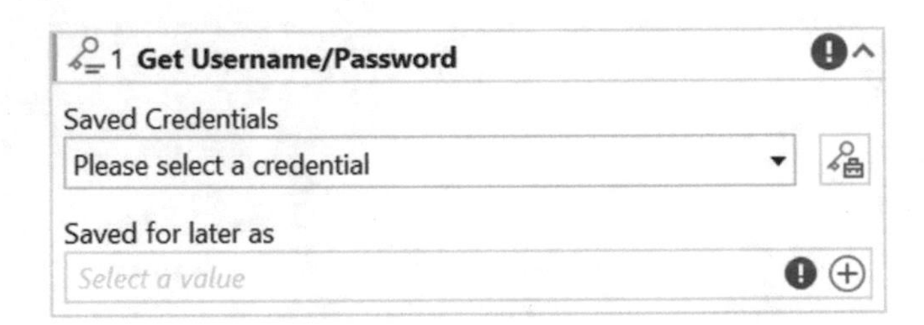

Figure 3-59. *Activity card for Get Username/Password*

Saved Credentials: This is a required configuration available on the activity card. This configuration allows you to select existing credentials or, if required, add new credentials.

To add credentials, click the `Key` icon, and it will open the `UiPath Credential Manager`, shown in Figure 3-60.

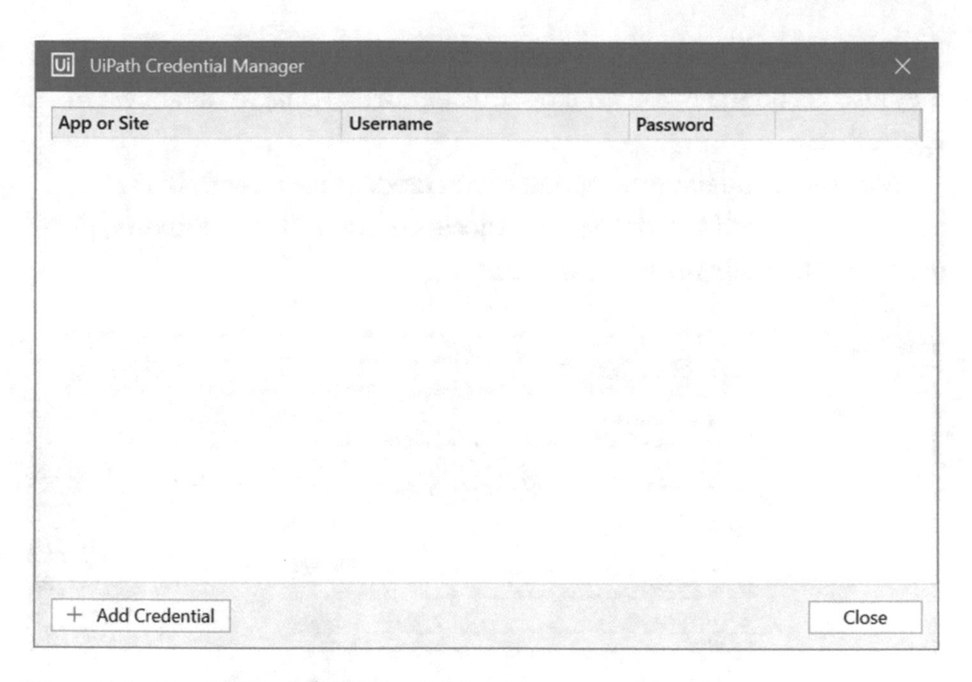

Figure 3-60. *UiPath Credential Manager*

Next, click + `Add Credential`, and enter credentials as shown in Figure 3-61.

Figure 3-61. *Add Credentials dialog*

Once you have added new credentials in `UiPath Credential Manager`, they also get added to the `Windows Credentials Manager`, as shown in Figure 3-62.

Moreover, you have the option to add credentials directly in the `Windows Credentials Manager` and access them in UiPath StudioX; just make sure to prefix the name with `UiPath`.

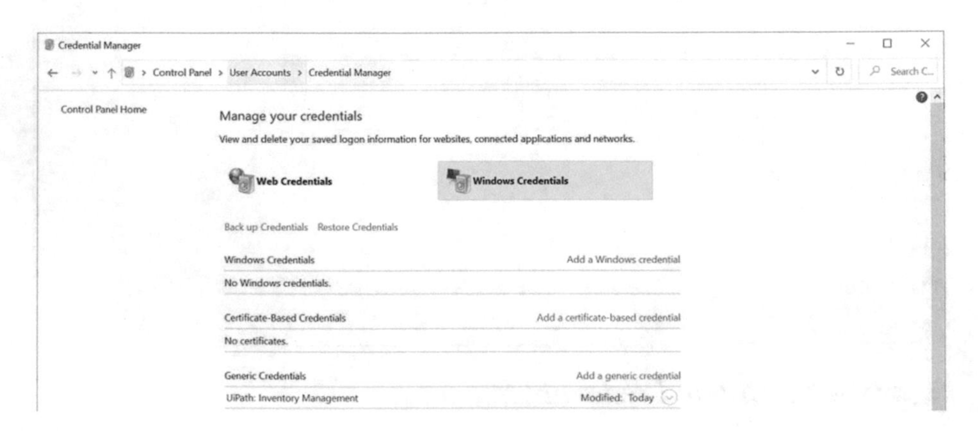

Figure 3-62. *Windows Credentials Manager*

Saved for later as: This is a required configuration available on the activity card. This configuration allows you to store credentials in a value for later use in the automation. You can later access both the `Username` and `Password` from the saved value, as demonstrated in the inventory management example in Figure 3-63.

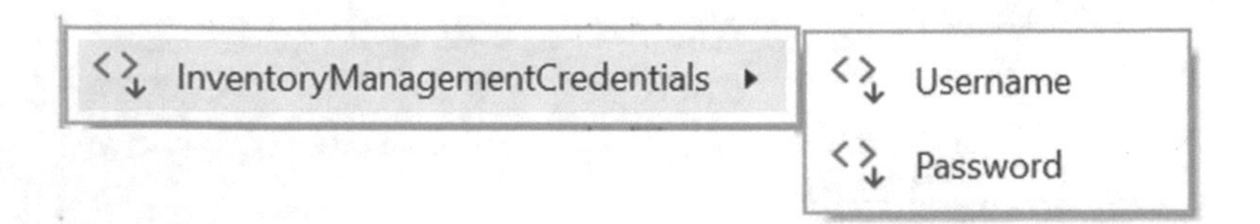

Figure 3-63. *Accessing username and password example*

Get Orchestrator Asset

The **Get Orchestrator Asset** activity allows you to retrieve credentials and other assets from Orchestrator for use in automation.

Tip For consistency and reusability, values that are shared across multiple automations should be stored in Orchestrator.

Configuration

This section provides instructions on how to configure a **Get Orchestrator Asset** activity, shown in Figure 3-64.

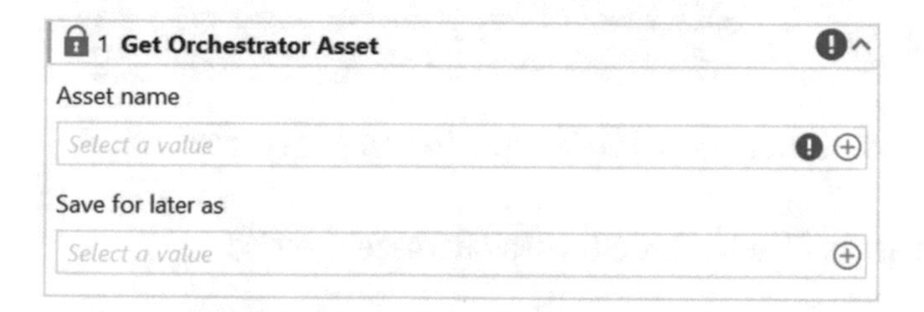

Figure 3-64. *Activity card for Get Orchestrator Asset*

Asset name: This is a required configuration available on the activity card. This configuration allows you to specify the asset that you want to retrieve from Orchestrator.

Orchestrator folder path: This is a required configuration available in the `Properties` panel. This configuration allows you to specify the folder path of the asset in Orchestrator.

Save for later as: This is an optional configuration available on the activity card. This configuration allows you to save the asset returned from Orchestrator for later use in automation. You can later reference the asset from the Use Saved Value menu.

EXERCISE

Goal: Use the `Get Orchestrator Asset` activity to retrieve a `DocumentPath` asset from Orchestrator, shown in Figure 3-65.

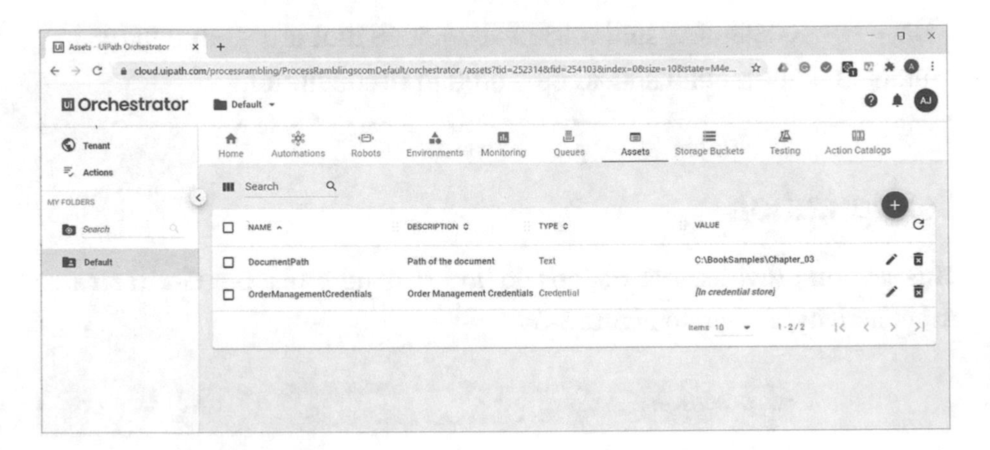

Figure 3-65. DocumentPath asset in Orchestrator

Source Code: Chapter_3-GetOrchestratorAssetExercise

Setup: Here are step-by-step implementation instructions:

1. In `StudioX`, add the `Get Orchestrator Asset` activity to a blank process.

2. In the `Asset name` field, click the `Plus` icon, select the `Text` option, and type `DocumentPath`.

3. In the `Orchestrator folder path` field in `Properties` panel, click the `Plus` icon, select the `Text` option, and type `/`. This specifies that we are looking for the asset in the parent folder.

4. Click the `Save for later as` field, click the `Plus` icon, and select `Save for Later Use`. Enter `OrchestratorAssetValue` as the value name.

5. Next, add the `Write Line` activity after the `Get Orchestrator Asset` activity.
6. In the `Text` field, click the `Plus` icon, and hover over `Use Saved Value` to select `OrchestratorAssetValue`.

Once you have completed the exercise, the final configuration of the **Get Orchestrator Asset** activity should resemble Figure 3-66.

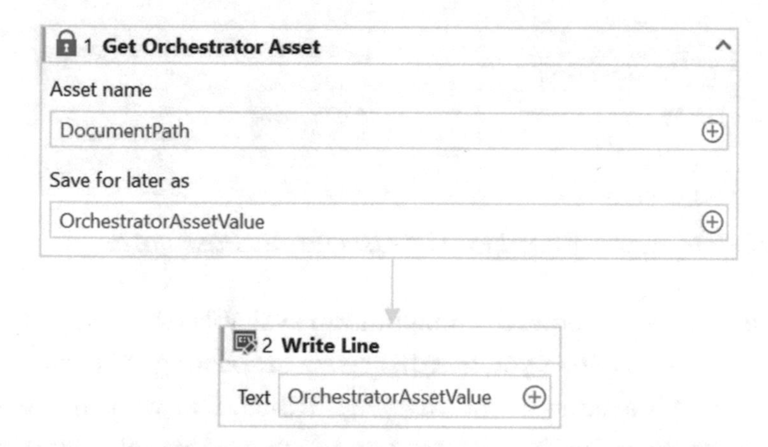

***Figure 3-66.** Configuration of Get Orchestrator Asset activity exercise*

Once you execute the activities shown in Figure 3-66, the automation will retrieve the asset from Orchestrator and print the value to the Output panel, as shown in Figure 3-67.

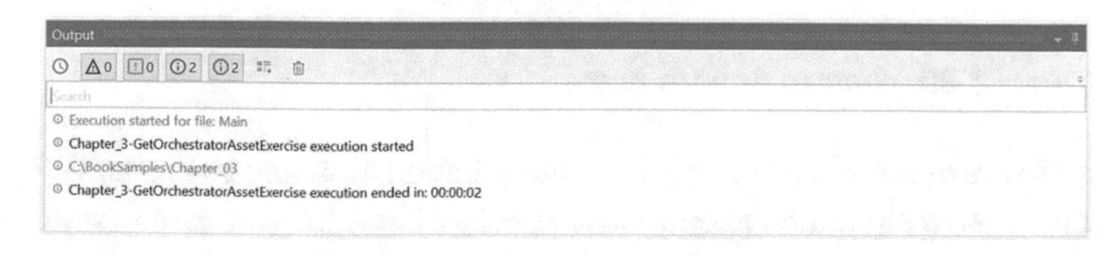

***Figure 3-67.** The output of the automation run*

Save For Later

The **Save For Later** activity allows you to save any value for use later.

Configuration

This section provides instructions on how to configure a **Save For Later** activity, shown in Figure 3-68.

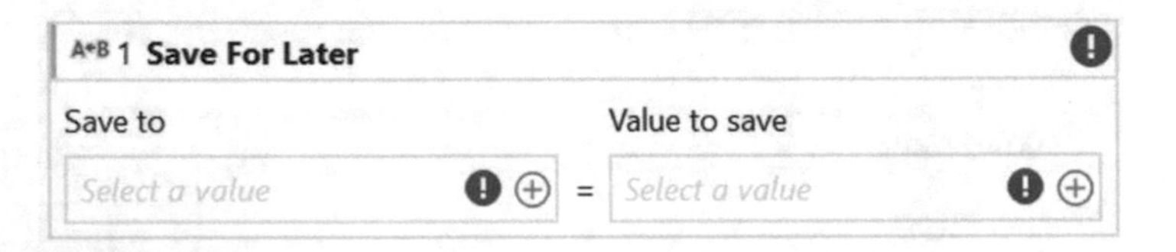

Figure 3-68. *Activity card for Get Username/Password*

Save to: This is a required configuration available on the activity card. This configuration allows you to either select an existing value name or create a new value name, as shown in Figure 3-69. This value name will start showing up in the `Data Manager` panel. To use this value later in an activity, click the `Plus` icon and select it from the `Use Saved Value` menu.

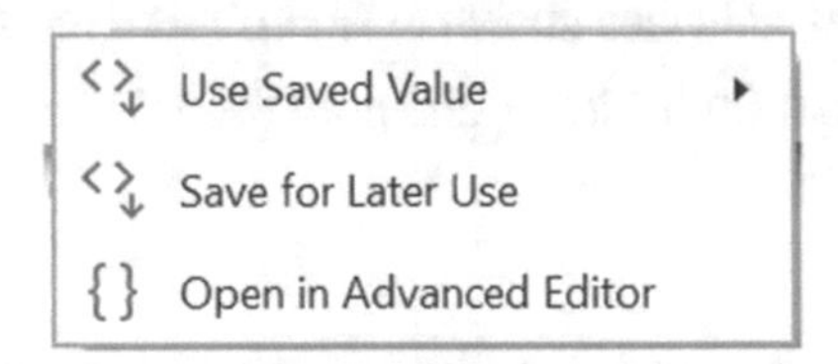

Figure 3-69. *Save to options menu*

Value to save: This is a required configuration available on the activity card. This configuration allows you to specify what value you want to save. Figure 3-70 shows different options available to specify the value.

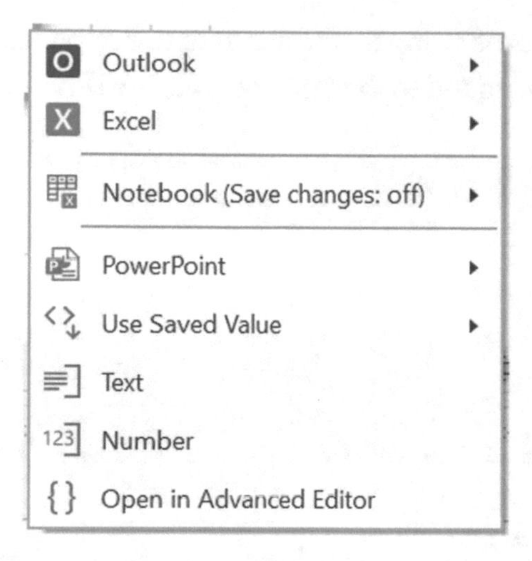

Figure 3-70. *Options to specify value to save*

Wait for Download

The **Wait for Download** activity allows you to monitor and wait for a file download to complete before proceeding.

Configuration

This section provides instructions on how to configure a **Wait for Download** activity, shown in Figure 3-71.

Any activity that is going to interact with the application to start file download must be added to the body of **Wait for Download** activity.

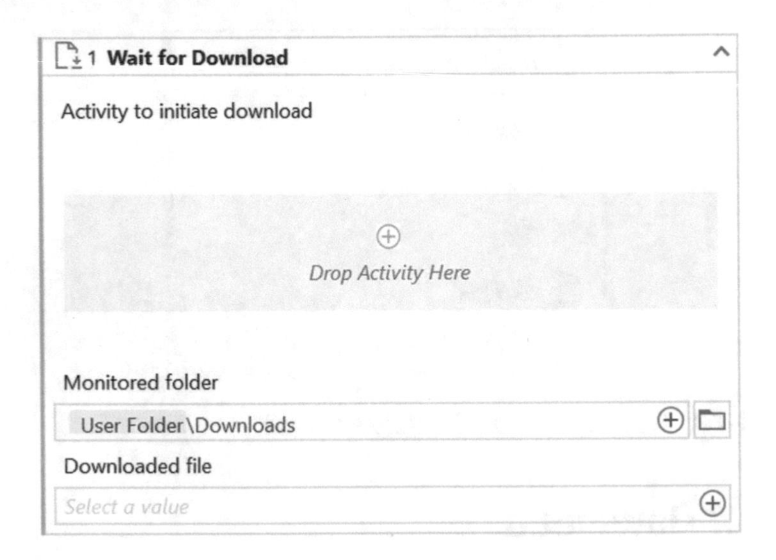

Figure 3-71. *Activity card for Wait for Download*

Monitored folder: This is a required configuration available on the activity card. This configuration allows you to specify a folder that the automation will monitor for the file download. By default, the value is `User Folder\Downloads`, that is, the automation monitors the default `Downloads` folder of your Windows profile or browser.

Downloaded file: This is an optional configuration available on the activity card. This configuration allows you to store downloaded file details for later use. For example, you could use the file path to copy it to another folder for processing.

Timeout: This is an optional configuration available on the `Properties` panel. This configuration allows you to specify how long (in seconds) should the activity wait for file download to complete before proceeding. By default, the timeout value is set to 300. This should be increased for files with larger size.

There is no exercise for this activity in this chapter. We will use the `Wait for Download` activity in the exercise for the `Hover` activity in Chapter 4. Figure 3-72 shows how this activity looks like when configured.

***Figure 3-72.** Configuration for Wait for Download activity*

Group

The **Group** activity allows you to logically group a set of activities. All the activities inside the body of the Group are executed in the same sequence. This is just a great way to logically split a large automation project.

Configuration

The **Group** activity, shown in Figure 3-73, does not have any configurations.

Figure 3-73. *Activity card for Group*

CHAPTER 4

UI Automation

Interacting with several desktop and web applications is an essential part of our daily work routine. We use these applications to accomplish tasks like logging timesheets in a time tracking system, storing, organizing, and retrieving documents in a document management system, or tracking projects in a project management software. To interact with these various applications, we use their user interface (UI). Similarly, UiPath StudioX UI automation interacts with the user interfaces to automate desired tasks by replicating the user interaction like checking boxes, selecting dropdowns, entering data, and clicking buttons.

Learning Objectives

At the end of this chapter, you will learn how to

- Load web and desktop apps
- Enter data in forms
- Extract data from forms and tables
- Perform miscellaneous activities to interact with user interfaces
- Use the App/Web Recorder to generate an automation

A. Javed et al., *Robotic Process Automation using UiPath StudioX*,
https://doi.org/10.1007/978-1-4842-6794-3_4

Sample Overview

The sample application used for all exercises of this chapter is Contacts Management. The application is accessible via a browser from `https://therpabook.com/samples/contactsmanagement/` location. This section provides a quick overview of the sample application.

Contacts List: This is the default page that loads when you open this web application. As shown in Figure 4-1, this page contains the following sections:

1. Header: To show the title of the application
2. Buttons: To open the add contact details dialog, view contact details dialog, and download to Excel and CSV
3. Table: With a list of contacts spanning multiple pages

Add Contact: This dialog box, shown in Figure 4-2, is displayed when you click the Add Contact button from the home page. This screen allows you to enter new contact details. The application is not operational, so this does not persist with any new contacts.

View Contact: This dialog box, shown in Figure 4-3, is displayed when you click the View Contact button from the home page. This screen allows you to view contact details. The application is not operational, so this always shows the same contact information.

Download: This button allows you to download the contacts list in Excel and CSV formats.

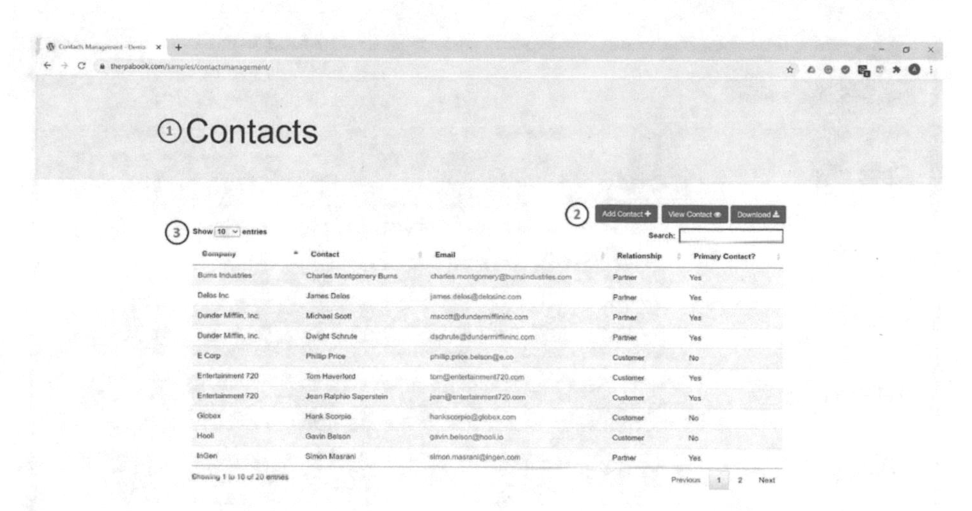

***Figure 4-1.** Contacts Management web application home page*

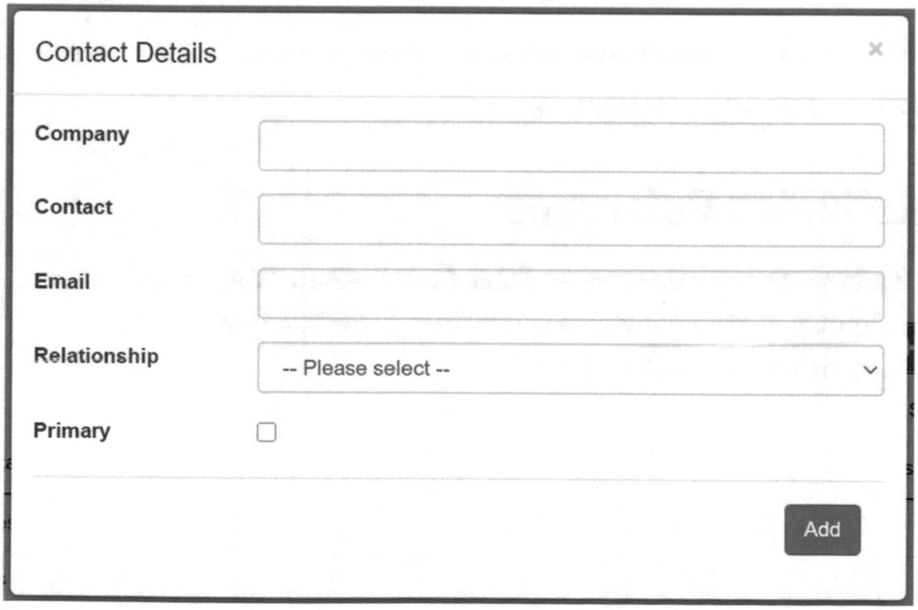

Contact Details ×

Company

Contact

Email

Relationship -- Please select --

Primary

Add

***Figure 4-2.** Add Contact dialog*

Figure 4-3. *View Contact dialog*

Activities Reference

UI automation activities are available from the App/Web category shown in Figure 4-4. The following sections provide instructions on how to configure and use each activity.

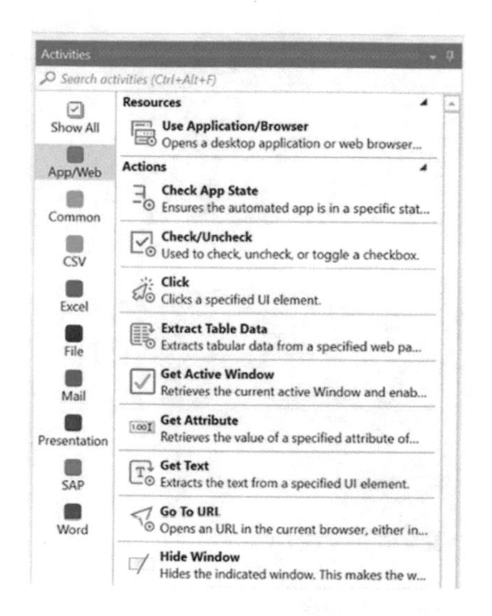

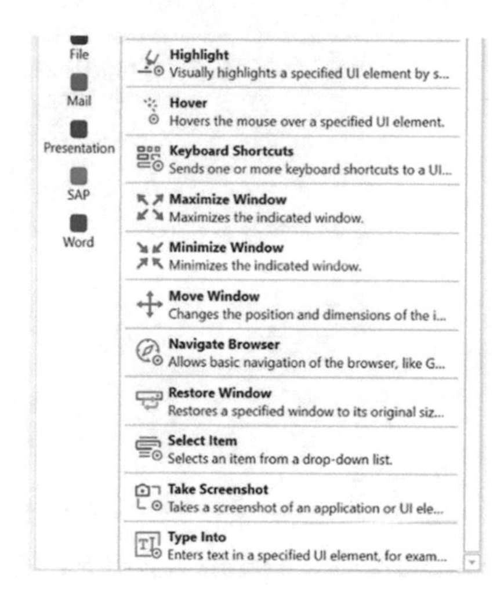

Figure 4-4. *Activities for App/Web automation*

Use Application/Browser

The **Use Application/Browser** activity allows you to open, close, and interact with a web or a desktop application.

Note The **Use Application/Browser** activity will contain all the actions that you want to perform on the UI of a target application. For example, if you want to click a button on the screen, the Click activity will have to be added to the body of this activity.

Configuration

This section provides instructions on how to configure a, shown in Figure 4-5.

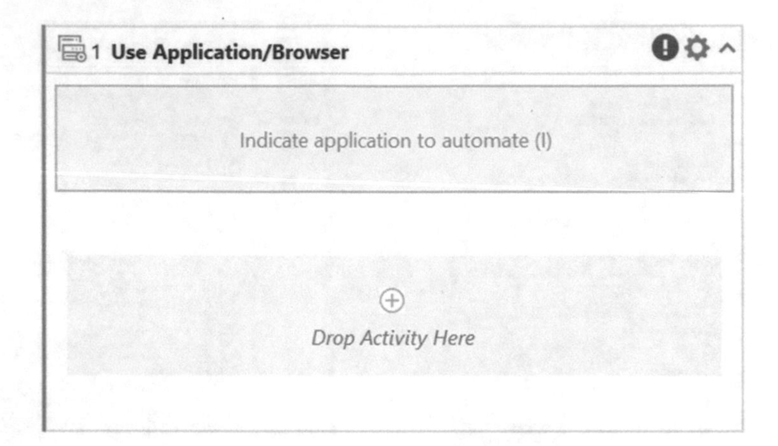

Figure 4-5. *Activity card for Use Application/Browser*

Note Before you can automate a web application, you will need to install the relevant web browser extension from Home ➤ `Tools` ➤ `UiPath Extensions`.

Indicate application to automate (I): This is a required configuration available on the activity card. This configuration allows you to indicate the target application that you are going to automate. To configure this activity, make sure your target application is already open. Once you have indicated your target application, StudioX will automatically identify if it is a desktop application or a web application. Figure 4-6 shows the activity card for a web application, while Figure 4-7 shows the activity card for a desktop application.

Browser URL: This is a required configuration available on the activity card. This configuration is only available when the target is a web application. This configuration allows you to specify the URL of the web application. By default, this is populated with the browser URL identified in the indicate application configuration.

Application path: This is a required configuration available on the activity card. This configuration is only available when the target is a desktop application. This configuration allows you to specify the complete path of the executable file on a local system. By default, this is populated with the application path identified in the indicate application configuration.

Application arguments: This is an optional configuration available on the activity card. This configuration is only available when the target is a desktop application. This configuration allows you to specify arguments for the application to execute. For example, when launching Notepad, you can specify what file to load when Notepad opens.

Match exact title: This is an optional configuration available on the activity card. This configuration is only available when the target is a desktop application. This configuration allows you to specify if the automation should exactly match the application title. By default, this option is not checked, that is, automation will not use exact title matching.

Open: This is an optional configuration available on the `Properties` panel. This configuration allows you to specify if the automation should open this application when executing. There are three options: `Never` (never open the application, this is useful when the application might already be open), `Always` (always open the application), and `IfNotOpen` (only open the application if it is not already open).

Close: This is an optional configuration available on the `Properties` panel. This configuration allows you to specify if the automation should close the target application after completion. There are three options: `Never` (never close the application, this is useful when a person or another program might have already opened the application), `Always` (always close the application), and `IfOpenedByAppBrowser` (only close the application if the automation opened it).

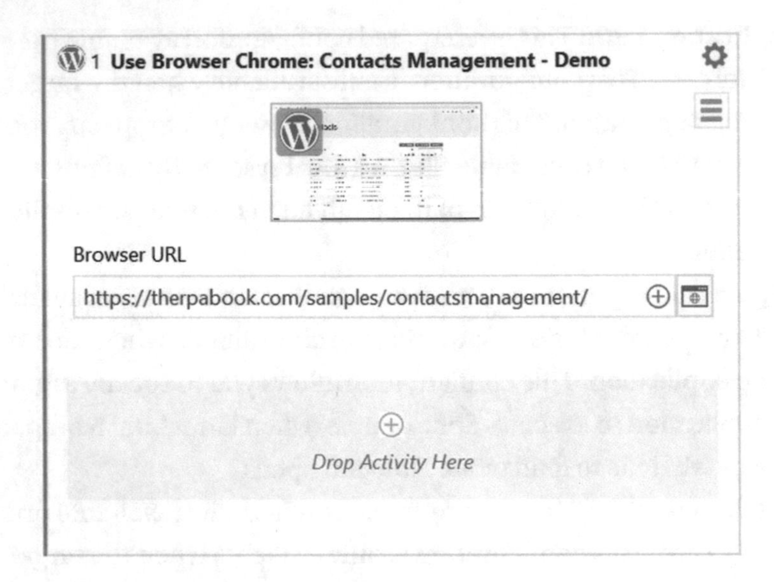

Figure 4-6. *Activity card for Use Browser*

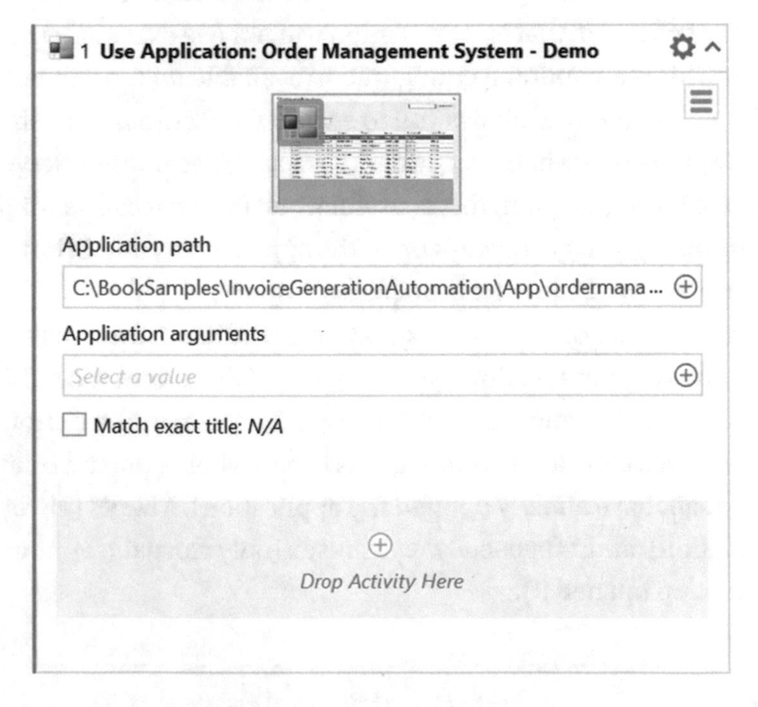

Figure 4-7. *Activity card for Use Application*

Go To URL

The **Go To URL** activity allows you to open a specified URL in a browser.

Configuration

This section provides instructions on how to configure a **Go To URL** activity, shown in Figure 4-8.

Figure 4-8. *Activity card for Go To URL*

URL: This is a required configuration available on the activity card. This configuration allows you to specify the URL that you want to open in the browser.

EXERCISE

Goal: Use the `Go To URL` activity to navigate to `https://therpabook.com/samples/contactsmanagement/`.

Source Code: Chapter_4_MiscellaneousActivitiesExercise

Setup: Here are step-by-step implementation instructions:

1. Open a browser of your choice and enter `about:blank` in the address bar. This will ensure that by default, no page is loaded.
2. In `StudioX`, add the `Use Application/Browser` activity to a blank process.

3. Next, click `Indicate application` in the `Use Application/Browser` activity card. You will notice a blue shade over the screen; this is to point your mouse to the browser to select the window. Clicking the browser will automatically populate the activity card.
4. Next, select the `Use Application/Browser` activity card, and from `Properties`, set the `Options` ➤ `Open` property to `Always`. This will ensure that the automation always opens a new browser.
5. Next, add a `Go To URL` activity in the body of `Use Application/Browser` activity.
6. Next, in the `URL` field of `Go To URL` activity, click the `Plus` icon, select the `Text` option, and type `https://therpabook.com/samples/contactsmanagement/`. Click `Save`.

Once you have completed the exercise, the final configuration of the **Go To URL** activity should resemble Figure 4-9. Figure 4-10 shows the state of the target web application once the automation has completed its run. In this case, the browser has been redirected to a new URL.

Figure 4-9. Final configuration of the Go To URL activity exercise

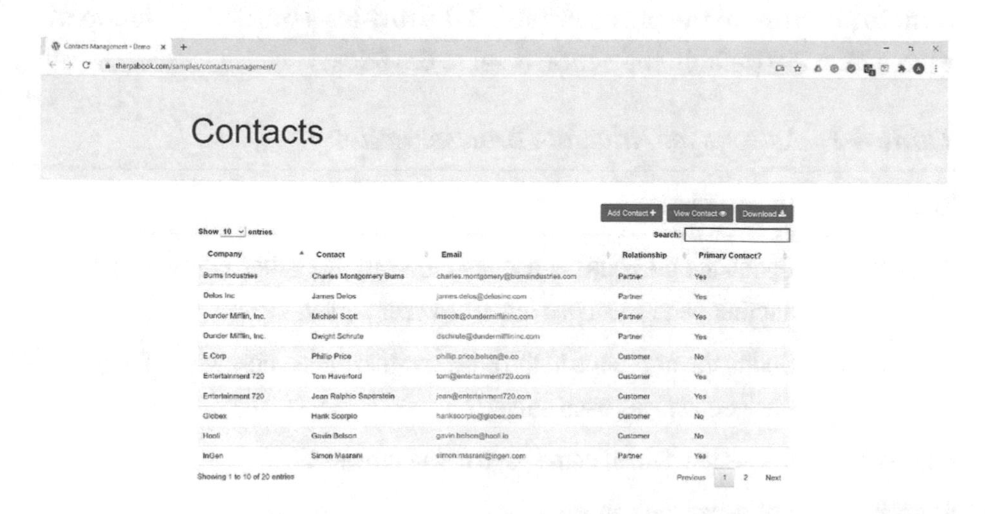

Figure 4-10. Result of the Go To URL activity exercise

Navigate Browser

The **Navigate Browser** activity allows you to perform basic browser navigations such as Go Back, Go Forward, Go Home, Refresh, and Close Tab.

Configuration

This section provides instructions on how to configure a **Navigate Browser** activity, shown in Figure 4-11.

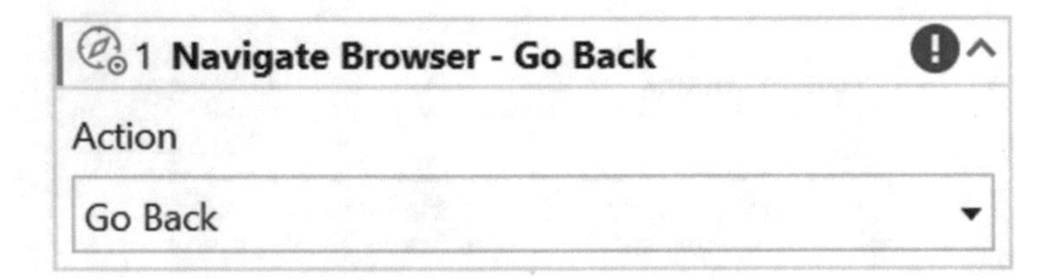

Figure 4-11. *Activity card for Navigate Browser*

Action: This is a required configuration available on the activity card. This configuration allows you to specify the navigation action that you want to perform on the browser. Table 4-1 provides a quick description of each action. By default, the action is set to `Go Back`.

Table 4-1. *Actions for Navigate Browser activity*

Action	Description
Go Back	Opens the previous page in the browser; this action only works if another page was open in the browser earlier.
Go Forward	Opens the next page in the browser; this action only works if the Go Back action was used earlier.
Go Home	Opens the default home page of the browser.
Refresh	Refreshes the page.
Close Tab	Closes the currently active tab.

EXERCISE

Goal: Use the `Navigate Browser` activity to refresh the already open web page. This exercise builds upon the `Go To URL` exercise. Figure 4-10 shows the state of the target web application before this exercise.

Source Code: Chapter_4_MiscellaneousActivitiesExercise

Setup: Here are step-by-step implementation instructions:

1. In `StudioX`, add the `Navigate Browser` activity in the body of `Use Application/Browser` activity after the `Go To URL` activity.
2. Next, from the `Action` dropdown, select `Refresh`.

Once you have completed the exercise, the final configuration of the **Navigate Browser** activity should resemble Figure 4-12.

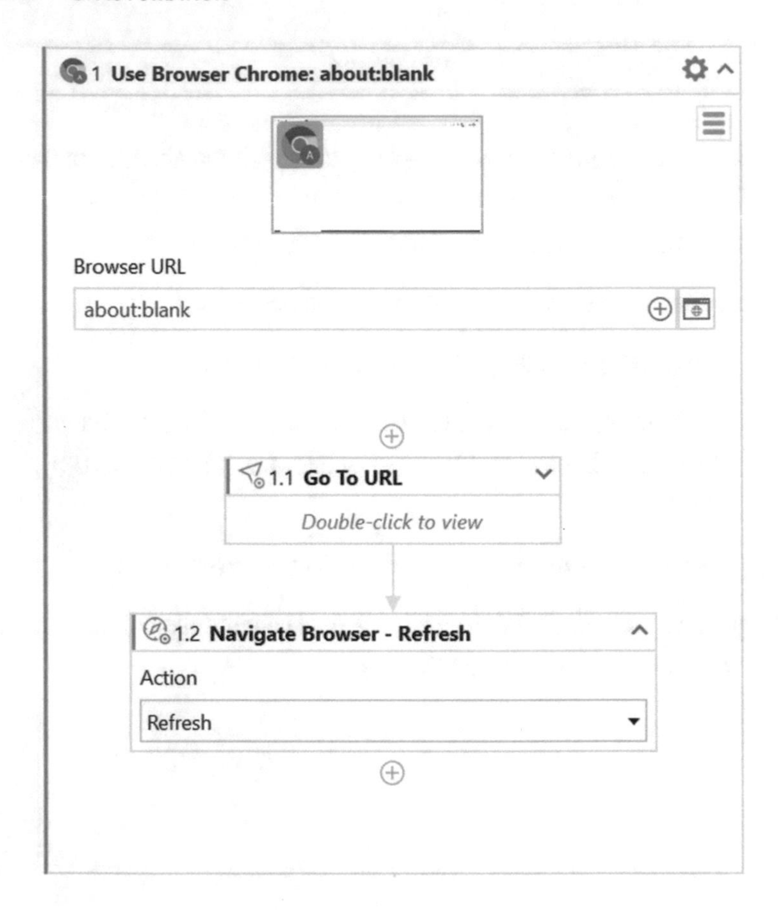

Figure 4-12. *Final configuration of the Navigate Browser activity exercise*

Highlight

The **Highlight** activity allows you to visually highlight a specified area while the automation is running. The automation creates a box around the specified element.

Tip For tasks where you want to visually emphasize specific UI elements while a process is executing, the Highlight activity is a great way to support such requirements.

Configuration

This section provides instructions on how to configure a **Highlight** activity, shown in Figure 4-13.

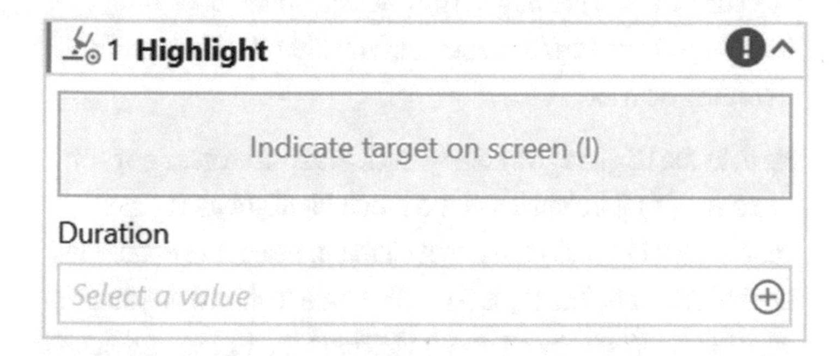

Figure 4-13. *Activity card for Highlight*

Indicate target on screen (I): This is a required configuration available on the activity card. This configuration allows you to specify the element or area on UI that you want to highlight.

Duration: This is an optional configuration available on the activity card. This configuration allows you to specify (in seconds) how long the automation should highlight the specified element.

Color: This is an optional configuration available on the `Properties` panel. This configuration allows you to specify the color of the highlight box and is set to gold by default.

EXERCISE

Goal: Use the `Highlight` activity to highlight the number of records in the Contacts table visually. This exercise builds upon the `Go To URL` and `Navigate Browser` exercises. Figure 4-10 shows the state of the target web application before this exercise.

Source Code: Chapter_4_MiscellaneousActivitiesExercise

Setup: Here are step-by-step implementation instructions:

1. In `StudioX`, add the `Highlight` activity in the body of `Use Application/Browser` activity after the `Navigate Browser` activity.

2. Next, in the `Highlight` activity, click `Indicate target on screen (I)` link. You'll notice a green highlight as you point your mouse to specify the target element; select the bottom-left area of the table. Then, you'll notice a blue highlight as you point your mouse to specify an anchor; indicate the `Previous` button. At this point, your selection should resemble Figure 4-14. Click the `Confirm` button.

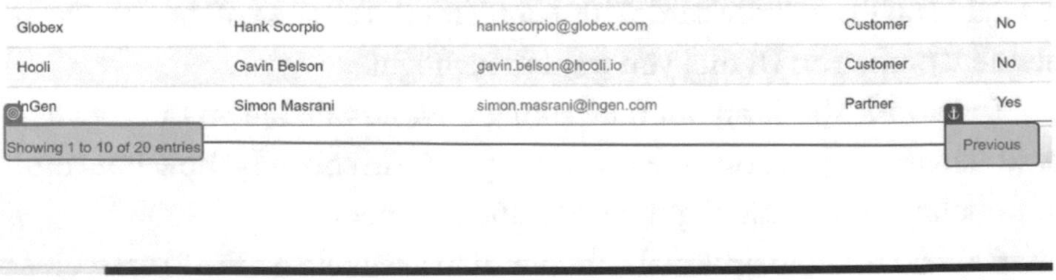

Figure 4-14. *Selection of element to be highlighted*

3. Next, in the `Duration` field of `Highlight` activity, click the `Plus` icon, select the `Number` option, and enter `5` seconds. This will highlight the specified UI element for 5 seconds.

Once you have completed the exercise, the final configuration of the **Highlight** activity should resemble Figure 4-15. Figure 4-16 shows the state of the target web application once the automation has completed its run. In this case, the total number of entries area is highlighted.

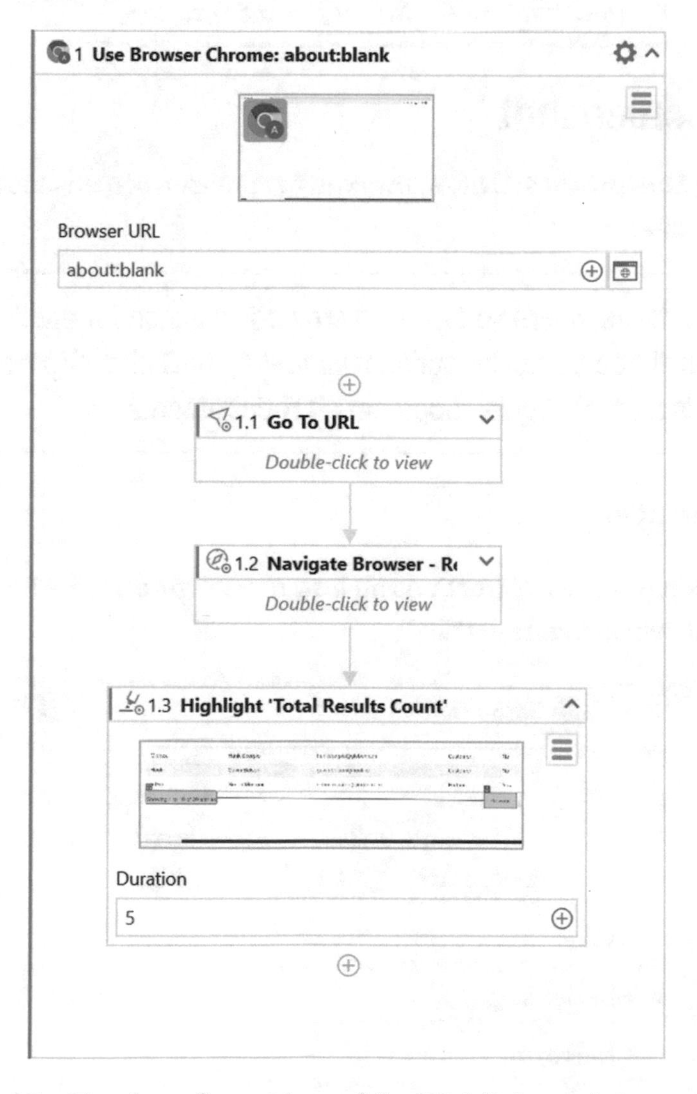

***Figure 4-15.** Final configuration of the Highlight activity exercise*

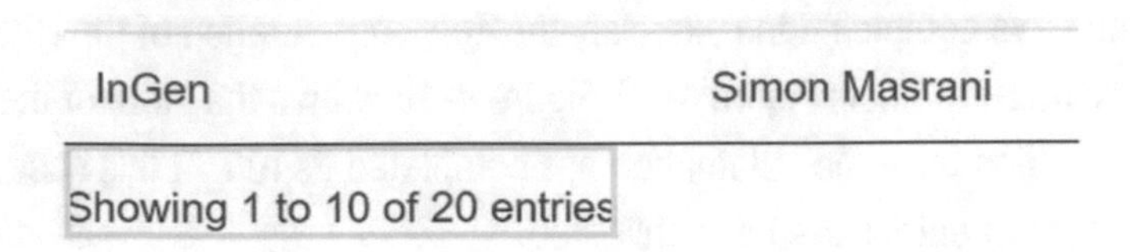

InGen
Simon Masrani
Showing 1 to 10 of 20 entries

Figure 4-16. *Result of the Highlight activity exercise*

Take Screenshot

The **Take Screenshot** activity allows you to capture a screenshot of a specified area.

Tip For tasks where you need to provide evidence for audit purposes that a particular action was taken, the `Take Screenshot` activity is a great way to support such requirements.

Configuration

This section provides instructions on how to configure a **Take Screenshot** activity, shown in Figure 4-17.

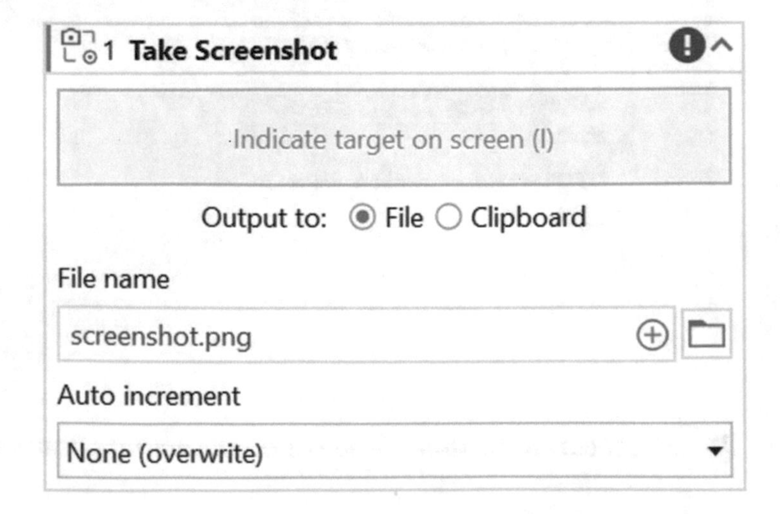

Figure 4-17. *Activity card for Take Screenshot*

Indicate target on screen (I): This is a required configuration available on the activity card. This configuration allows you to specify the element on UI that you want to capture in your screenshot.

Output to: This is a required configuration available on the activity card. This configuration allows you to specify if the screenshot should be physically saved as a file or just saved in the Clipboard memory.

File name: This is a required configuration available on the activity card. This configuration is only available when you select `File` from the `Output to` field. This configuration allows you to specify the folder location where screenshots should be stored and the name of the screenshot.

Auto increment: This is an optional configuration available on the activity card. This configuration is only available when you select `File` from the `Output to` field. This configuration automatically appends an index or a timestamp, shown in Figure 4-18, to the file name only in case the same file already exists in the `Save to` location. This field is helpful in avoiding overwriting over an existing file.

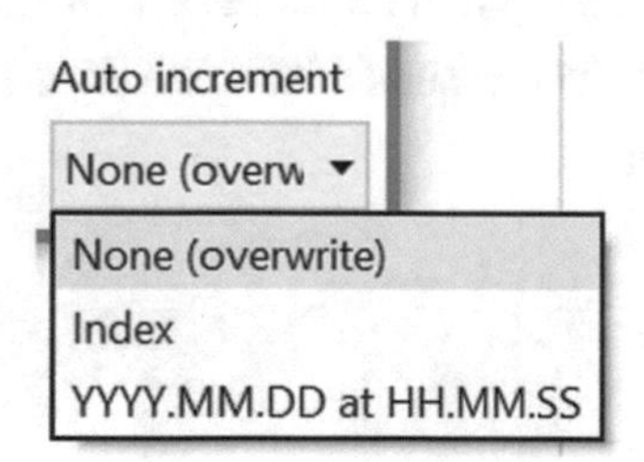

Figure 4-18. *Available options for Auto increment*

EXERCISE

Goal: Use the `Take Screenshot` activity to take a screenshot of the Contacts table. This exercise builds upon the `Go To URL`, `Navigate Browser`, and `Highlight` exercises. Figure 4-10 shows the state of the target web application before this exercise.

Source Code: Chapter_4_MiscellaneousActivitiesExercise

Setup: Here are step-by-step implementation instructions:

1. In `StudioX`, add the `Take Screenshot` activity in the body of `Use Application/Browser` activity after the `Highlight` activity.

2. Next, in the `Take Screenshot` activity, click `Indicate target on screen (I)` link and point your mouse to the contacts list table. Once you have specified the target element, you'll notice a red highlight over the table. This is to prompt you to specify an anchor; in this case, use the page header. At this point, your selection should look like Figure 4-19. Click the `Confirm` button.

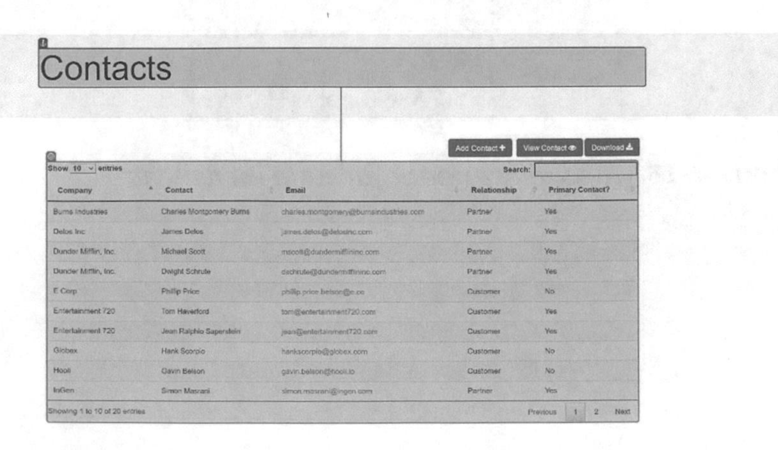

Figure 4-19. *Selection of element to be captured in the screenshot*

3. Next, in the `Save to folder` field of the `Take Screenshot` activity, click the `Browse for folder` icon and select `C:\BookSamples\Chapter_04` folder.

4. The name of the screenshot file will be automatically appended to the folder path. The default name is `screenshot.png`.

5. From the `Auto increment` dropdown, select the `YYYY.MM.DD at HH.MM.SS` option. This will append a timestamp at the end of the screenshot file name.

Once you have completed the exercise, the final configuration of the **Take Screenshot** activity should resemble Figure 4-20. Figure 4-21 shows the folder with multiple screenshots. If you run the automation multiple times, it will keep adding screenshots to the specified folder with a timestamp appended to the name.

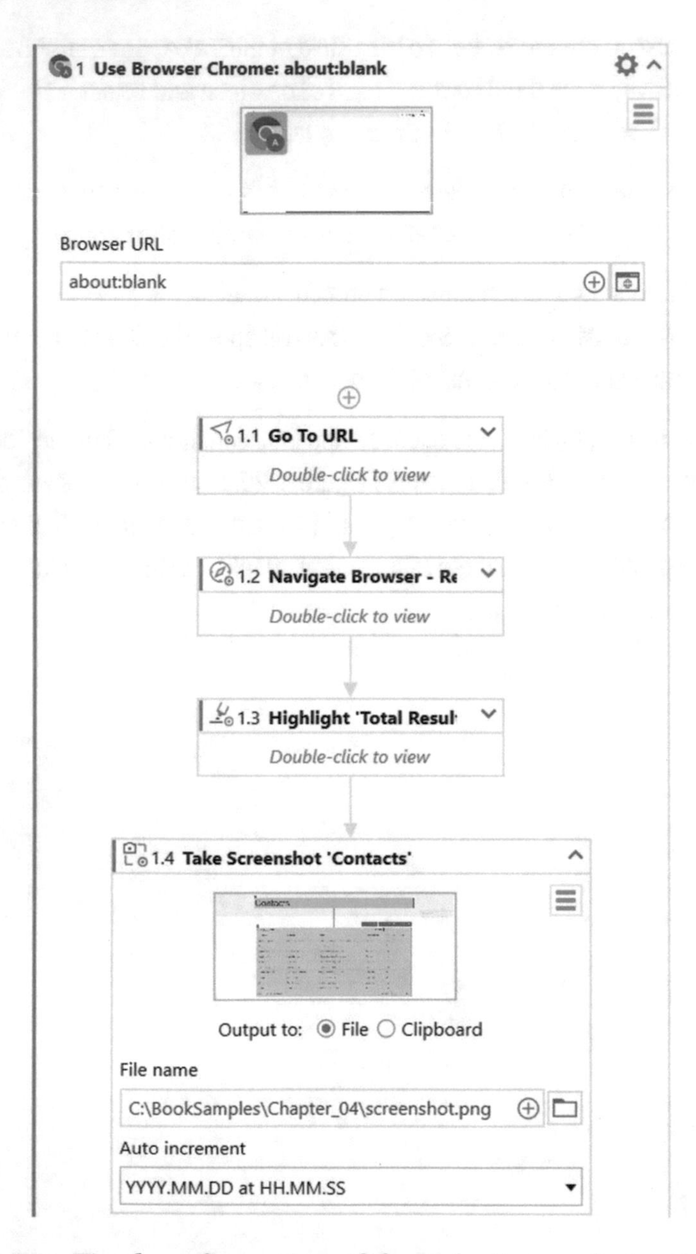

Figure 4-20. *Final configuration of the Take Screenshot activity exercise*

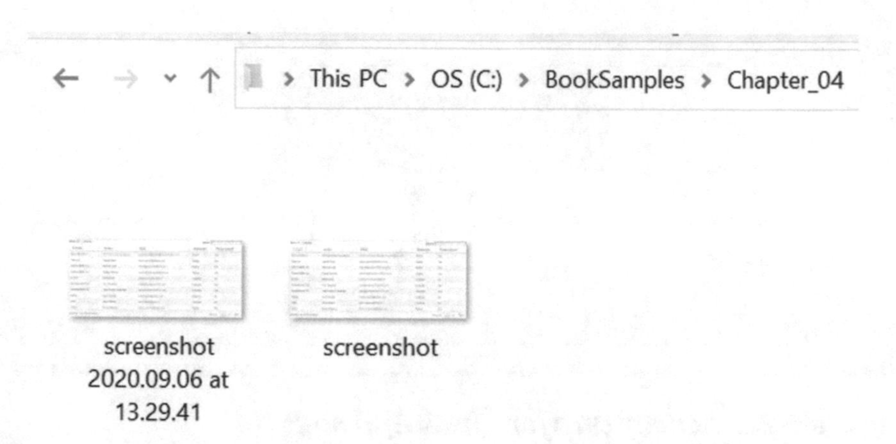

***Figure 4-21.** Result of the Take Screenshot activity exercise*

Check App State

When needing to perform an action based on a particular UI element such as a loading icon on an application, the Check App State activity can be used. We, as humans, understand that this means the screen is still loading, but the automation does not. The **Check App State** activity is a great way to ensure that a screen has completely loaded before the automation starts interacting with it.

Tip For tasks where you need to perform different actions based on UI elements such as a message box or error pop-up window, the Check App State activity is a great way to support such requirements.

Configuration

This section provides instructions on how to configure a **Check App State** activity, shown in Figure 4-22.

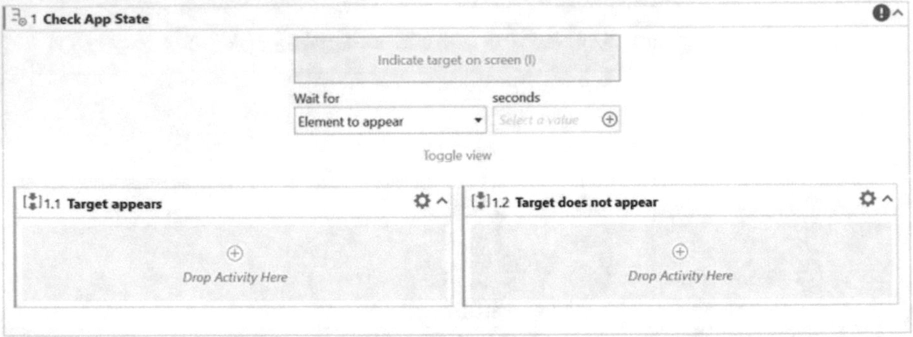

***Figure 4-22.** Activity card for Check App State*

Indicate target on screen (I): This is a required configuration available on the activity card. This configuration allows you to specify a UI element that the automation will wait for either to appear or to disappear before proceeding.

Wait for: This is a required configuration available on the activity card. This configuration allows you to specify if the automation needs to wait for the target element to appear or disappear before proceeding. By default, this is set to `Element to appear`.

Seconds: This is an optional configuration available on the activity card. This configuration allows you to specify the amount of time (in seconds) that the automation needs to wait for the target element to appear or disappear before proceeding. By default, this is set to `5` seconds.

Target appears/disappears: A set of activities you want to run if the target successfully appears or disappears will be added to this block.

Target does not appear/disappear: A set of activities you want to run if the target does not appear or disappear will be added to this block.

Result: This is an optional configuration available on the `Properties` panel. This configuration allows you to save the result (`True` or `False`) for later use.

EXERCISE

Goal: Use the `Check App State` activity to wait for the spinner to disappear before proceeding.

Source Code: Chapter_4_CheckAppStateExercise

Setup: Here are step-by-step implementation instructions:

1. Open a browser of your choice and enter `https://therpabook.com/samples/contactsmanagement/home_load.html` in the URL field.
2. In `StudioX`, add the `Use Application/Browser` activity to a blank process.
3. Next, in the `Use Application/Browser` activity card, click `Indicate application` and point your mouse to the browser. This will automatically populate the activity card.

Note Because the Use Application/Browser card is capturing the URL and it's not changing while it's loading, you can indicate the browser while the page is loading or has loaded.

---

4. Next, select `Use Application/Browser` activity card, and in the `Properties`, set the `Options` ➤ `Open` property to `Always`. This will ensure that the automation always opens a new browser.
5. Next, add a `Check App State` activity in the body of `Use Application/Browser` activity.

6. Next, in the `Check App State` activity, click the `Indicate target on screen (I)` link and point your mouse to the spinner. The sample app shows the spinner for 10 seconds, so you will need to indicate the element within that timeframe. If that timeframe is too short, use the pause option (pressing F2 from your keyboard) while indicating the target. Additionally, you can hit refresh (F5 or Ctrl + R) as the pause is about to end, so that you have enough time to indicate the spinner.
7. Next, from the `Wait for` dropdown, select the `Element to disappear` option.
8. In the `Seconds` field, click the `Plus` icon, select the `Number` option, and type `15` seconds (this is adding a 5-second buffer).

Tip When there is a potential lag of a UI element appearing or disappearing, adding a buffer to the seconds field of the Check App State activity is a great way to support latency.

---

9. Next, in the `Target disappears` block, add a `Message Box` activity.
10. Next, in the `Message Box` activity, click the `Plus` icon, select the `Text` option, and type `Target disappeared`. Click `Save`.
11. Next, in the `Target does not disappear` block, add a `Message Box` activity.
12. Next, in the `Message Box` activity, click the `Plus` icon, select the `Text` option, and type `Target did not disappear`. Click `Save`.

Once you have completed the exercise, the final configuration of the **Check App State** activity should resemble Figure 4-23. Once you run the automation, it will wait for 15 seconds and then check if the spinner has disappeared. Depending on the state of the target application, a message box will appear with the appropriate message.

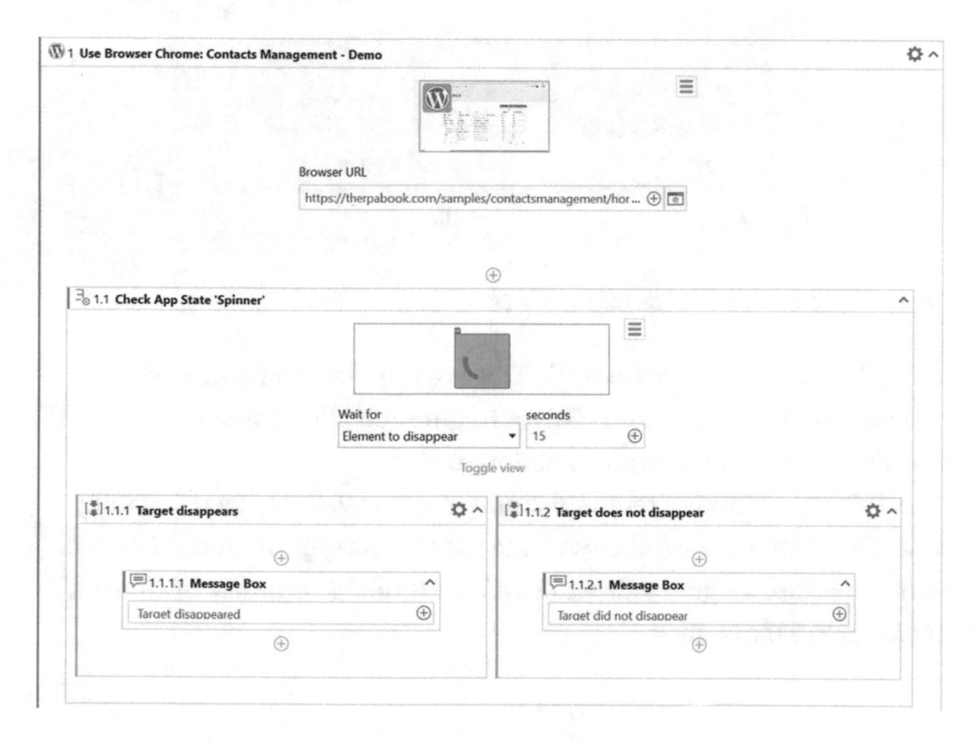

Figure 4-23. *Final configuration of the Check App State activity to wait for spinner to disappear*

Click

The **Click** activity allows you to perform a mouse click on a specified UI element.

Configuration

This section provides instructions on how to configure a **Click** activity, shown in Figure 4-24.

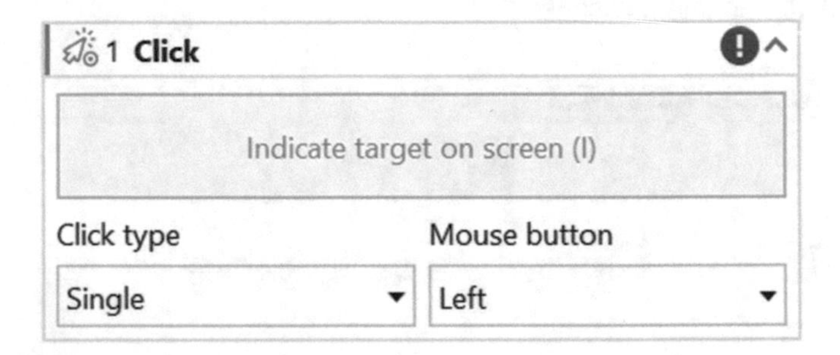

Figure 4-24. *Activity card for Click*

Indicate target on screen (I): This is a required configuration available on the activity card. This configuration allows you to specify a UI element that you want the automation to click.

Click type: This is a required configuration available on the activity card. This configuration allows you to specify the type of mouse click you want to perform. Figure 4-25 shows all the available options. By default, this is set to `Single` click.

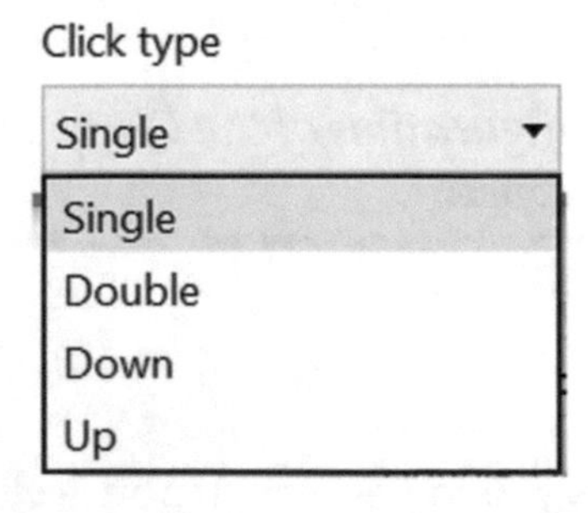

Figure 4-25. *Available options for click types*

Mouse button: This is a required configuration available on the activity card. This configuration allows you to specify which mouse button to click. Figure 4-26 shows all the available options. By default, this is set to `Left` click.

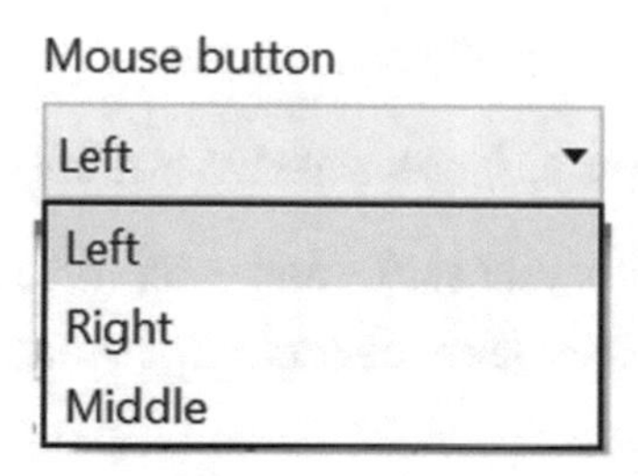

***Figure 4-26.** Available options for mouse button click*

Key modifiers: This is an optional configuration available on the `Properties` panel. This configuration allows you to use keys in combination with the mouse click action. Figure 4-27 shows all available options. By default, this is set to `None`.

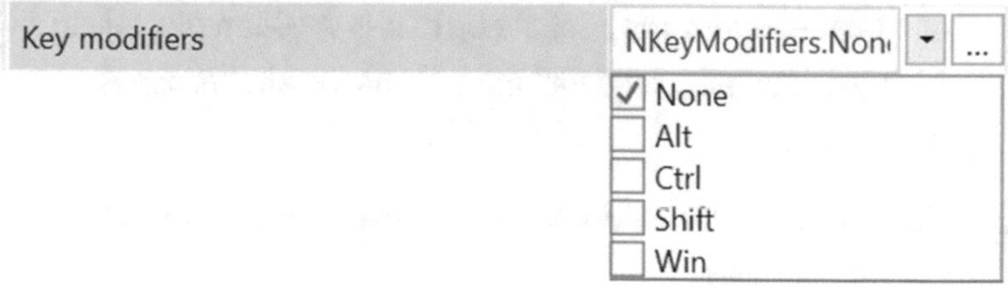

***Figure 4-27.** Available options for key modifiers*

EXERCISE

Goal: Use the `Click` activity to click the `Add Contact +` button to open the Contact Details dialog.

Source Code: Chapter_4_FormDataEntryExercise

Setup: Here are step-by-step implementation instructions:

1. Open a browser of your choice and enter `https://therpabook.com/samples/contactsmanagement/` in the address bar.
2. In `StudioX`, add the `Use Application/Browser` activity to a blank process.
3. Next, click `Indicate application` in the `Use Application/Browser` activity card, and point your mouse to the browser. This will automatically populate the activity card.
4. Next, select the `Use Application/Browser` activity card, and from `Properties`, set the `Options` ➤ Open property to `Always`. This will ensure that the automation always opens a new browser.
5. Next, add a `Click` activity within the `Use Application/Browser` activity.
6. Next, in the `Click` activity, click the `Indicate target on screen (I)` link and you'll notice a green highlight as you point your mouse to the `Add Contact` button. Then, you'll see a blue highlight as you point your mouse to specify an anchor; indicate the `View Contact` button. If you are unable to detect any elements, click `F4` to change the detection mechanism. Once you have selected the target element, use the `View Contact` button as an anchor. At this point, your selection should look like Figure 4-28. Click the `Confirm` button.

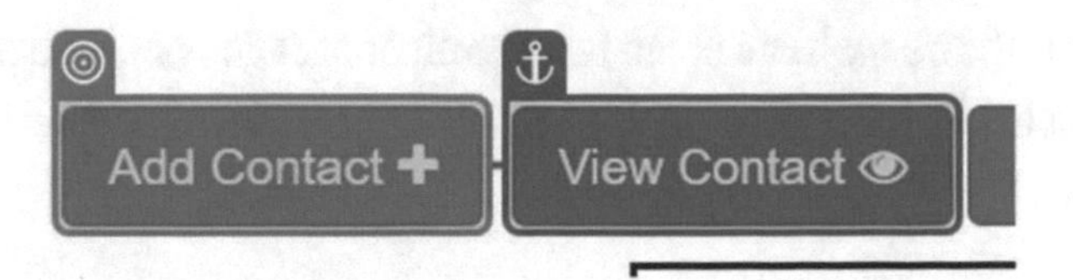

Figure 4-28. Target and anchor for the Add Contact button

Once you have completed the exercise, the final configuration of the **Click** activity should resemble Figure 4-29.

Tip When needing to click within an application that is minimized, using the Maximize Window activity can be used to maximize the application for the bot to click the correct UI element. Additionally, setting the Input mode to Simulate within the properties pane of the Click activity can have the bot click the application even when minimized.

Figure 4-29. Final configuration of the Click activity exercise

Figure 4-30 shows the state of the target web application once the automation has completed its run. In this case, an empty Contact Details dialog is displayed.

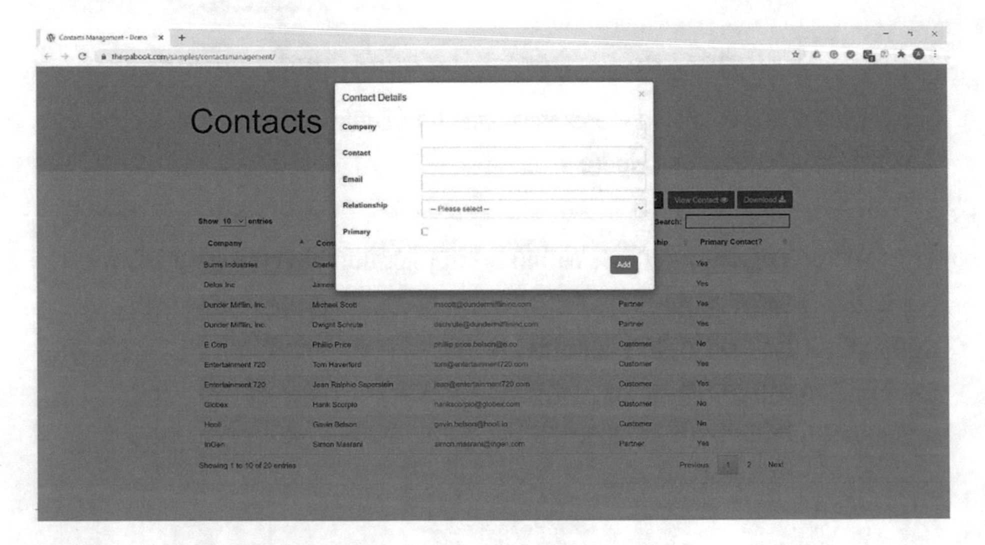

Figure 4-30. Result of the Click activity exercise

Type Into

The **Type Into** activity allows you to enter text in a specified element on the UI.

Tip When a dropdown menu of an application does not support the Select Item activity (discussed next), using a Type Into activity is a great way to support selecting a specific dropdown option.

Configuration

This section provides instructions on how to configure a **Type Into** activity, shown in Figure 4-31.

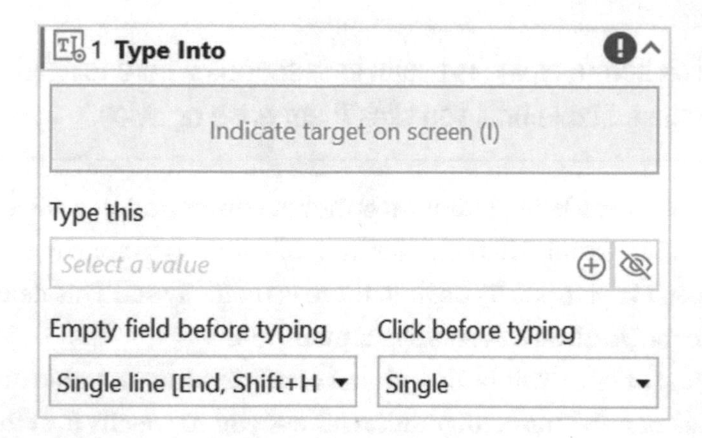

Figure 4-31. Activity card for Type Into

Indicate target on screen (I): This is a required configuration available on the activity card. This configuration allows you to specify which UI element you want to enter data in.

Type this: This is an optional configuration available on the activity card. This configuration allows you to specify the text that should be entered in the target element. You also have the option to include special keys in the text or toggle the password mode to type the text securely. This text can be static or dynamic.

Empty field before typing: This is an optional configuration available on the activity card. This configuration allows you to specify if you want to keep the existing text in the target element or clear it. Set to `None` if you want to leave the existing text as is. If you want to clear the existing text, then use either the single line or multiline options. If you are dealing with a single line Text field, then using `Single line` will clear the text, and if you are dealing with a multiline component like a Text Area, then the `Multi line` option will work the best. As a note, field data is cleared using keyboard shortcuts.

Click before typing: This is an optional configuration available on the activity card. This configuration allows you to specify if the automation should click the target element before it starts typing. You can either do a `Single` click or a `Double` click. By default, this is set to `Single`.

Tip Click before typing is useful in scenarios where text fields do not become editable until you click them once or twice.

Delay between keys: This is an optional configuration available on the `Properties` panel. This configuration allows you to specify a delay between each keystroke. By default, there is no delay set. This field is only applicable for Hardware Events input methods.

Deselect at end: This is an optional configuration available on the `Properties` panel. This configuration allows you to specify if, at the end of the activity, the automation should remove move from the UI element. This field is only applicable for Simulate input methods.

EXERCISE

Goal: Use the `Type Into` activity to fill out text fields of the Contact Details dialog. This exercise builds upon the previous exercise of the `Click` activity. Figure 4-30 shows the state of the target web application prior to this exercise.

Source Code: Chapter_4_FormDataEntryExercise

Setup: Here are step-by-step implementation instructions:

1. In `StudioX`, add the `Type Into` activity within the `Use Application/Browser` activity right after the `Click 'Add Contact'` activity.
2. Next, in the `Type Into` activity, click `Indicate target on screen (I)` link and point your mouse to the `Company` field. The field label is auto-detected as an anchor. At this point, your selection should look like Figure 4-32. Click the `Confirm` button.

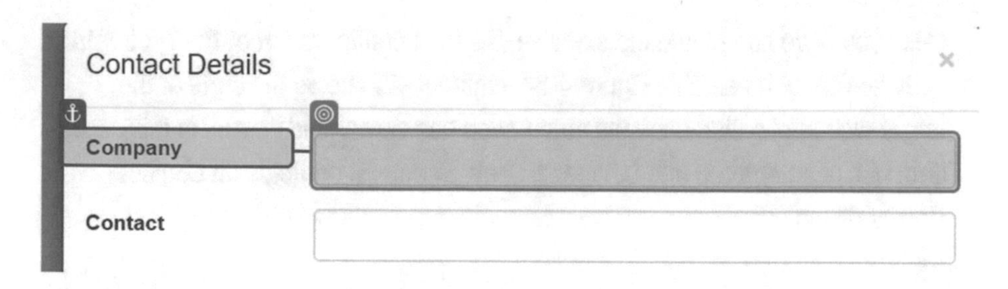

Figure 4-32. *Target and anchor for the Company text field*

3. Next, in the `Type this` field, click the `Plus` icon, select the `Text` option, and type `Stark Industries`. At this point, your `Type Into` activity configuration should resemble Figure 4-33.

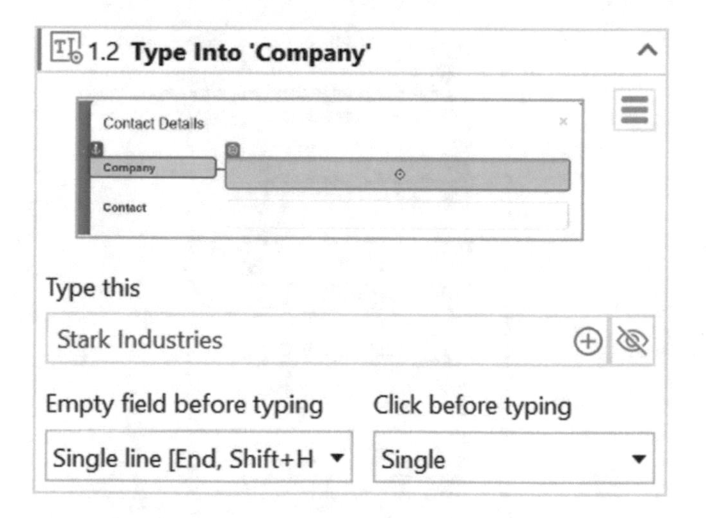

Figure 4-33. *Configuration of Type Into activity*

4. Next, repeat steps 2–4 for the `Contact` field.

5. Next, In the `Type this` field, click the `Plus` icon, select the `Text` option, and type `Tony Stark`.

6. Next, repeat steps 2–4 for the `Email` field.

7. Next, in the `Type this` field, click the `Plus` icon, select the `Text` option, and type tstark@starkindustries.com.

Once you have completed the exercise, the final configuration of the **Type Into** activities should resemble Figure 4-34. Figure 4-35 shows the state of the target web application once the automation has completed its run. In this case, data has been entered into Company, Contact, and Email fields on Contact Details dialog.

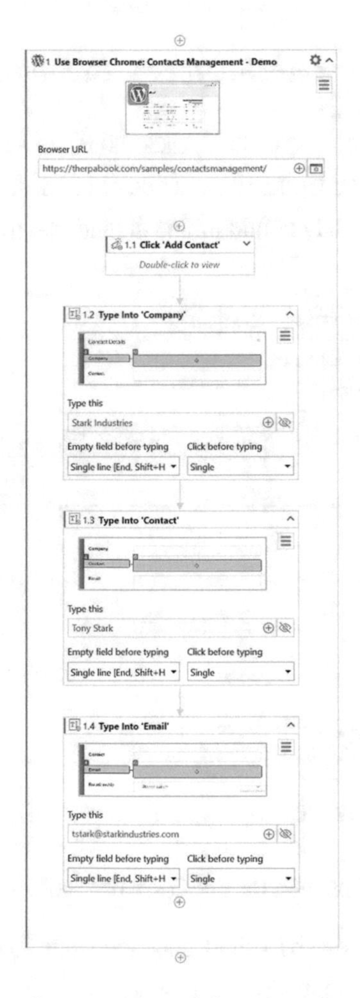

Figure 4-34. *Final configuration of the Type Into activity exercise*

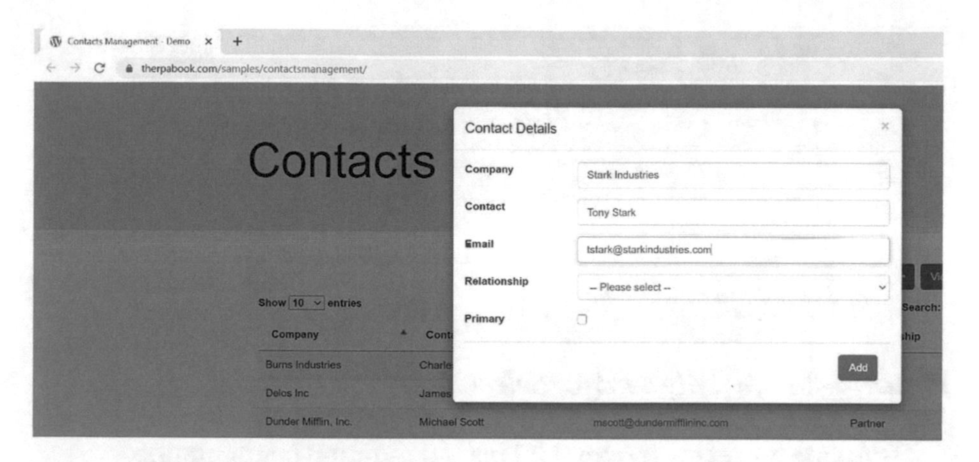

***Figure 4-35.** Result of the Type Into activity exercise*

Select Item

The **Select Item** activity allows you to select an item in a specified dropdown list.

Note If the application dropdown is not a combo box or list box, the Select Item activity may not be supported. In this case, it's advised to use a Type Into activity.

Configuration

This section provides instructions on how to configure a **Select Item** activity, shown in Figure 4-36.

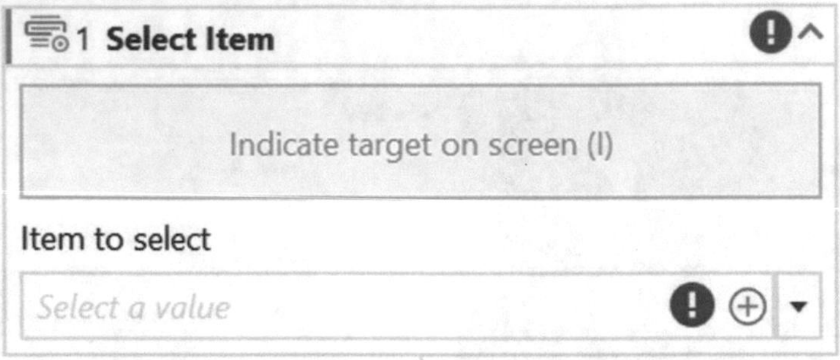

1 Select Item

Indicate target on screen (I)

Item to select

Select a value

Figure 4-36. Activity card for Select Item

Indicate target on screen (I): This is a required configuration available on the activity card. This configuration allows you to specify a dropdown on the UI in which you want to select an item.

Item to select: This is an optional configuration available on the activity card. By default, this list is empty. Once you have indicated the dropdown element on the UI, this list will be populated with all options available in the dropdown. Figure 4-37 shows an example when this list is populated.

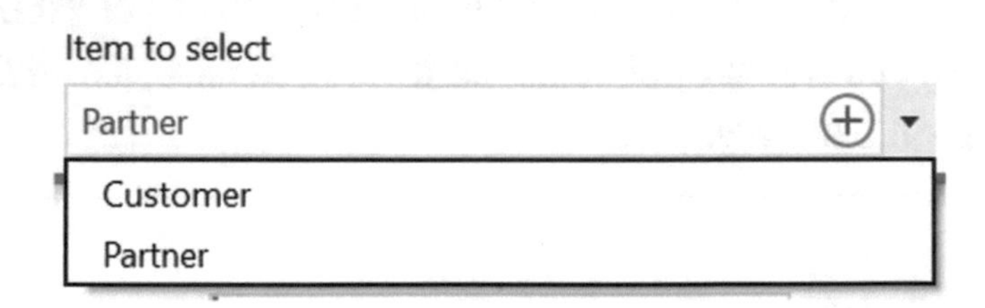

Figure 4-37. Available options in the dropdown

EXERCISE

Goal: Use the `Select Item` activity to choose Partner from Relationship dropdown on the Contact Details dialog. This exercise builds upon the previous exercises for `Click` and `Type Into` activities. Figure 4-35 shows the state of the target web application before this exercise.

Source Code: Chapter_4_FormDataEntryExercise

Setup: Here are step-by-step implementation instructions:

1. In `StudioX`, add the `Select Item` activity in the body of `Use Application/Browser` activity after the three `Type Into` activities.

2. In the `Select Item` activity, click the `Indicate target on screen (I)` link and point your mouse to the `Relationship` dropdown. The field label, Relationship, is auto-detected as an anchor. At this point, your selection should resemble Figure 4-38. Click the `Confirm` button.

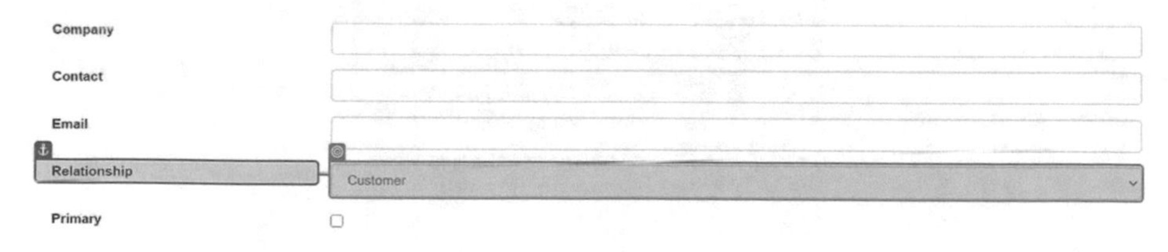

Figure 4-38. *Target and anchor for Relationship dropdown*

3. At this point, the `Item to select` field will be populated with all options available in the `Relationship` dropdown. Select `Partner` as the relationship from the list.

Once you have completed the exercise, the final configuration of the **Select Item** activity should resemble Figure 4-39. Figure 4-40 shows the state of the target web application once the automation has completed its run. In this case, an item has been selected from the Relationship field.

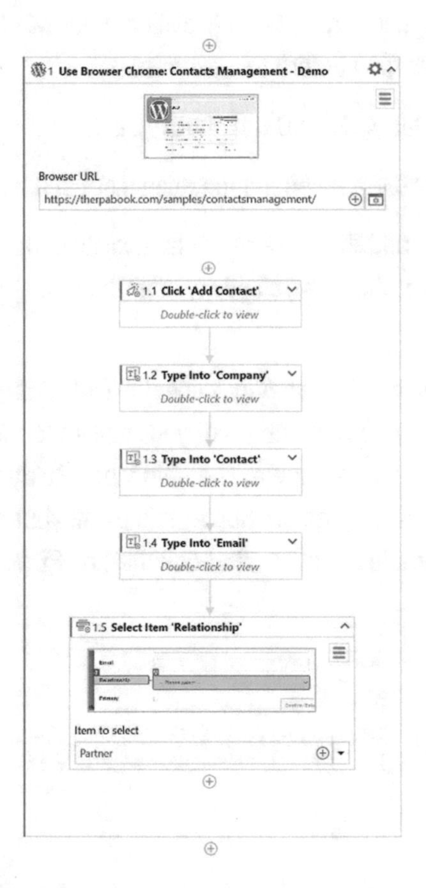

***Figure 4-39.** Final configuration of the Select Item activity exercise*

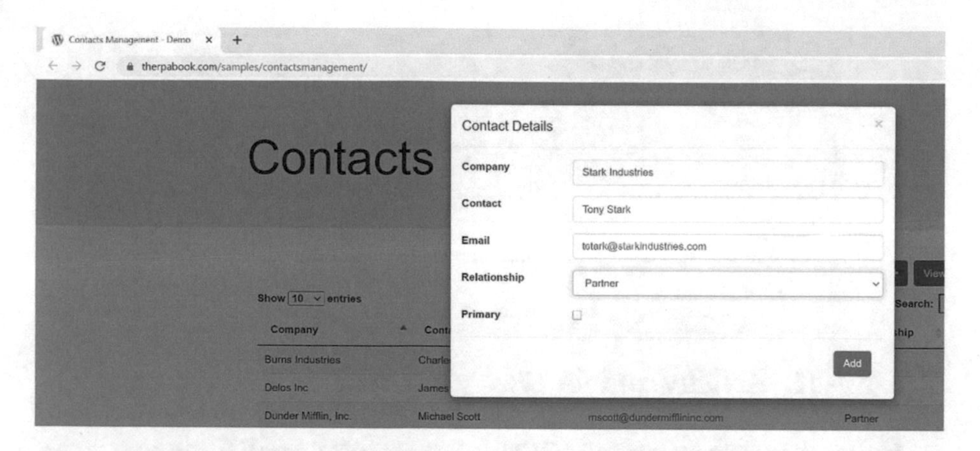

Figure 4-40. Result of the Select Item activity exercise

Check/Uncheck

The **Check/Uncheck** activity allows you to interact with checkboxes on a user interface.

Tip In place of using a Click activity to interact with a checkbox within an application, using the Check/Uncheck activity gives the option to toggle, check, or uncheck a field. This way, if the application may already have a checkmark marked or unmarked, it won't unnecessarily change it.

Configuration

This section provides instructions on how to configure a **Check/Uncheck** activity, shown in Figure 4-41.

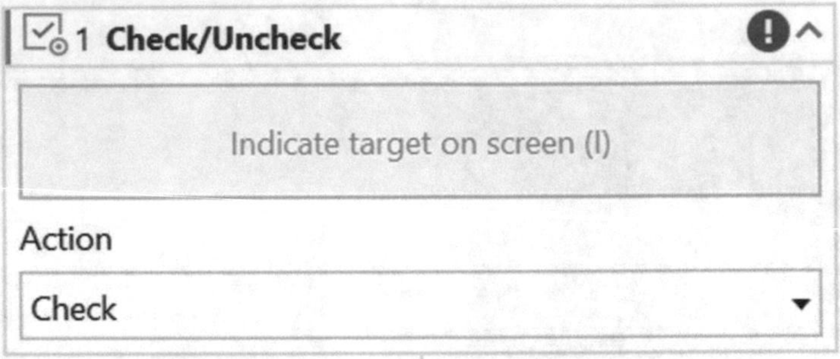

1 **Check/Uncheck**

Indicate target on screen (I)

Action

Check

Figure 4-41. *Activity card for Check/Uncheck*

Indicate target on screen (I): This is a required configuration available on the activity card. This configuration allows you to specify a checkbox on the UI which you want to check or uncheck.

Action: This is a required configuration available on the activity card. This configuration allows you to specify the state of the checkbox. Figure 4-42 shows all available actions. By default, this is set to Check.

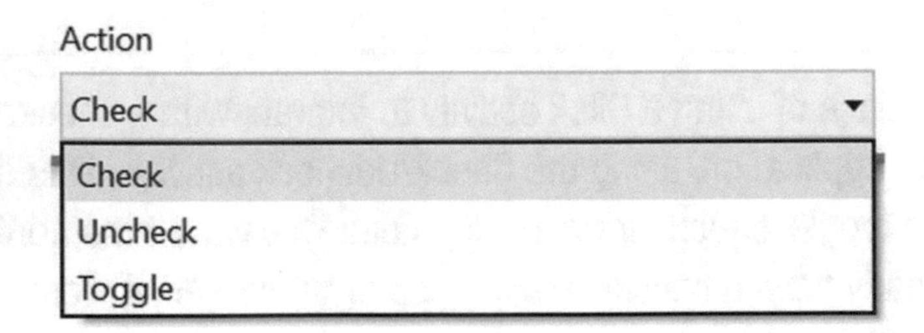

Figure 4-42. *Available action for Check/Uncheck activity*

EXERCISE

Goal: Use the Check/Uncheck activity to specify that the contact being added is primary by checking the Primary option in the Contact Details dialog. This exercise builds upon the previous exercises for Click, Type Into, and Select Item activities. Figure 4-40 shows the state of the target web application before this exercise.

Source Code: Chapter_4_FormDataEntryExercise

Setup: Here are step-by-step implementation instructions:

1. In StudioX, add the Check/Uncheck activity in the body of Use Application/Browser activity after the Select Item activity.
2. Next, in the Check/Uncheck activity, click Indicate target on screen (I) link and point your mouse to the Primary checkbox. Select the field label, Primary, as an anchor. At this point, your selection should look like Figure 4-43. Click the Confirm button.

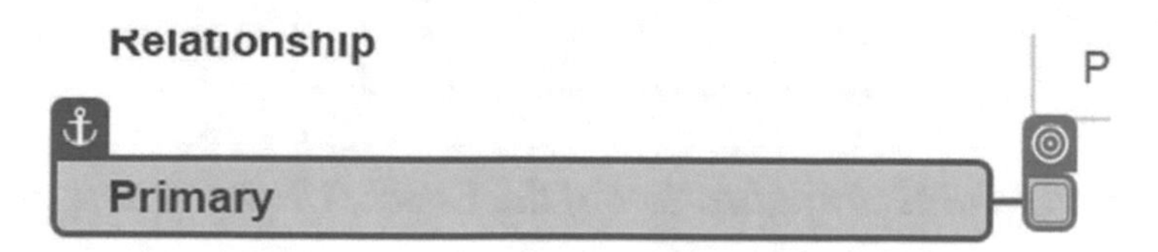

***Figure 4-43.** Target and anchor for Primary checkbox*

3. Next, select Check from the Action dropdown.

Once you have completed the exercise, the final configuration of the **Check/Uncheck** activity should resemble Figure 4-44. Figure 4-45 shows the state of the target web application once the automation has completed its run. In this case, the Primary checkbox has been checked.

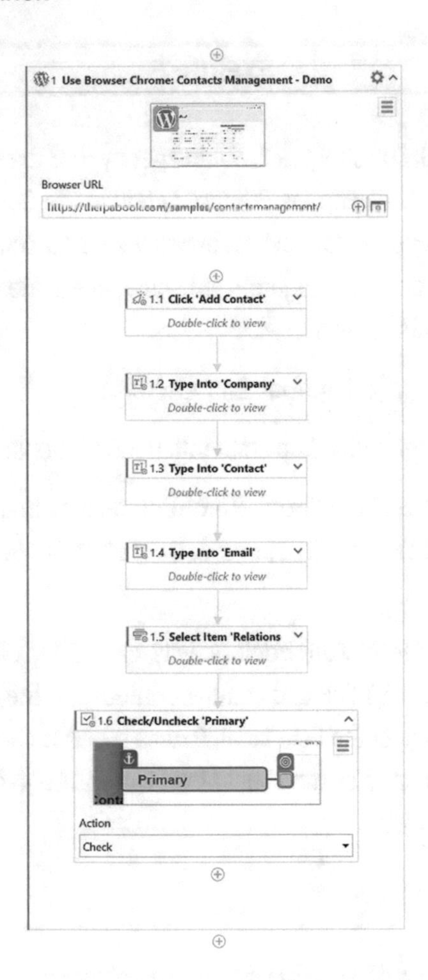

Figure 4-44. *Final configuration of the Check/Uncheck activity exercise*

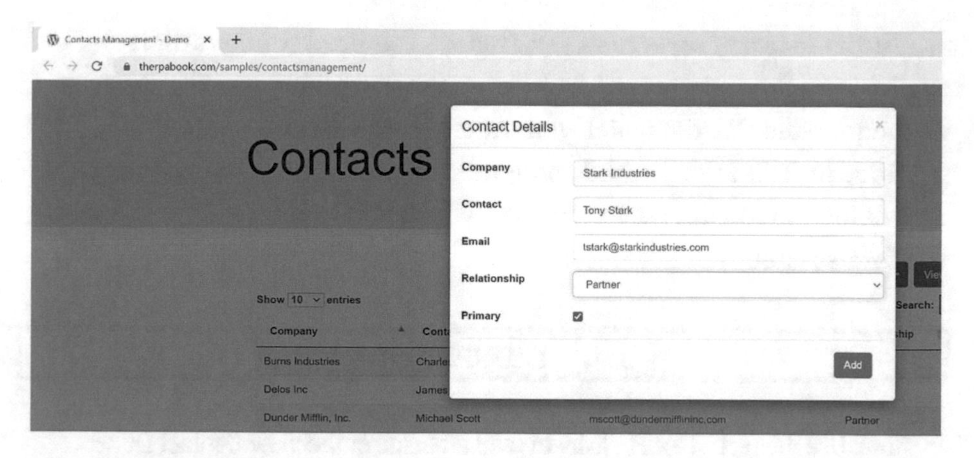

***Figure 4-45.** Result of the Check/Uncheck activity exercise*

Get Text

The **Get Text** activity allows you to extract text from a specified element on the UI.

Configuration

This section provides instructions on how to configure a **Get Text** activity, shown in Figure 4-46.

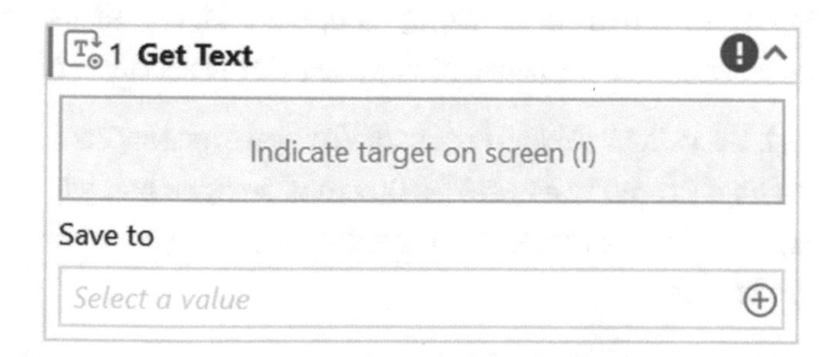

***Figure 4-46.** Activity card for Get Text*

Indicate target on screen (I): This is a required configuration available on the activity card. This configuration allows you to specify an element on the UI from which you want to retrieve text.

Save to: This is a required configuration available on the activity card. This configuration allows you to specify how you want to store retrieved text for use later.

EXERCISE

Goal: Use the `Get Text` activity to read all fields of the View Contact Details dialog.

Source Code: Chapter_4_FormDataExtractionExercise

Setup: Here are step-by-step implementation instructions:

1. Open a browser of your choice and enter `https://therpabook.com/samples/contactsmanagement/` in the address bar.
2. In `StudioX`, add the `Use Application/Browser` activity to a blank process.
3. Next, click `Indicate application` in the `Use Application/Browser` activity card, and point your mouse to the browser. This will automatically populate the activity card.
4. Next, select the `Use Application/Browser` activity card, and from `Properties`, set the `Options` ➤ `Open` property to `Always`. This will ensure that the automation always opens a new browser.
5. Add a `Click` activity within the `Use Application/Browser` activity and configure it to click the `View Contact` button.
6. Next, make sure the `View Contact` dialog is open.

7. Add the `Get Text` activity within the `Use Application/Browser` activity right after the `Click` activity.

8. In the `Get Text` activity, click the `Indicate target on screen (I)` link and point your mouse to the `Company` field. The field label is auto-detected as an anchor. At this point, your selection should resemble Figure 4-47. Click the `Confirm` button.

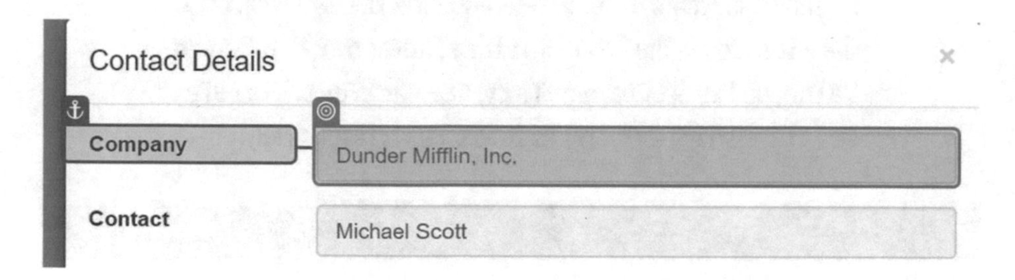

Figure 4-47. *Target and anchor for the Company field*

9. Next, click the `Plus` icon in the `Save to` the field. Select the `Save for Later Use` option and name your saved value as `CompanyText`. Click `Ok`. At this point, your activity `Get Text` activity configuration should resemble Figure 4-48.

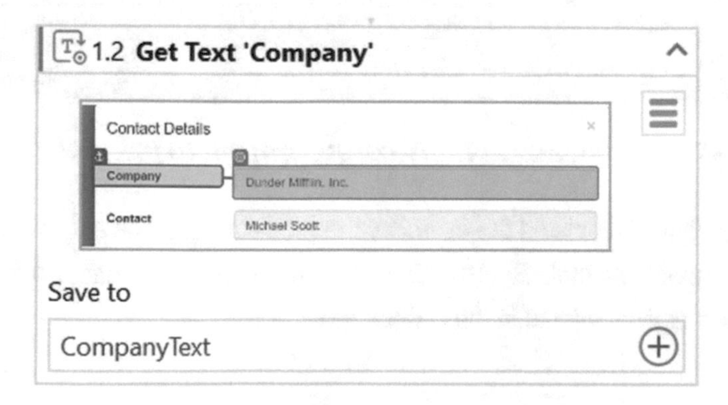

Figure 4-48. *Configuration of Get Text activity*

10. Next, repeat steps 7–9 for the remaining fields, that is, `Contact`, `Email`, `Relationship`, and `Primary`. Name the saved values as `ContactText`, `EmailText`, `RelationshipText`, and `PrimaryText`, respectively.

11. Next, add a `Write Line` activity at the end. Click the `Plus` icon in the `Text` field, select the `Text` option, and enter the text as shown in Figure 4-49. To reference the values saved, click the `Plus` icon in the `Text Builder`, hover over `Use Saved Value`, and select `CompanyText`, `ContactText`, `EmailText`, `RelationshipText`, and `PrimaryText`, respectively.

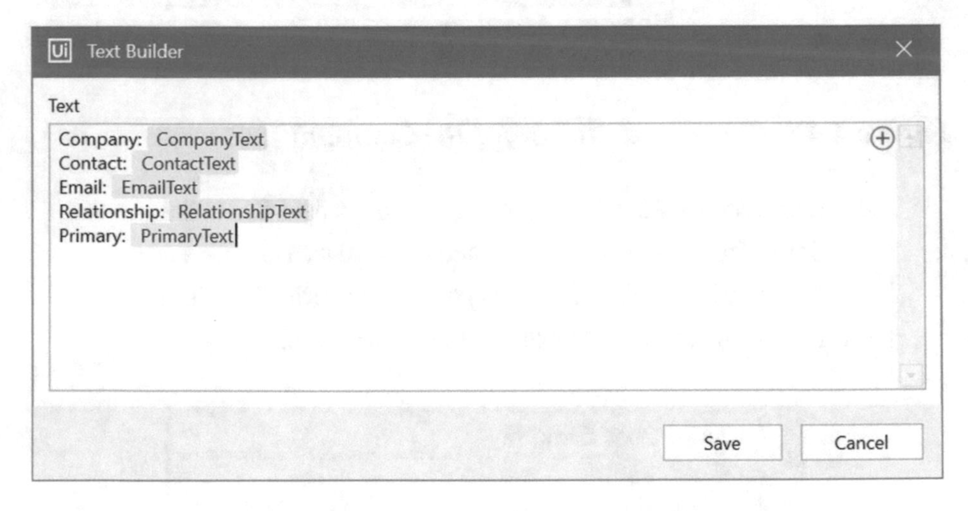

Figure 4-49. *Write Line activity configuration to print all form data*

Once you have completed the exercise, the final configuration of the **Get Text** activity should resemble Figure 4-50. Figure 4-51 shows the data from the Write Line activity printed in the Output panel.

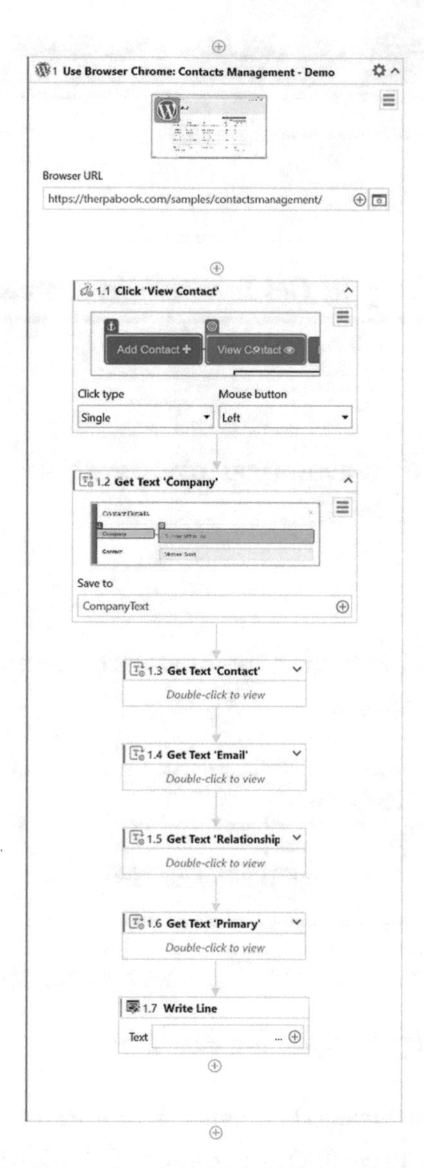

Figure 4-50. *Final configuration of automation with Get Text activities*

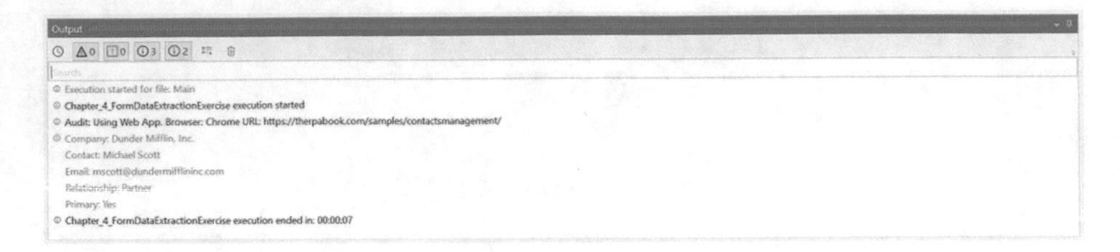

Figure 4-51. *Output of the Get Text activity exercise*

Get Attribute

The **Get Attribute** activity allows you to retrieve the value of an attribute of a specified element on the UI.

Configuration

This section provides instructions on how to configure a **Get Attribute** activity, shown in Figure 4-52.

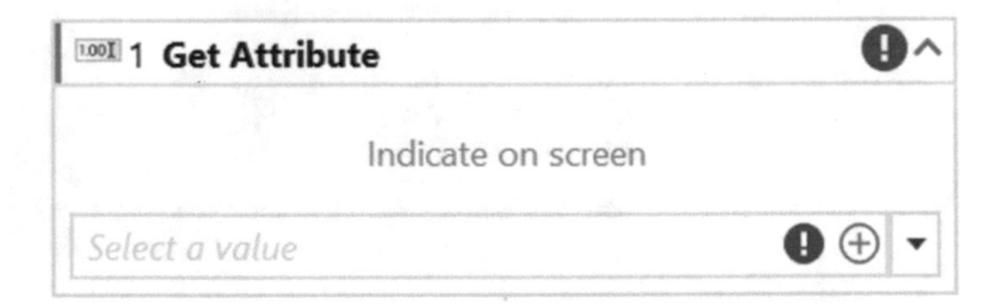

Figure 4-52. *Activity card for Get Attribute*

Indicate target on screen (I): This is a required configuration available on the activity card. This configuration allows you to specify an element on the UI from which you want to retrieve the attribute value.

Attribute: This is required configuration available on the activity card. This configuration allows you to specify the attribute whose value you want to retrieve.

Result: This is an optional configuration available on the activity card. This configuration allows you to specify how you want to store retrieved value of the attribute for use later.

EXERCISE

Goal: Use the `Get Attribute` activity to read value of the Company field from the View Contact Details dialog. This exercise builds upon the previous exercise for the `Get Text` activity.

Source Code: Chapter_4_FormDataExtractionExercise

Setup: Here are step-by-step implementation instructions:

1. In `StudioX`, add the `Get Attribute` activity within the `Use Application/Browser` activity right after the `Write Line` activity.
2. In the `Get Attribute` activity, click the `Indicate on screen` link and point your mouse to the `Company` field.
3. Next, click the `Plus` icon in the `Attribute` field, select the `Text` option, and type value. This is the attribute that we are going to retrieve. Click `Save`.
4. Next, from the `Properties` panel, click the `Plus` icon in the `Result` field. Select the `Save for Later Use` option and name your saved value as `AttributeValue`. Click `Ok`.
5. Next, add the `Write Line` activity at the end. Click the `Plus` icon in the `Text` field, select the `Text` option, and enter the text as shown in Figure 4-53.

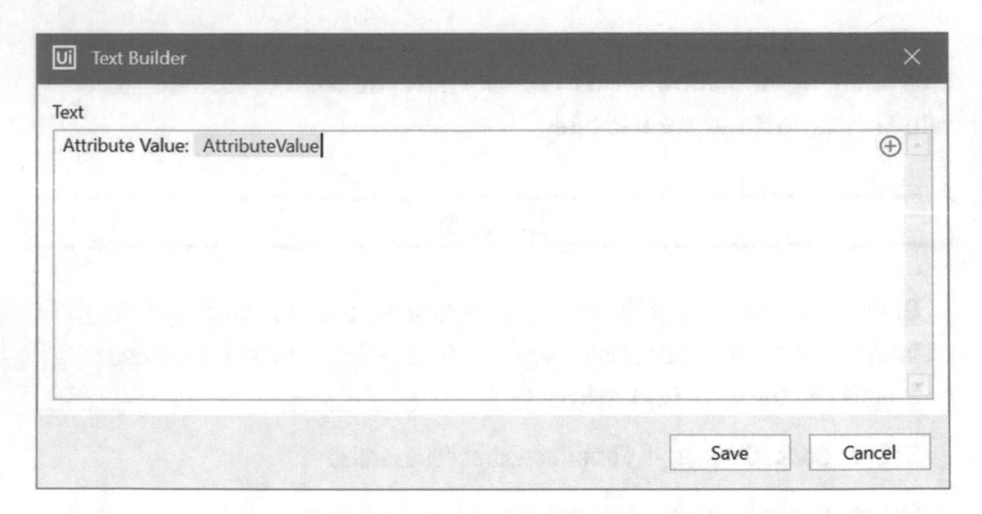

***Figure 4-53.** Write Line activity configuration to print all form data*

Once you have completed the exercise, the final configuration of the **Get Attribute** activity should resemble Figure 4-54. Figure 4-55 shows the data from the Write Line activity printed in the Output panel.

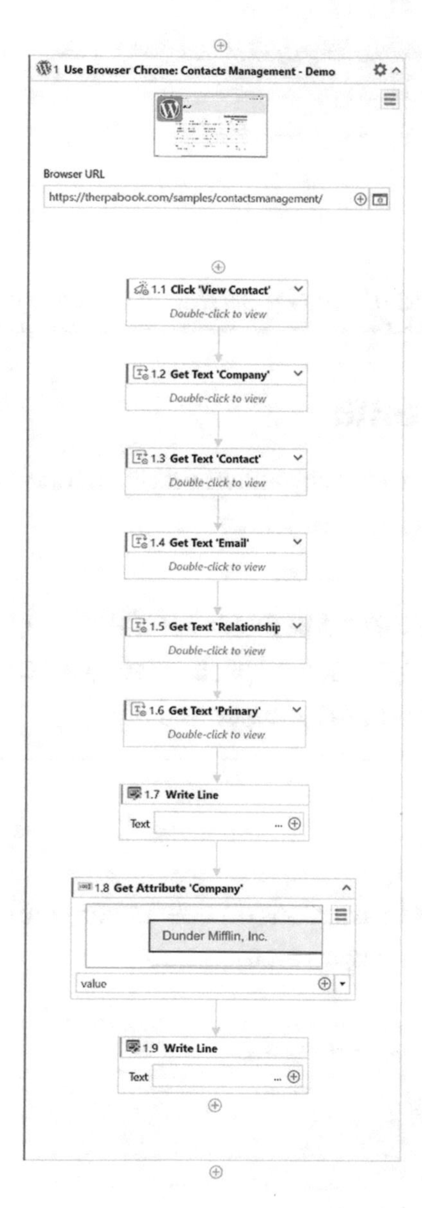

Figure 4-54. *Final configuration of automation with Get Attribute activities*

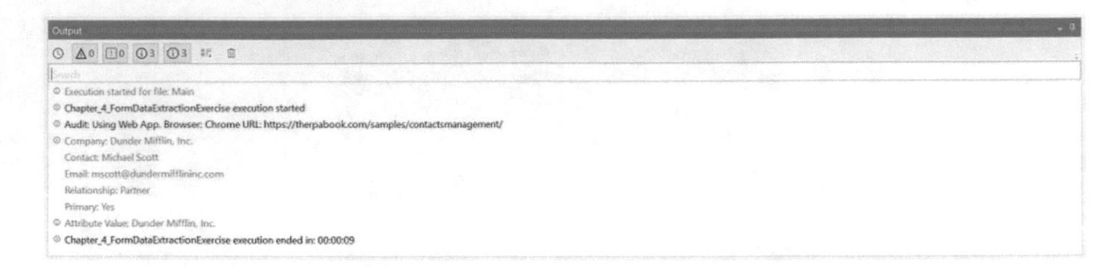

Figure 4-55. *Output of the Get Attribute activity exercise*

Extract Table Data

The **Extract Table Data** activity allows you to extract data in tabular format from web and desktop applications.

Note The **Table Extraction** menu item in the top ribbon also follows similar configuration steps. It creates a separate container inside a **Use Application/Browser** activity.

Configuration

This section provides instructions on how to configure an **Extract Table Data** activity, shown in Figure 4-56.

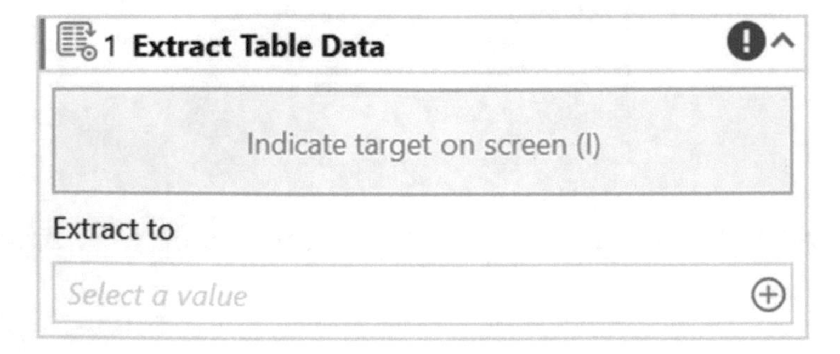

Figure 4-56. *Activity card for Extract Table Data*

Indicate target on screen (I): This is a required configuration available on the activity card. This configuration allows you to specify the table from which you want to extract data. If the table data spans multiple pages, this configuration will also allow you to specify the element that automation can use to navigate between pages.

Extract to: This is an optional configuration available on the activity card. This configuration allows you to specify how you want to store extracted data.

Append results: This is an optional configuration available on the `Properties` panel. This configuration allows you to specify if the automation should append any new extractions at the end of the previously extracted data or overwrite it.

Edit Next Link: This configuration becomes available from the activity menu only after you have indicated the target on the screen. This configuration, shown in Figure 4-57, allows you to edit the UI element you have specified to navigate to the next page of data.

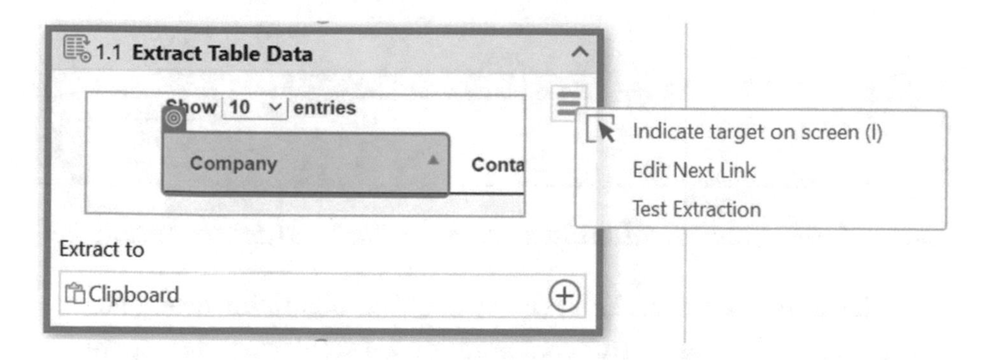

***Figure 4-57.** Options in the activity menu*

Test Extraction: This configuration becomes available from the activity menu only after you have indicated the target on the screen. This configuration, shown in Figure 4-57, allows you to test your data extraction. Once you click the `Test Extraction` option from the menu options, a popup shown in Figure 4-58 will appear containing extracted data from first page indicated.

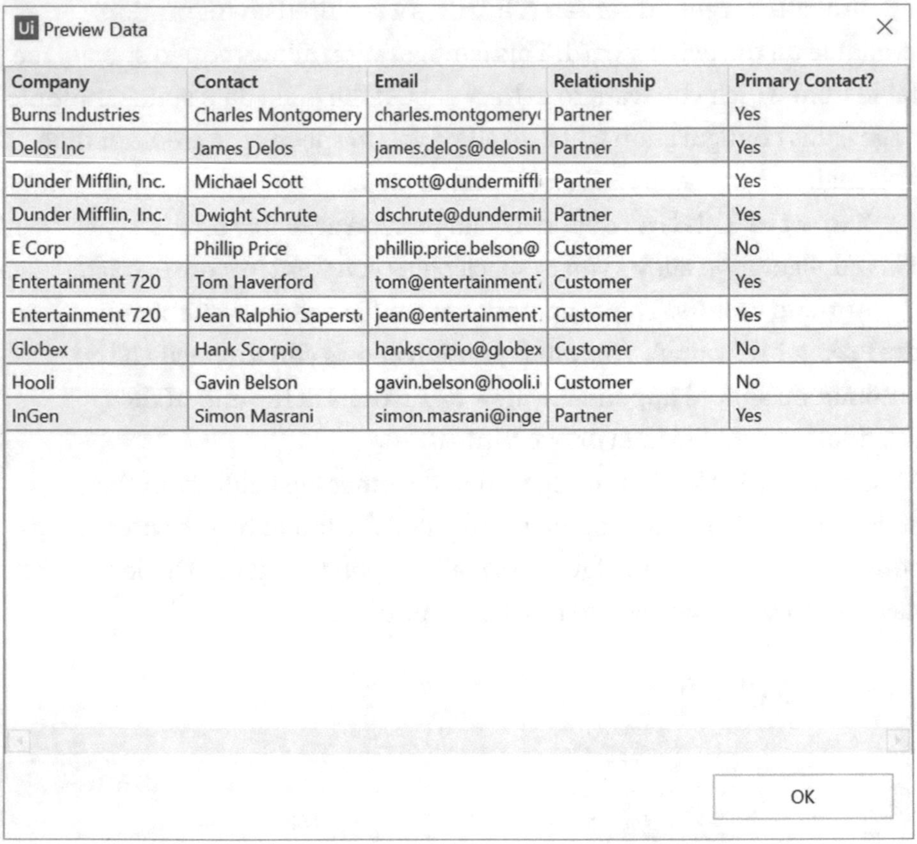

Preview Data

Company	Contact	Email	Relationship	Primary Contact?
Burns Industries	Charles Montgomery	charles.montgomery	Partner	Yes
Delos Inc	James Delos	james.delos@delosin	Partner	Yes
Dunder Mifflin, Inc.	Michael Scott	mscott@dundermiffl	Partner	Yes
Dunder Mifflin, Inc.	Dwight Schrute	dschrute@dundermi	Partner	Yes
E Corp	Phillip Price	phillip.price.belson@	Customer	No
Entertainment 720	Tom Haverford	tom@entertainment	Customer	Yes
Entertainment 720	Jean Ralphio Saperst	jean@entertainment	Customer	Yes
Globex	Hank Scorpio	hankscorpio@globe	Customer	No
Hooli	Gavin Belson	gavin.belson@hooli.i	Customer	No
InGen	Simon Masrani	simon.masrani@inge	Partner	Yes

OK

***Figure 4-58.** Preview table data using the Test Extraction option*

Delay between pages: This is an optional configuration available on the `Properties` panel. This configuration is useful when data spans multiple pages, and after opening the next page, data takes some time to load. You can specify how long (in seconds) you want to wait before extracting data from the next page.

Maximum results: This is an optional configuration available on the `Properties` panel. This configuration allows you to specify the maximum number of results you want to extract. By default, this is `100`.

EXERCISE

Goal: Use the Extract Table Data activity to read all contacts from all pages of the table.

Source Code: Chapter_4_TableDataExtractionExercise

Setup: Here are step-by-step implementation instructions:

1. Open a browser of your choice and enter https://therpabook.com/samples/contactsmanagement/ in the address bar.
2. In StudioX, add the Use Application/Browser activity to a blank process.
3. Next, click Indicate application in the Use Application/Browser activity card, and point your mouse to the browser. This will automatically populate the activity card.
4. Next, select the Use Application/Browser activity card, and from Properties, set the Options ➤ Open property to Always. This will ensure that the automation always opens a new browser.
5. Next, add an Extract Table Data activity to the Use Application/Browser activity.
6. Next, in the Extract Table Data activity, click the Indicate target on screen (I) link. This will start the table data extraction wizard, shown in Figure 4-59.

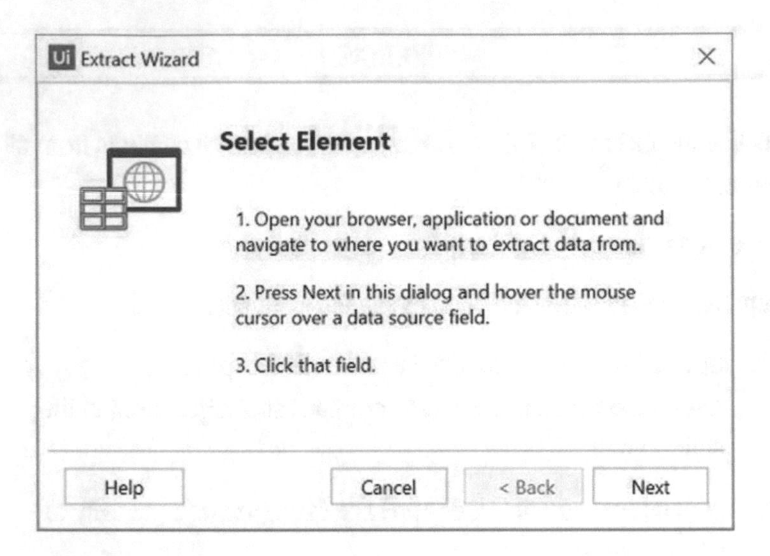

Figure 4-59. *Extract Wizard – initial window*

7. Next, on the `Extract Wizard` dialog, click the `Next` button.
8. At this point, the `Extract Wizard` will prompt you to select a column of the table. Click the `Company` column, as shown in Figure 4-60.

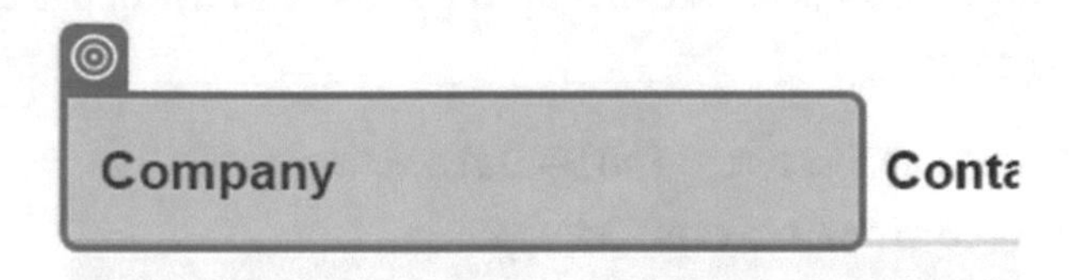

Figure 4-60. *Extract Wizard – column selection*

9. Next, the `Extract Wizard` will prompt you to confirm if you want to extract all columns of the specified table or just the column that you selected, shown in Figure 4-61. Click `Yes`.

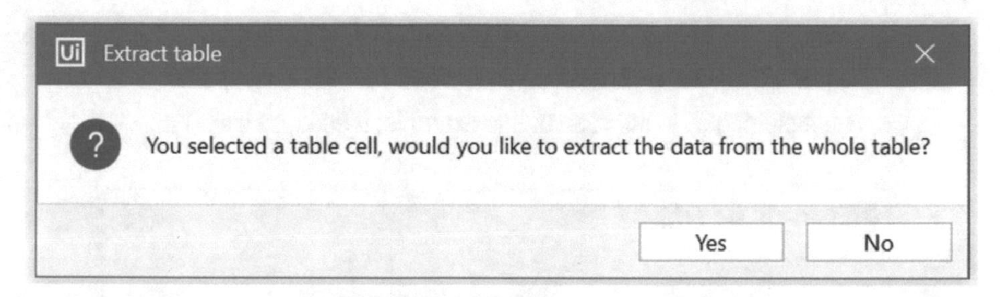

***Figure 4-61.** Extract Wizard – prompt to confirm data extraction from a single column or the entire table*

10. Next, the `Extract Wizard` will show you the data it was able to extract, shown in Figure 4-62. If the data looks accurate, that is, you are not missing any columns, then click `Finish`.

***Figure 4-62.** Extract Wizard – preview extracted data*

11. Next, the `Extract Wizard` will prompt another message, shown in Figure 4-63, asking you to confirm if the data spans multiple pages. In the case of this example, it does, so click `Yes`.

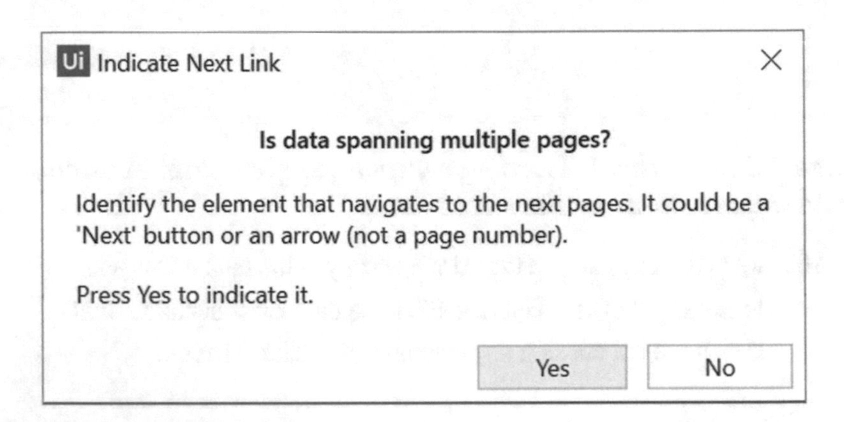

Figure 4-63. *Extract Wizard – confirm if data spans multiple pages*

12. At this point, you will select the `Next` button on the Contacts List screen, shown in Figure 4-64. This button will allow the automation to move to the next page of data. This completes the extraction configuration.

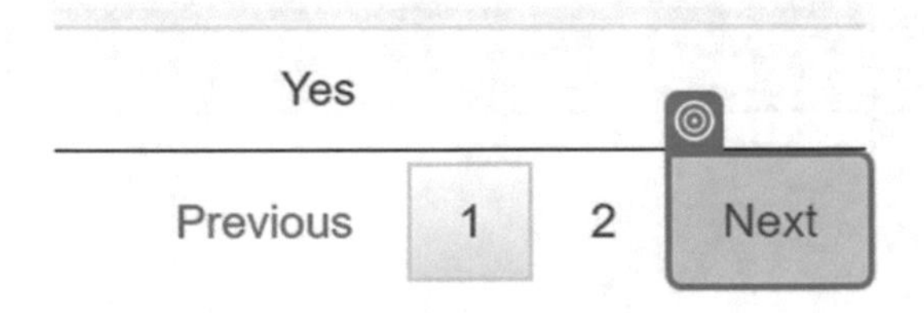

Figure 4-64. *Extract Wizard – select the data navigation UI element*

13. Next, in the `Extract to` field, select `Copy to clipboard` option.

14. Next, add a `Write Line` activity at the end. In the `Text` field, click the `Plus` icon, and select `Paste from clipboard` option.

Once you have completed the exercise, the final configuration of the **Extract Table Data** activity should resemble Figure 4-65. Figure 4-66 shows the data extracted from the table by using the Write Line activity to print to the Output panel.

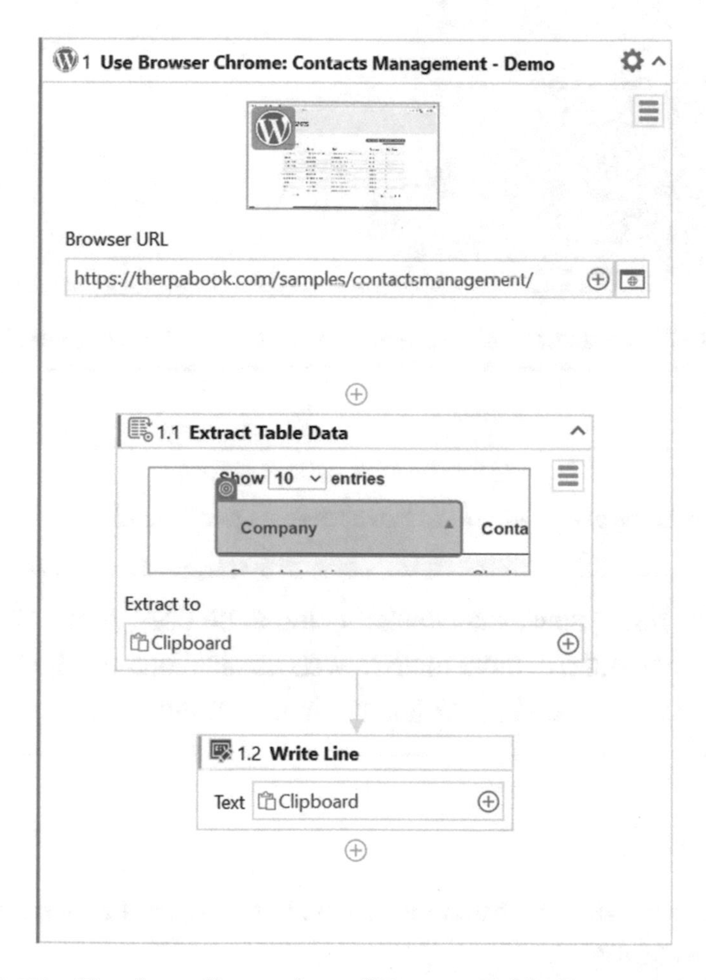

***Figure 4-65.** Final configuration of Extract Table Data activity exercise*

```
Output
Execution started for file: Main
Chapter_4_TableDataExtractionExercise execution started
Audit: Using Web App. Browser: Chrome URL: https://therpabook.com/samples/contactsmanagement/
Audit: Extract Data. From: <html app='chrome.exe' title='Contacts Management - Demo' />
Company	Contact	Email	Relationship	Primary Contact?
Burns Industries	Charles Montgomery Burns	charles.montgomery@burnsindustries.com	Partner	Yes
Delos Inc	James Delos	james.delos@delosinc.com	Partner	Yes
Dunder Mifflin, Inc.	Michael Scott	mscott@dundermifflininc.com	Partner	Yes
Dunder Mifflin, Inc.	Dwight Schrute	dschrute@dundermifflininc.com	Partner	Yes
E Corp	Phillip Price	phillip.price.belson@e.co	Customer	No
Entertainment 720	Tom Haverford	tom@entertainment720.com	Customer	Yes
Entertainment 720	Jean Ralphio Saperstein	jean@entertainment720.com	Customer	Yes
Globex	Hank Scorpio	hankscorpio@globex.com	Customer	No
Hooli	Gavin Belson	gavin.belson@hooli.io	Customer	No
InGen	Simon Masrani	simon.masrani@ingen.com	Partner	Yes
InGen	John Hammond	john.hammond@ingen.com	Partner	No
MomCorp	Carol Miller	carol.miller@mom.co	Partner	Yes
Monsters, Inc.	James P. Sullivan	jpsullivan@monstersinc.com	Customer	Yes
Monsters, Inc.	Mike Wazowski	mwazowski@monstersinc.com	Customer	No
Rainholm Industries	Douglas Reynholm	douglas@rainholmindustries.com	Partner	Yes
Rearden Steel	Hank Rearden	hankrearden@reardensteel.com	Customer	Yes
Stark Industries	Tony Stark	tstark@starkindustries.com	Customer	No
Umbrella Corp	Chris Valentine	chris.valentine@umbrella.co	Partner	Yes
Wayne Enterprises, Inc.	Bruce Wayne	bruce.wayne@wayneenterprises.inc	Customer	Yes
Wayne Enterprises, Inc.	Lucius Fox	lucius.fox@wayneenterprises.inc	Customer	No
Chapter_4_TableDataExtractionExercise execution ended in: 00:00:38
```

Figure 4-66. *Output of the Extract Table Data activity exercise*

Hover

The **Hover** activity allows you to hover over a specified element on the UI.

Tip The Hover activity is typically going to be used with a Click activity. For example, most modern web applications use hover menus, but you need to click a menu item to open it.

Configuration

This section provides instructions on how to configure a **Hover** activity, shown in Figure 4-67.

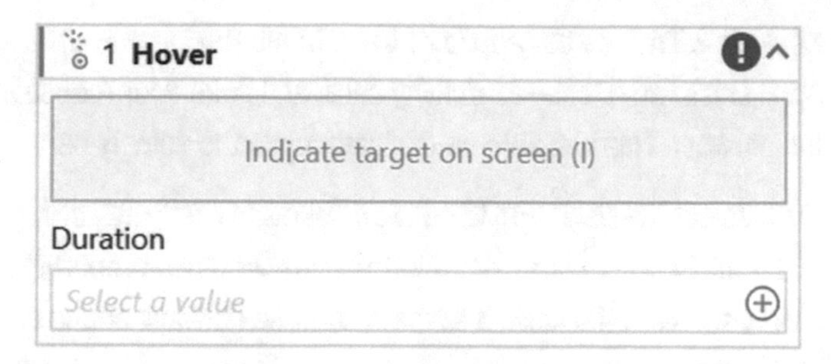

Figure 4-67. Activity card for Hover

Indicate target on screen (I): This is a mandatory configuration available from the activity card. This configuration allows you to specify an element on the UI that you want to hover over.

Duration: This is an optional configuration available from the activity card. This configuration allows you to specify how long (in seconds) you want to hover over the specified element.

EXERCISE

Goal: Use the `Hover` activity to open the `Download` menu, and use the `Click` activity to initiate the download of contacts data in Excel format. Use a `Wait for Download` activity to monitor file download.

Source Code: Chapter_4_MenuHoverExercise

Setup: Here are step-by-step implementation instructions:

1. Open a browser of your choice and enter `https://therpabook.com/samples/contactsmanagement/` in the address bar.
2. In `StudioX`, add the `Use Application/Browser` activity to a blank process.

3. Next, click `Indicate application` in the `Use Application/Browser` activity card, and point your mouse to the browser. This will automatically populate the activity card.

4. Next, select the `Use Application/Browser` activity card, and from `Properties`, set the `Options` ➤ `Open` property to `Always`. This will ensure that the automation always opens a new browser.

5. Next, add the `Hover` activity within the `Use Application/Browser` activity.

6. Next, in the `Hover` activity, click `Indicate target on screen (I)` link to select the `Download` button. Once you have selected the target element, use the `View Contact` button as an anchor. At this point, your selection should resemble Figure 4-68. Click the `Confirm` button.

Figure 4-68. *Target and anchor for the Company field*

7. Next, in the `Duration` field, click the `Plus` icon, select the `Number` option, and type `3` seconds.

8. Next, add a `Wait for Download` activity to the body of `Use Application/Browser` activity right after the `Hover` activity.

9. Next, in the `Downloaded file` field, click the `Plus` icon, select `Save for Later Use`, and name your saved value as `DownloadedFile`.

10. Leave the `Monitored folder` field as is.

11. Next, add a `Click` activity to the body of `Wait for Download` activity.

12. Next, in the `Click` activity, click `Indicate target on screen (I)` link to specify the `Excel` option from the `Download` menu. This is not going to be straightforward, because when you are in element selection mode, hover will not work, and you will not be able to select the `Excel` download option. To make this work, after you click Indicate, you must pause the selection process by pressing `F2`. While the selection process is paused, hover over the `Download` button, and make sure you are pointing your mouse on the `Excel` option. Once the selection process resumes, you will be able to select the `Excel` option and use the `CSV` option as an anchor. At this point, your selection should look like Figure 4-69. Click the `Confirm` button.

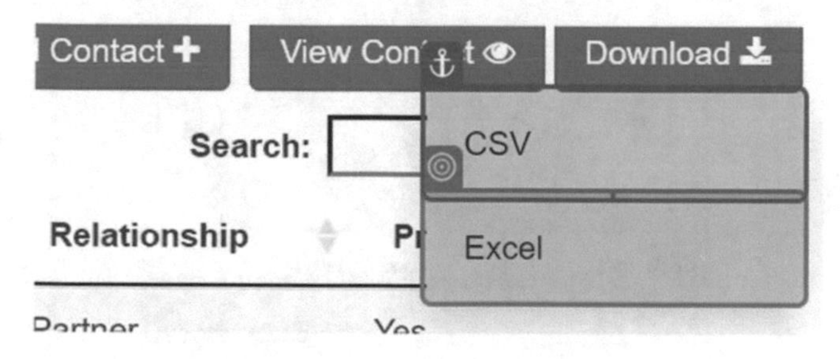

Figure 4-69. *Hover menu selection*

13. Next, add a `Write Line` activity to the body of `Use Application/Browser` activity after the `Wait for Download` activity.

14. Next, in the `Text` field of the `Write Line` activity, click the `Plus` icon, select `Text` option, and type `File downloaded at path: DownloadedFile` ➤ `Full Name`. You will need to use the saved value `DownloadedFile` to display the path of the downloaded file.

Once you have completed the exercise, the final configuration of all activities should resemble Figure 4-70. Figure 4-71 shows the full path of the downloaded file in the Output panel.

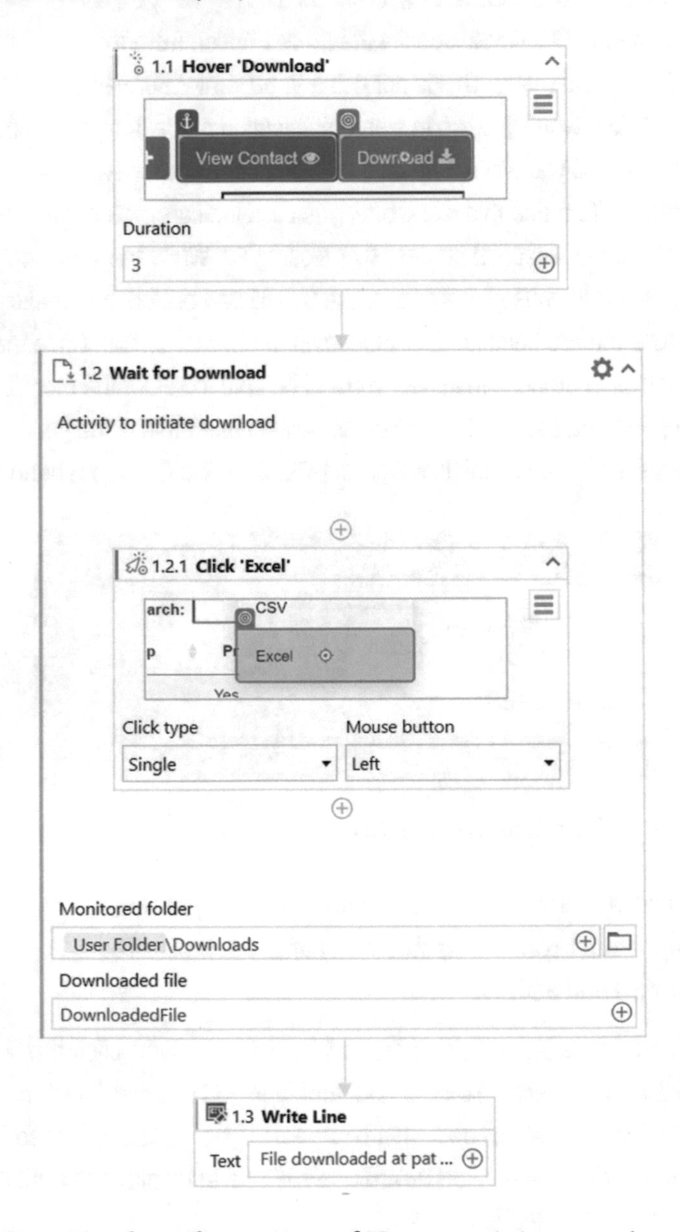

Figure 4-70. *Final configuration of Hover activity exercise*

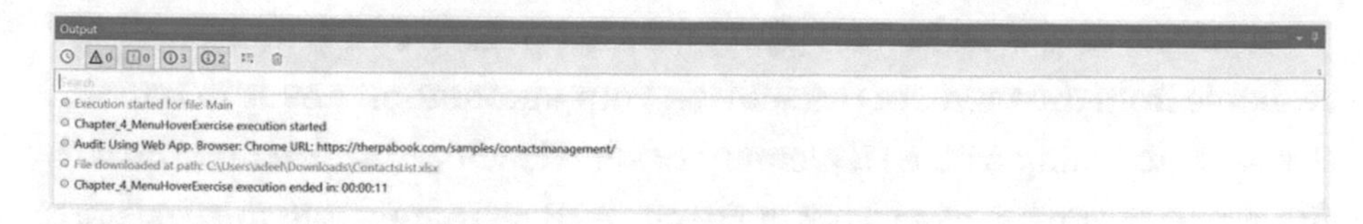

Figure 4-71. *Output of the Hover activity exercise*

Keyboard Shortcuts

The **Keyboard Shortcuts** activity allows you to send one or multiple keyboard shortcuts to a UI element.

Tip Keyboard shortcuts can be used while automating legacy applications or in scenarios where you are unable to click a UI element. This activity should not be used as the default navigation mechanism.

Configuration

This section provides instructions on how to configure a **Keyboard Shortcuts** activity, shown in Figure 4-72.

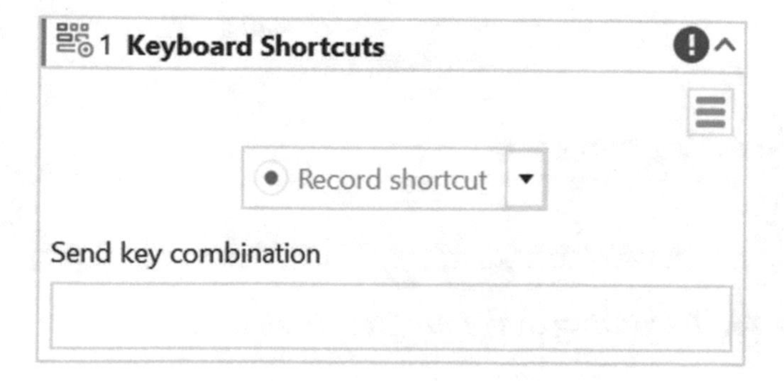

Figure 4-72. *Activity card for Keyboard Shortcuts*

Indicate target on screen (I): This is an optional configuration available from the menu of the activity. Your automation needs to be already interacting with a UI element to send keyboard shortcuts. You can either use an activity like `Click` or specify a UI element using this `Indicate target on screen (I)` link.

Record shortcut: This is a required configuration available on the activity card. This configuration allows you to specify the actual shortcuts that you want to send. Figure 4-73 shows the three options for specifying keyboard shortcuts.

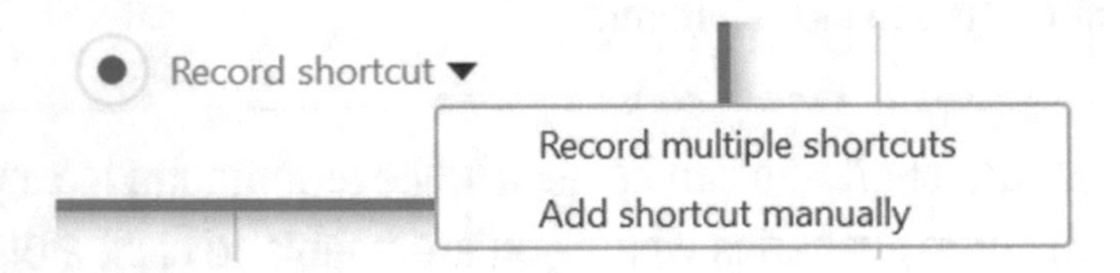

***Figure 4-73.** Available options for recording shortcuts*

First, you can use the `Record shortcut` option to record a single keyboard shortcut. As soon as you click this option, the recording starts, as shown in Figure 4-74.

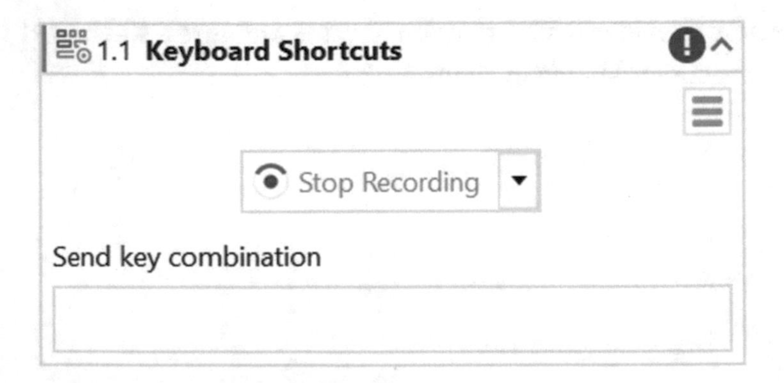

***Figure 4-74.** Recording in the progress indicator*

The next keystroke you type will be recorded as your shortcut. Figure 4-75 shows the result of the recording.

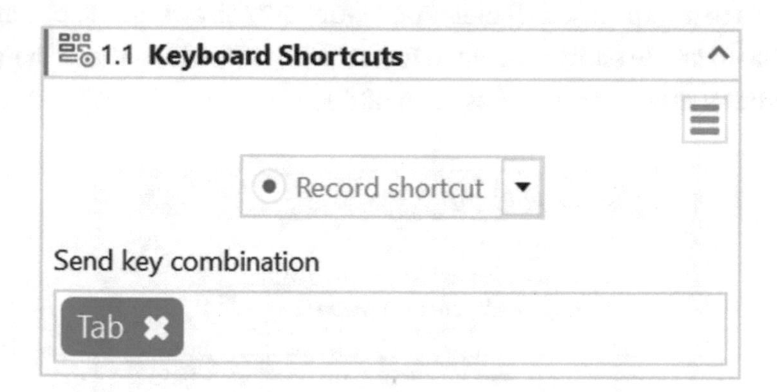

Figure 4-75. *Results of single keyboard shortcut recording*

Second, you can use `Record multiple shortcuts` to record one or multiple keyboard shortcuts. The recording process is the same as the previous option. In this option, the recording does not stop after a single keyboard shortcut. Figure 4-76 shows how the activity looks after multiple shortcuts are captured.

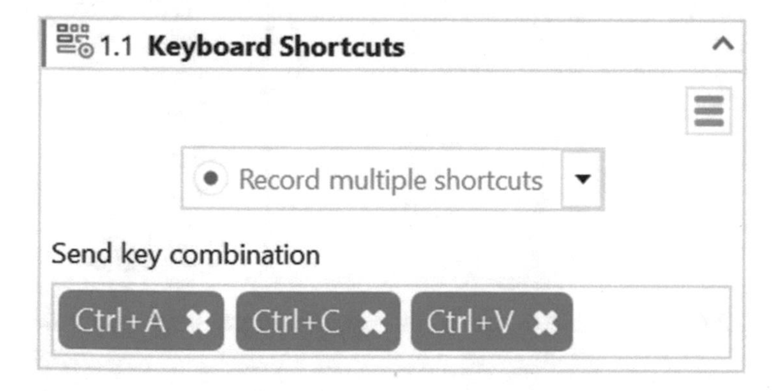

Figure 4-76. *Results of multiple keyboard shortcuts recording*

Finally, if you are having difficulty recording your shortcuts, you can use the `Add a shortcut manually` option to send one or multiple shortcuts to the specified UI element. Figure 4-77 shows all the options available to create shortcuts, while Figure 4-78 shows how the activity looks after a single shortcut has been added.

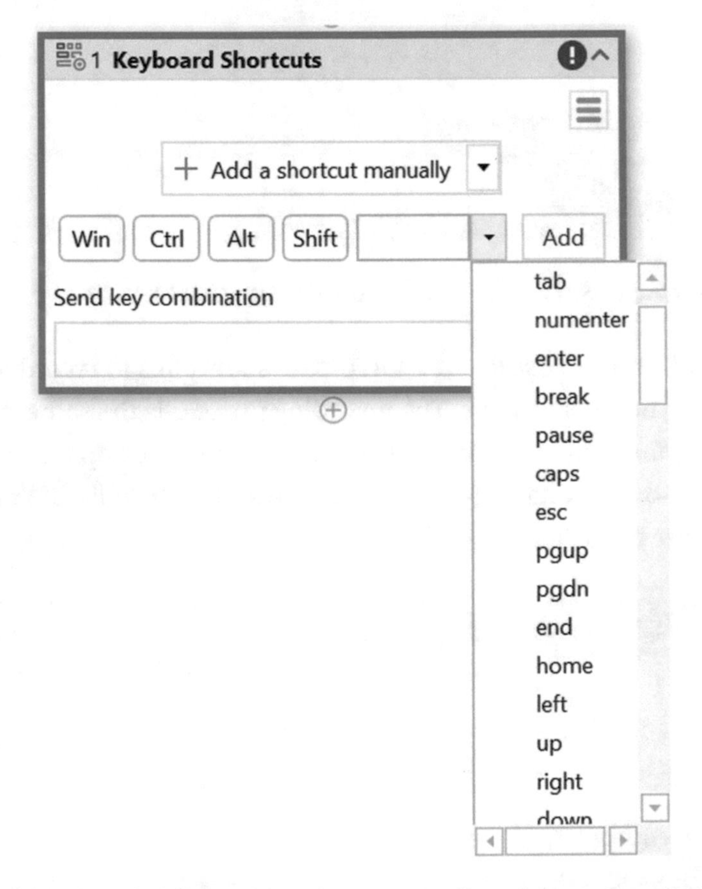

Figure 4-77. *Available options for manually adding shortcuts*

Figure 4-78. *Results of manually adding keyboard shortcuts*

Click before typing: This is an optional configuration available on the `Properties` panel. This configuration allows you to perform a `Single` or `Double` click on the UI element before sending keyboard shortcuts. By default, this is set to `None`.

Delay between shortcuts: This is an optional configuration available on the `Properties` panel. This configuration allows you to specify the delay (in seconds) between multiple keyboard shortcuts. By default, this is empty, that is, no delay is set.

EXERCISE

Goal: Use the `Keyboard Shortcuts` activity to navigate between text fields of Contact Details dialog and enter data.

Source Code: Chapter_4_FormDataEntryKeyShortcutsExercise

Setup: Here are step-by-step implementation instructions:

1. Open a browser of your choice and enter `https://therpabook.com/samples/contactsmanagement/` in the address bar.

2. In StudioX, add the Use Application/Browser activity to a blank process.
3. Next, click Indicate application in the Use Application/Browser activity card, and point your mouse to the browser. This will automatically populate the activity card.
4. Next, select the Use Application/Browser activity card, and from Properties, set the Options ➤ Open property to Always. This will ensure that the automation always opens a new browser.
5. Next, add a Click activity within the Use Application/Browser activity and configure it to click the Add Contact button (see Click activity exercise).
6. Before proceeding, make sure the Add Contact dialog is open.
7. When the dialog is opened, by default, none of the elements are active, so you cannot be sure to which element the automation will send keyboard shortcuts. To ensure that there is a known starting point, add another Click activity within the Use Application/Browser activity. Configure it to click the title of the dialog, that is, Contact Details.
8. Next, add a Keyboard Shortcuts activity right after the Click activity. This will make the automation move to the first field in the form. Use the Record shortcut feature to record a single Tab.
9. Next, add a Type Into activity. Click the Plus icon, select the Text option, and type Delos Inc. Click Save.
10. Next, add another Keyboard Shortcuts activity. This will make the automation move to the next field in the form. Use the Record shortcut feature to record a single Tab.

11. Next, add a `Type Into` activity. Click the `Plus` icon, select the `Text` option, and type `James Delos`. Click `Save`.

Once you have completed the exercise, the final configuration of the **Keyboard Shortcuts** activity should resemble Figure 4-79. Figure 4-80 shows the state of the target web application once the automation has completed its run. In this case, data in the Company and Contact fields have been entered.

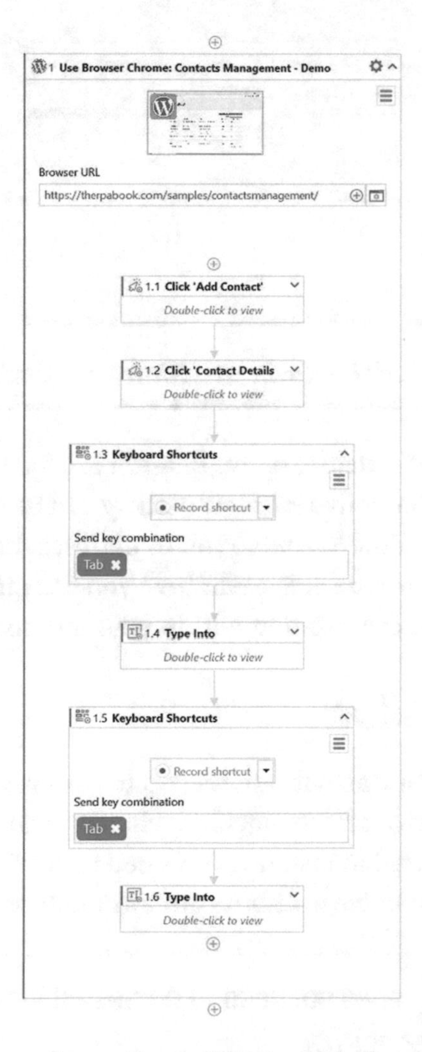

Figure 4-79. *Final configuration of Keyboard Shortcuts activity exercise*

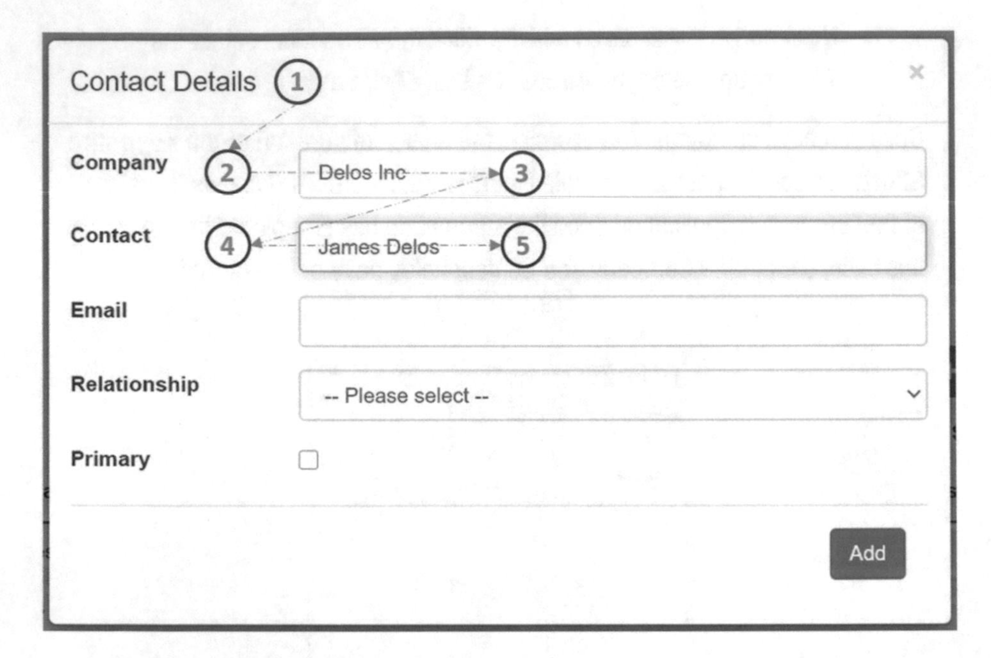

Figure 4-80. *Result of the Keyboard Shortcuts activity exercise*

The next group of activities of this chapter are focused on window manipulation. The Get Active Window activity can be used to reference the current active window to enable you to perform multiple actions. For example, once you have the active window, you can choose to maximize, hide, restore, or move that window within your automation.

Get Active Window

The **Get Active Window** activity allows you to get a reference to the window currently active on the machine where the automation is running. This activity can be used in instance you need to get the active window to later minimize or maximize it within your automation.

Note This activity does not need to be nested inside a Use Application/Browser activity.

Configuration

This section provides instructions on how to configure a **Get Active Window** activity, shown in Figure 4-81.

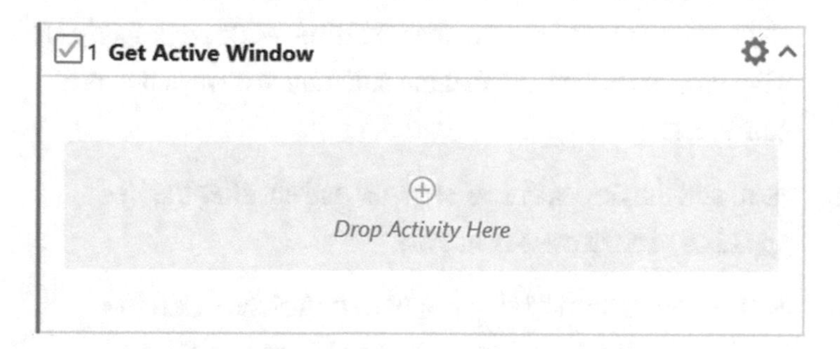

Figure 4-81. *Activity card for Get Active Window*

ApplicationWindow: This is an optional configuration available on the `Properties` panel. This configuration allows you to save a reference to the active window for use later.

EXERCISE

Goal: Use the `Get Active Window` activity to get a reference of the Contacts Management application window.

Source Code: Chapter_4_WindowOperationsExercise

Setup: Here are step-by-step implementation instructions:

1. Open a browser of your choice and enter `https://therpabook.com/samples/contactsmanagement/` in the address bar.
2. In `StudioX`, add the `Use Application/Browser` activity to the `Designer` panel.

3. Next, click Indicate application in the Use Application/Browser activity card, and point your mouse to the browser. This will automatically populate the activity card.

4. Next, select the Use Application/Browser activity card, and from Properties, set the Options ➤ Open property to Always. This will ensure that the automation always opens a new browser.

5. Next, add the Get Active Window activity after the Use Application/Browser activity.

6. Next, in the Properties panel of Get Active Window activity, click the Plus icon in the ApplicationWindow field. Click Save for Later Use and name the variable as ActiveWindow.

Once you have completed the exercise, the final configuration of the **Get Active Window** activity should resemble Figure 4-82.

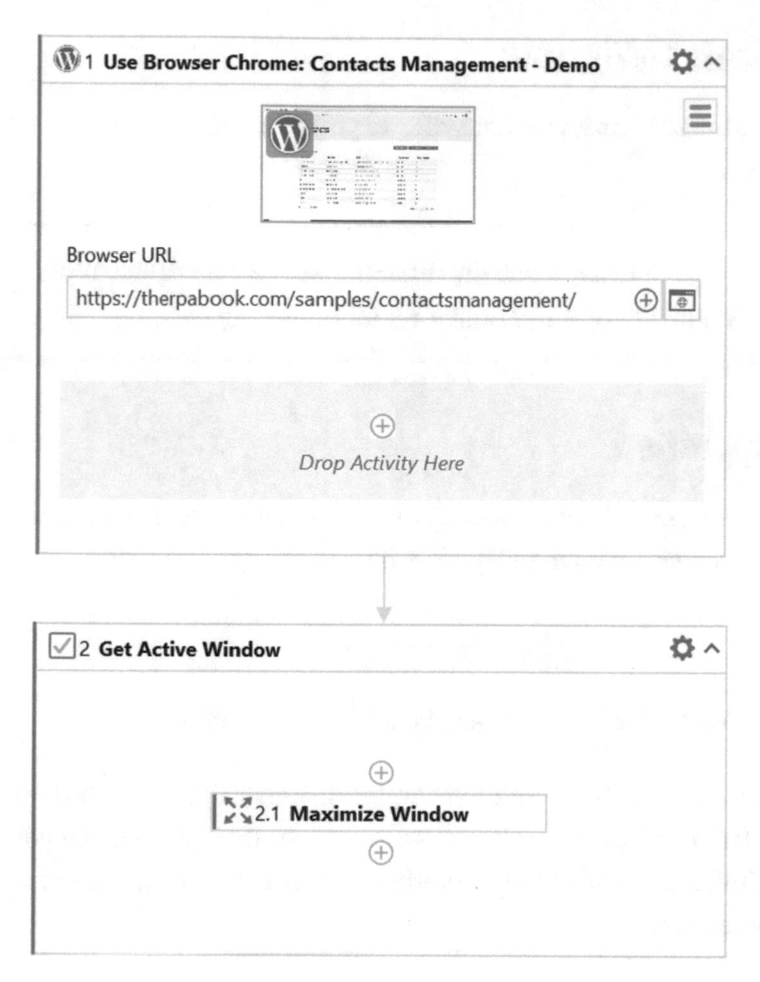

***Figure 4-82.** Final configuration of the Get Active Window exercise*

Maximize Window

The **Maximize Window** activity allows you to maximize the specified window.

Note This activity is usually nested inside a Use Application/ Browser activity or a Get Active Window activity.

Configuration

This section provides instructions on how to configure a **Maximize Window** activity, shown in Figure 4-83.

Figure 4-83. *Activity card for Maximize Window*

Window: This is an optional configuration available on the `Properties` panel. This configuration allows you to specify the window that you want to maximize. The reference to window can be obtained using `Get Active Window` activity.

EXERCISE

Goal: Use the `Maximize Window` activity to maximize the Contacts Management application window. This exercise builds upon the previous exercise for `Get Active Window` activity.

Source Code: Chapter_4_WindowOperationsExercise

Setup: Here are step-by-step implementation instructions:

1. In `StudioX`, add the `Maximize Window` activity inside the `Get Active Window` activity.
2. Next, in the `Properties` panel of `Maximize Window` activity, click the `Plus` icon in the `Window` field. Hover over `Use Saved Value` and select `ActiveWindow`.

Once you have completed the exercise, the final configuration of the **Maximize Window** activity should resemble Figure 4-84.

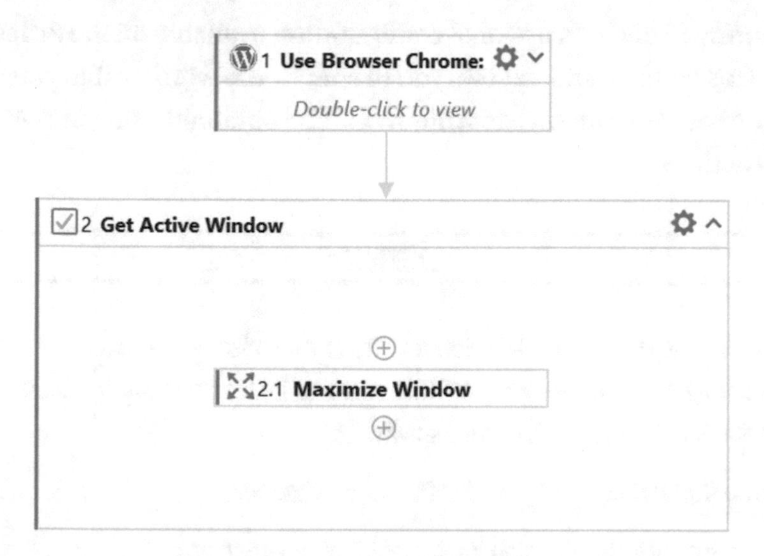

Figure 4-84. *Final configuration of the Maximize Window activity exercise*

Minimize Window

The **Minimize Window** activity allows you to minimize the specified window.

Note This activity does not need to be nested inside a Use Application/Browser activity or a Get Active Window activity.

Configuration

This section provides instructions on how to configure a **Minimize Window** activity, shown in Figure 4-85.

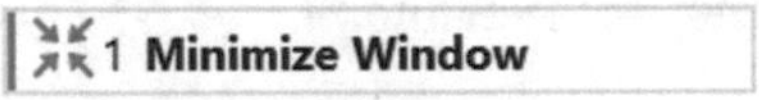

1 Minimize Window

Figure 4-85. *Activity card for Minimize Window*

Window: This is an optional configuration available on the `Properties` panel. This configuration allows you to specify the window that you want to minimize. The reference to window can be obtained using `Get Active Window` activity.

EXERCISE

Goal: Use the `Minimize Window` activity to minimize the Contacts Management application window. This exercise builds upon the previous exercise for `Maximize Window` activity.

Source Code: Chapter_4_WindowOperationsExercise

Setup: Here are step-by-step implementation instructions:

1. In `StudioX`, add the `Delay` activity inside the `Get Active Window` activity after the `Maximize Window` activity. Update the `Duration` field to `3` seconds. We are adding a slight delay just so that we can see the operations happening; otherwise, this is not needed.
2. Next, add the `Minimize Window` activity inside the `Get Active Window` activity after the `Delay` activity.

3. In the Properties panel of Minimize Window activity, click the Plus icon in the Window field. Hover over Use Saved Value and select ActiveWindow.

Once you have completed the exercise, the final configuration of the **Minimize Window** activity should resemble Figure 4-86.

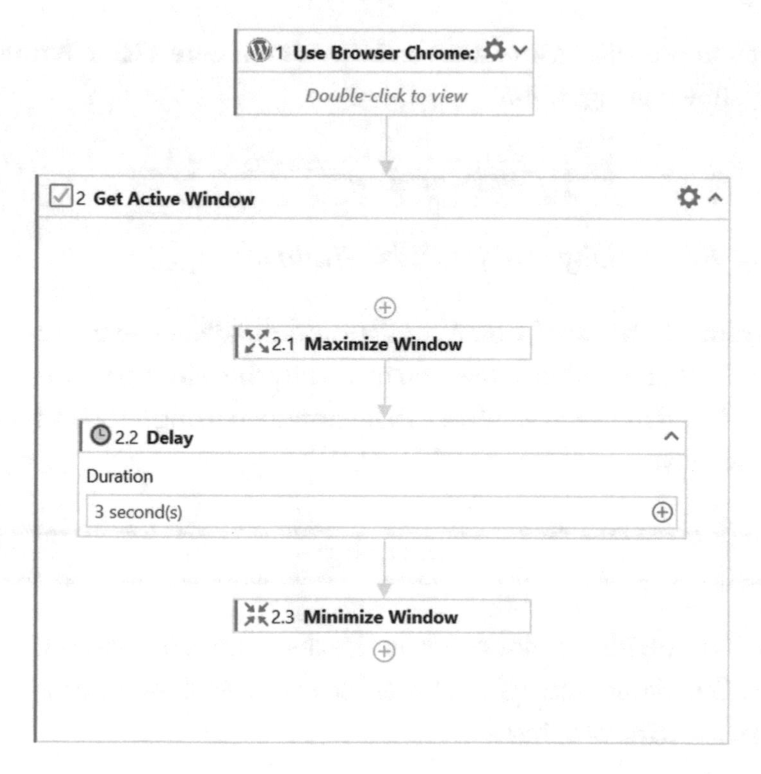

Figure 4-86. *Final configuration of the Minimize Window activity exercise*

Hide Window

The **Hide Window** activity allows you to hide the specified window.

Note This activity does not need to be nested inside a Use Application/Browser activity or a Get Active Window activity.

Configuration

This section provides instructions on how to configure a **Hide Window** activity, shown in Figure 4-87.

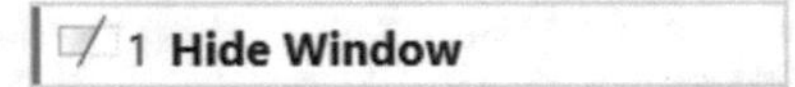

Figure 4-87. *Activity card for Hide Window*

Window: This is an optional configuration available on the `Properties` panel. This configuration allows you to specify the window that you want to hide. The reference to window can be obtained using `Get Active Window` activity.

EXERCISE

Goal: Use the `Hide Window` activity to hide the Contacts Management application window. This exercise builds upon the previous exercise for `Minimize Window` activity.

Source Code: Chapter_4_WindowOperationsExercise

Setup: Here are step-by-step implementation instructions:

1. In `StudioX`, add the `Delay` activity inside the `Get Active Window` activity after the `Minimize Window` activity. Update the `Duration` field to `3` seconds. We are adding a slight delay just so that we can see the operations happening; otherwise, this is not needed.

2. Next, add the Hide Window activity inside the Get Active Window activity after the Delay activity.
3. In the Properties panel of Hide Window activity, click the Plus icon in the Window field. Hover over Use Saved Value and select ActiveWindow.

Once you have completed the exercise, the final configuration of the **Hide Window** activity should resemble Figure 4-88.

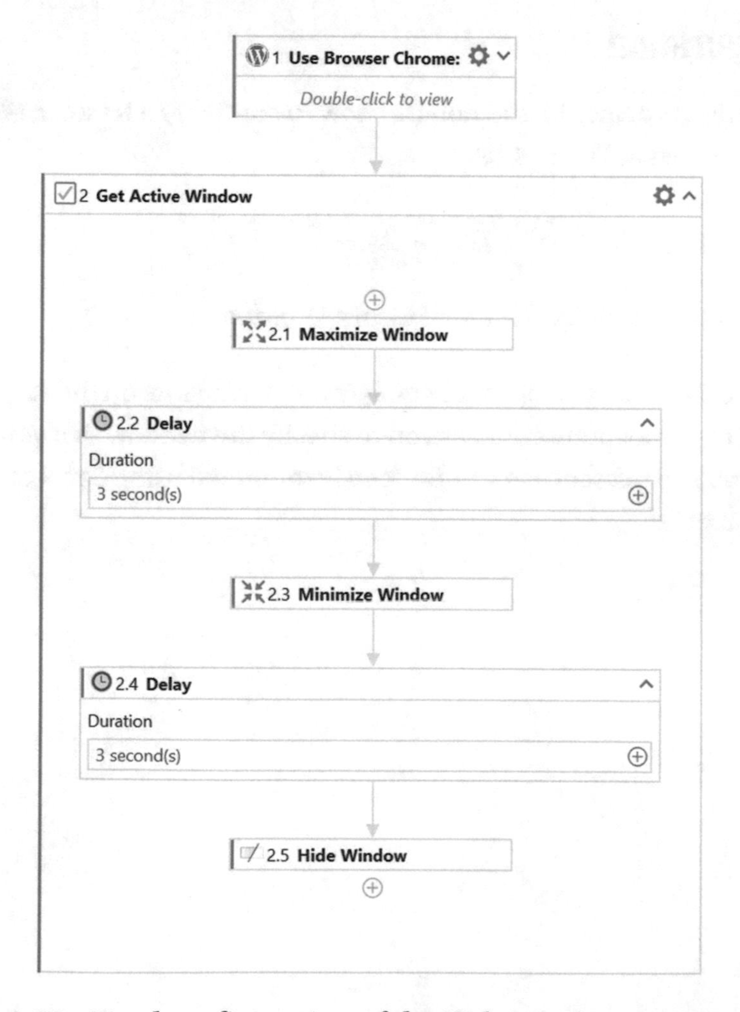

***Figure 4-88.** Final configuration of the Hide Window activity exercise*

Restore Window

The **Restore Window** activity allows you to restore the specified window.

Note This activity does not need to be nested inside a Use Application/Browser activity or a Get Active Window activity.

Configuration

This section provides instructions on how to configure a **Restore Window** activity, shown in Figure 4-89.

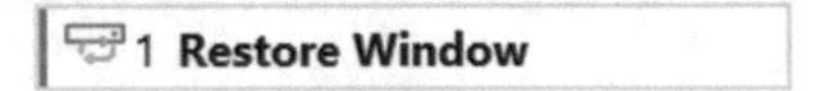

Figure 4-89. *Activity card for Restore Window*

Window: This is an optional configuration available on the `Properties` panel. This configuration allows you to specify the window that you want to restore. The reference to window can be obtained using `Get Active Window` activity.

EXERCISE

Goal: Use the Restore Window activity to restore the Contacts Management application window. This exercise builds upon the previous exercise for Hide Window activity.

Source Code: Chapter_4_WindowOperationsExercise

Setup: Here are step-by-step implementation instructions:

1. In StudioX, add the Delay activity inside the Get Active Window activity after the Minimize Window activity. Update the Duration field to 3 seconds. We are adding a slight delay just so that we can see the operations happening; otherwise, this is not needed.

2. Next, add the Restore Window activity inside the Get Active Window activity after the Delay activity.

3. In the Properties panel of Minimize Window activity, click the Plus icon in the Window field. Hover over Use Saved Value and select ActiveWindow.

Once you have completed the exercise, the final configuration of the **Restore Window** activity should resemble Figure 4-90.

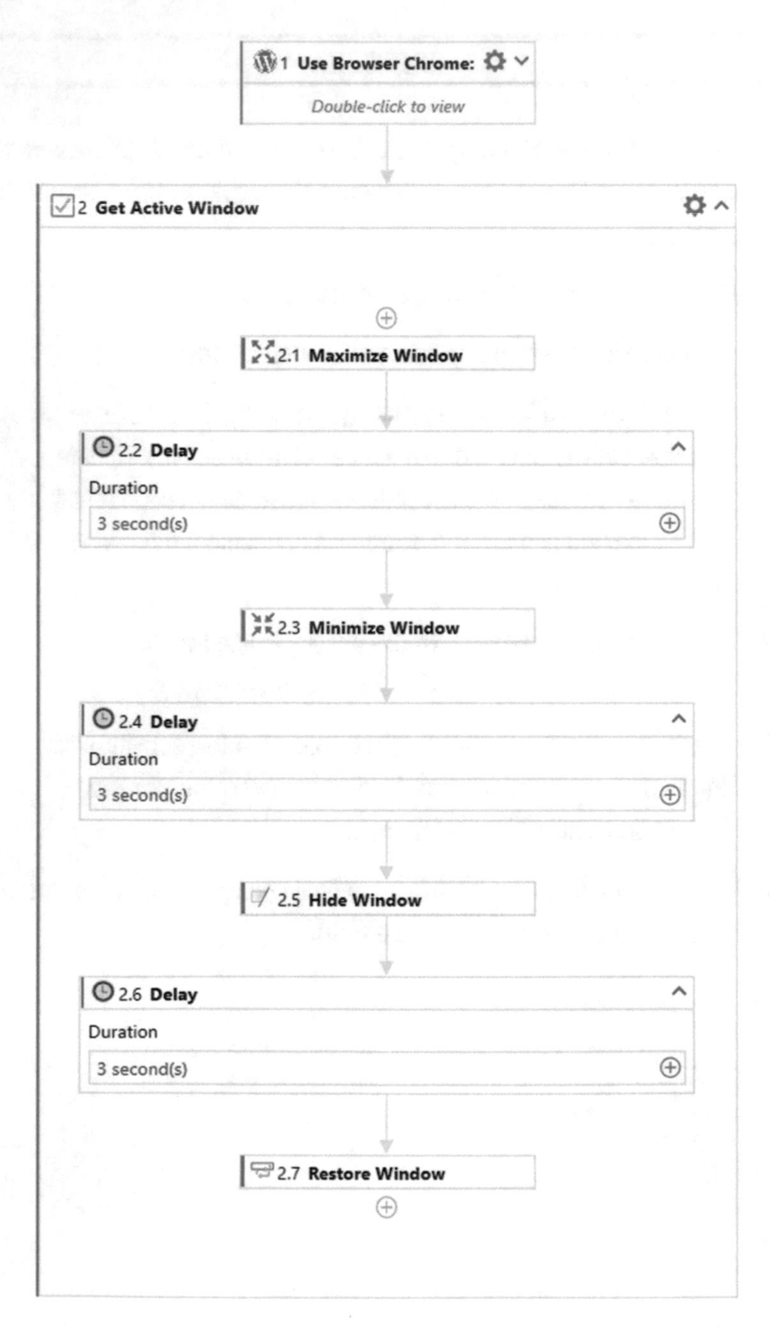

Figure 4-90. *Final configuration of the Restore Window activity exercise*

Move Window

The **Mode Window** activity allows you to change the size of the specified and move the specified window to a new position on the screen.

Note This activity does not need to be nested inside a Use Application/Browser activity or a Get Active Window activity.

Configuration

This section provides instructions on how to configure a **Move Window** activity, shown in Figure 4-91.

1 Move Window

Figure 4-91. *Activity card for Move Window*

Window: This is an optional configuration available on the `Properties` panel. This configuration allows you to specify the window that you want to minimize. The reference to window can be obtained using `Get Active Window` activity.

Height: This is an optional configuration available on the `Properties` panel. This configuration allows you to specify the new height of the window.

Width: This is an optional configuration available on the `Properties` panel. This configuration allows you to specify the new width of the window.

X: This is an optional configuration available on the `Properties` panel. This configuration allows you to specify the new X position of the window.

Y: This is an optional configuration available on the `Properties` panel. This configuration allows you to specify the new Y position of the window.

EXERCISE

Goal: Use the Move Window activity to reduce the screen size of the Contacts Management application to 500 x 500 and move it to X: 100 and Y: 100 location.

Source Code: Chapter_4_WindowOperationsExercise

Setup: Here are step-by-step implementation instructions:

1. In StudioX, add the Delay activity inside the Get Active Window activity after the Restore Window activity. Update the Duration field to 3 seconds. We are adding a slight delay just so that we can see the operations happening; otherwise, this is not needed.
2. Next, add the Move Window activity inside the Get Active Window activity after the Delay activity.
3. In the Properties panel of Move Window activity, click the Plus icon in the Window field. Hover over Use Saved Value and select ActiveWindow.
4. In the Height field, click the Plus icon, select the Number option, and type 500.
5. In the Width field, click the Plus icon, select the Number option, and type 500.
6. In the X field, click the Plus icon, select the Number option, and type 100.
7. In the Y field, click the Plus icon, select the Number option, and type 100.

Once you have completed the exercise, the final configuration of the **Move Window** activity should resemble Figure 4-92.

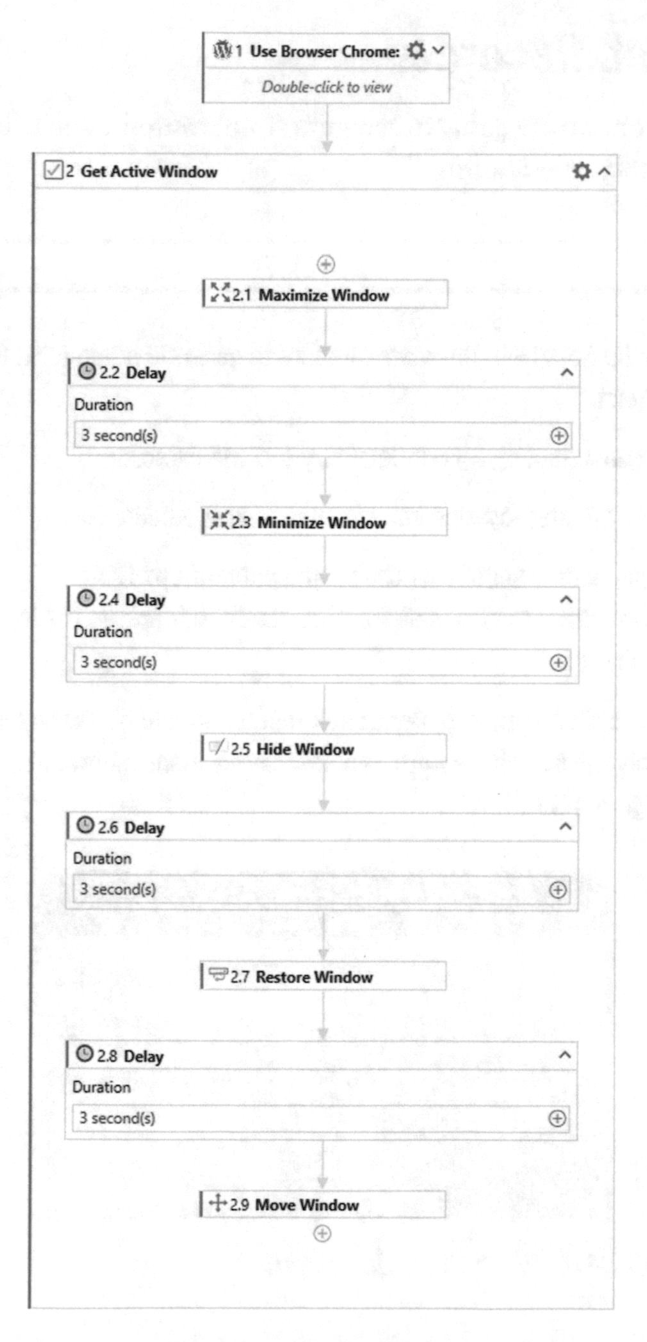

Figure 4-92. *Final configuration of the Move Window activity exercise*

App/Web Recorder

In this section, we are going to generate simple automation using the **App/Web Recorder** feature.

EXERCISE

Goal: Use the `App/Web Recorder` feature to generate automation for adding a new contact.

Source Code: Chapter_4_FormDataEntryRecorderExercise

Setup: Here are step-by-step implementation instructions:

1. Open a browser of your choice and enter `https://therpabook.com/samples/contactsmanagement/` in the address bar.
2. Next, click `App/Web Recorder` menu from the ribbon on top. This will launch the `App/Web Recorder` menu, shown in Figure 4-93.

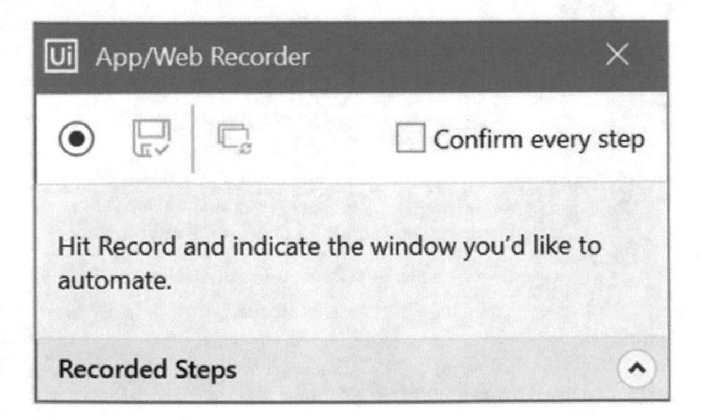

***Figure 4-93.** App/Web Recorder menu*

3. Next, click the `Start recording` button and point your mouse to the browser that has the Contacts Management application open.
4. Next, click `Add Contact` button.
5. Next, select the `Company` field, and using the prompt, type `E Corp`.
6. Next, select the `Contact` field, and using the prompt, type `Phillip Price`.
7. Next, select the `Email` field, and using the prompt, type phillip.price.belson@e.co.
8. Next, select the `Relationship` dropdown, and using the prompt, choose `Customer`.
9. Next, check the `Primary` checkbox.
10. Next, click the `Add` button.
11. We have completed data entry for a new contact, so click the `Pause` icon on the `App/Web Recorder` menu and click the `Save` icon.

Once you have completed the exercise, the final configuration of the automation generated by the **App/Web Recorder** should resemble Figures 4-94 and 4-95.

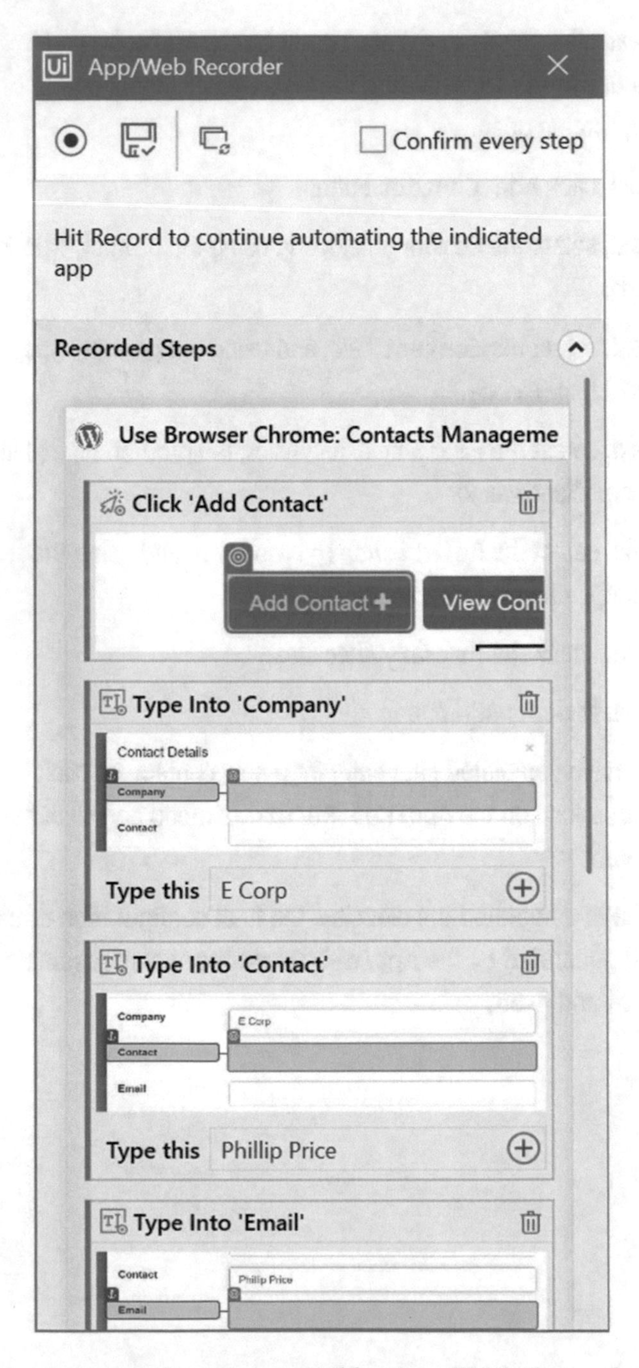

Figure 4-94. *Automation generated by App/Web Recorder*

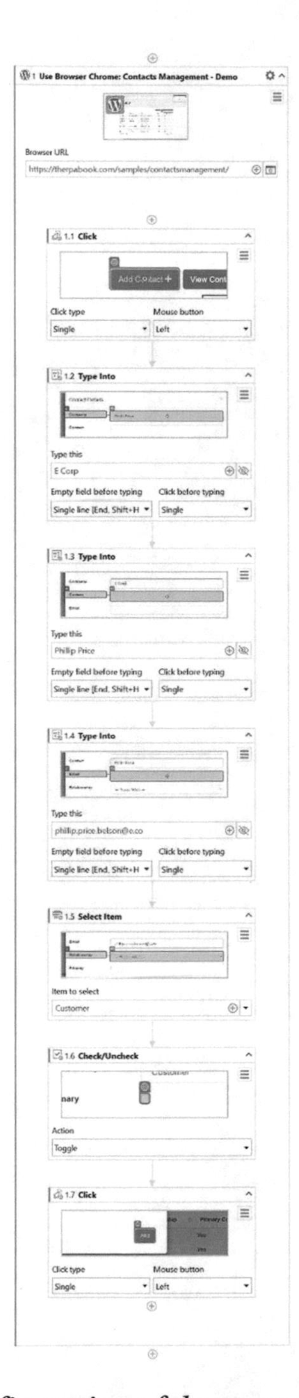

***Figure 4-95.** Final configuration of the generated automation*

CHAPTER 5

Mail Automation

Email is essential for organizing, communicating, and calendaring in our day-to-day. UiPath StudioX Mail automation enables business users to be able to automate those tasks to focus on the more creative high-value tasks. Often, a mailbox is the initial input of automation, and moving emails across mailbox folders can be a way to indicate the progress of a process, or the mailbox folder can serve as an exception folder for human review. It is also vital in providing a way for the robot to notify a user there may be a pending task or a way to automate communication by sending emails based on given business rules.

Learning Objectives

At the end of this chapter, you will learn how to

- Define a Desktop Outlook, Outlook 365, and Gmail account scope to automate
- Move emails to another folder
- Mark email as read or unread
- Forward emails
- Save attachments and emails
- Archive emails

A. Javed et al., *Robotic Process Automation using UiPath StudioX*,
https://doi.org/10.1007/978-1-4842-6794-3_5

- Send emails and calendar invites
- Reply to an email
- Delete emails

Sample Overview

Throughout this chapter, we will be using a simplified version of the initial recruitment screening process to showcase the usage of all Mail automation activities. Concepts discussed in this chapter can be applied across various areas of business, for example, vendor management, procurement, sales, and other processes within human resources.

This section will familiarize you with the prerequisites for all exercises in this chapter.

Desktop Outlook Setup

First, create a new **Initial Screen** folder inside the **Inbox** of the email account that you are going to use for automation. Figure 5-1 shows the Outlook folder structure.

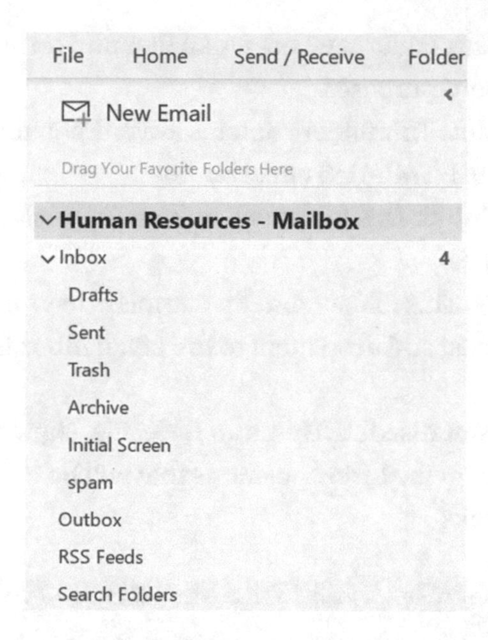

Figure 5-1. *Human Resources department Outlook folder structure*

File System Structure

Next, download the source code from the book's site, and make sure you move the entire BookSamples folder to your C:\ drive. All exercises in this chapter assume the folder paths will be **C:\BookSamples\Chapter_05**. Figure 5-2 shows the physical folder structure required for this sample. This folder structure comes with the source code.

This PC > OS (C:) > BookSamples > Chapter_05

Name	Date modified	Type	Size
Downloads	9/13/2020 9:29 AM	File folder	
EmailTemplates	9/13/2020 10:14 PM	File folder	
PreparationGuide	9/13/2020 9:58 AM	File folder	
SampleEmails	9/13/2020 10:14 PM	File folder	
Job_Posting_Details	9/13/2020 9:59 PM	Microsoft Excel Worksheet	11 KB

Figure 5-2. *Folder structure used for the Initial Recruitment Screening exercise*

Downloads: This folder is where email files and resumes will be saved for review and record purposes.

EmailTemplates: This folder contains a Word document that will be used to incorporate HTML-rich email bodies.

PreparationGuide: This folder will be sent as preparation material for interview candidates.

SampleEmails: This folder contains sample emails used in this chapter. You can drag and drop them to the email inbox that you are using for automation.

Job_Posting_Details.xlsx: This is an Excel file, shown in Figure 5-3, with details related to the hiring positions that will be referenced in Outlook email details.

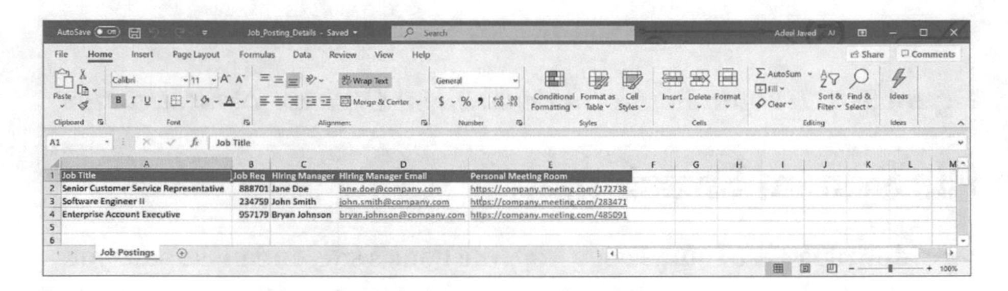

Job Title	Job Req	Hiring Manager	Hiring Manager Email	Personal Meeting Room
Senior Customer Service Representative	888701	Jane Doe	jane.doe@company.com	https://company.meeting.com/172738
Software Engineer II	234759	John Smith	john.smith@company.com	https://company.meeting.com/283471
Enterprise Account Executive	957179	Bryan Johnson	bryan.johnson@company.com	https://company.meeting.com/485091

Figure 5-3. *Job_Posting_Details Excel file with posting details*

Activities Reference

As shown in Figure 5-4, all Mail automation activities can be found under the Mail category. The following sections will provide instructions on how to configure and use each activity.

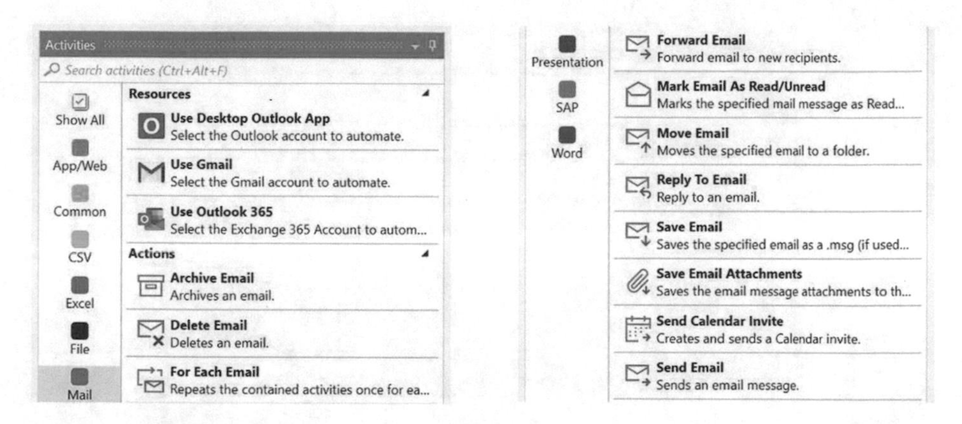

Figure 5-4. *Activities for Mail automation*

Use Desktop Outlook App

The **Use Desktop Outlook App** activity allows you to select the Outlook account to automate that is already configured in the Outlook desktop application.

This activity will contain all the actions that you want to take on the Outlook account. For example, suppose you want to send an email. In that case, the **Send Email** activity will be nested in the body of **Use Desktop Outlook App** activity.

Configuration

This section provides instructions on how to configure a **Use Desktop Outlook App** activity, shown in Figure 5-5.

Note To configure this activity, make sure your Microsoft Outlook desktop application is configured and signed in.

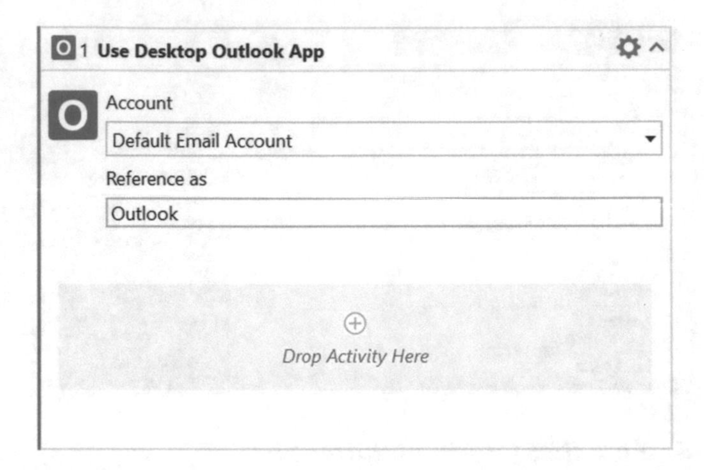

Figure 5-5. *Activity card for Use Desktop Outlook App*

Account: This is a required configuration available on the activity card. This configuration allows you to specify the Outlook account email address you plan to automate. By default, your `Default Email Account` is selected. If you have multiple Outlook accounts, then you can use the dropdown to select the appropriate account.

Tip Leaving the account as the Default Email Account option makes it easy to share the automation with other users. This way, automation is designed to run on the Outlook account of the workstation the process is executing from.

Reference as: This is a required configuration available on the activity card. This configuration allows you to provide a name for your Outlook account. All the activities that need to use the selected Outlook account will reference it by this name.

Use Outlook 365

The **Use Outlook 365** activity allows you to select the Exchange account to automate.

This activity will contain all the actions that you want to take on the Exchange account. For example, suppose you want to send an email. In that case, the **Send Email** activity will be nested in the body of **Use Outlook 365** activity.

Configuration

This section provides instructions on how to configure a **Use Outlook 365** activity, shown in Figure 5-6.

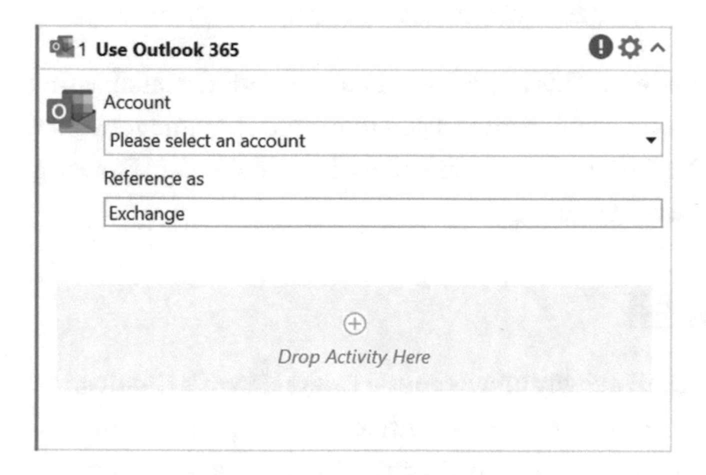

Figure 5-6. *Activity card for Use Outlook 365*

Account: This is a required configuration available on the activity card. This configuration allows you to specify or add the Outlook 365 account email address you plan to automate. Figure 5-7 shows the screen to add a new Exchange account. The organization's IT department will most likely provide this information.

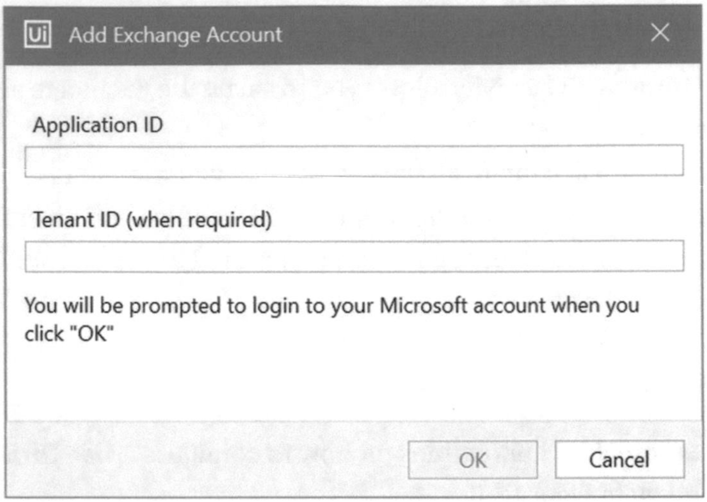

Figure 5-7. Add Exchange Account screen

Reference as: This is a required configuration available on the activity card. This configuration allows you to provide a name for your email account. All the activities that need to use the selected Exchange account will reference it by this name.

Use Gmail

The **Use Gmail** activity allows you to integrate with Gmail and Google Calendar to automate email activities for the Gmail account.

This activity will contain all the actions that you want to take on the Gmail account. For example, suppose you want to send an email. In that case, the **Send Email** activity will be nested in the body of **Use Gmail** activity.

Configuration

This section provides instructions on how to configure a **Use Gmail**, shown in Figure 5-8.

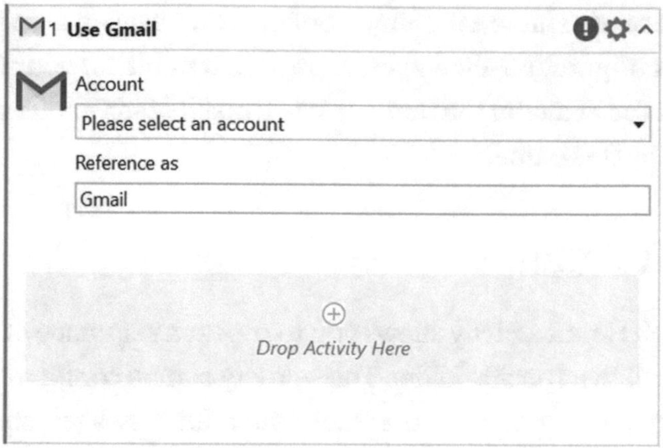

Figure 5-8. Activity card for Use Gmail

Account: This is a required configuration available on the activity card. This configuration allows you to specify the Gmail account email address you plan to automate. Figure 5-9 shows the screen to add a new Exchange account. The organization's IT department will most likely provide this information.

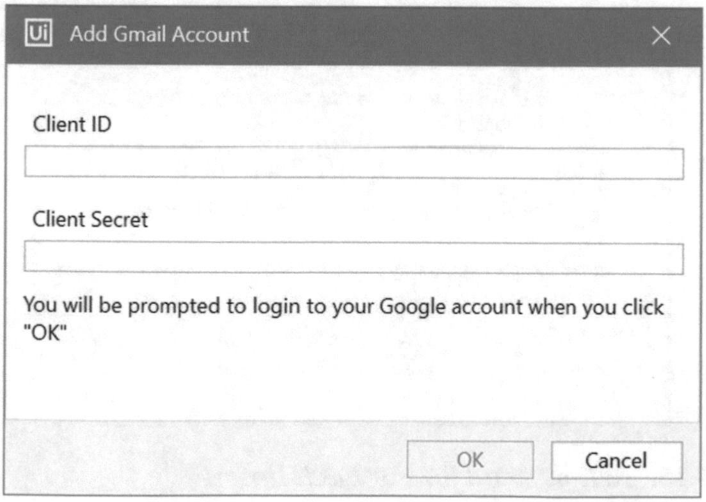

Figure 5-9. Add Gmail Account screen

Reference as: This is a required configuration available on the activity card. This configuration allows you to provide a name for your Gmail account. All the activities that need to use the selected Gmail account will reference it by this name.

For Each Email

The **For Each Email** activity allows you to repeat a sequence of activities for each email in a specific folder. This activity can be configured to filter for a specific number of emails, unread mail messages, emails with attachments only, and additional advanced custom filters.

Configuration

This section provides instructions on how to configure a **For Each Email** activity, shown in Figure 5-10.

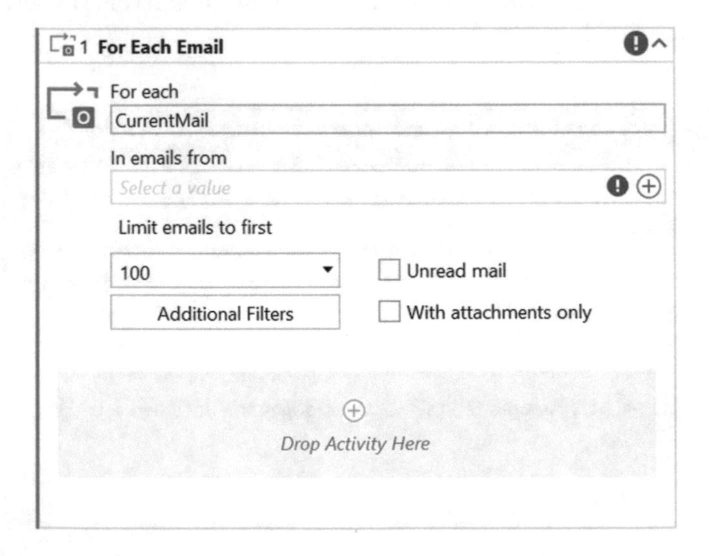

Figure 5-10. *Activity card for For Each Email*

For each: This is a required configuration available on the activity card. This configuration allows you to specify a name to reference each email within the activity card. By default, the value is set to `CurrentMail`.

In emails from: This is a required configuration available on the activity card. When placed within one of the `Use Desktop Outlook App`, `Use Outlook 365`, or `Use Gmail` activities, the folders within the specified account are auto-populated as options. You can select the mail folder from which you want this activity to read emails.

Limit emails to first: This is a required configuration available on the activity card. This configuration allows you to specify the number of emails to read from the specified folder. By default, this is set to `100`.

Unread mail: This is an optional configuration available on the activity card. When checked, only unread messages will be iterated. By default, this option is unchecked.

With attachments only: This is an optional configuration available on the activity card. When checked, only emails with attachments will be iterated. By default, this option is unchecked.

Additional Filters: This is an optional configuration available on the activity card. This configuration allows you to create additional mail filters. As shown in Figure 5-11, you can build conditional statements to filter emails by specific attributes.

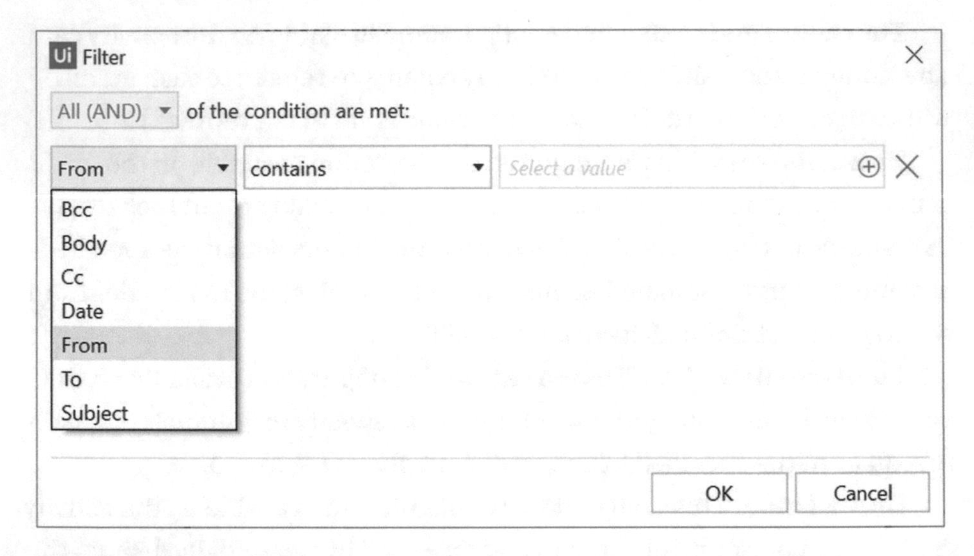

***Figure 5-11.** Additional Filters of the For Each Email activity*

EXERCISE

Goal: Utilize the `Use Desktop Outlook App` and `For Each Email` activities to read all unread emails from the Human Resources email inbox and print the subjects in the Output panel. Figure 5-12 shows the current state of the inbox.

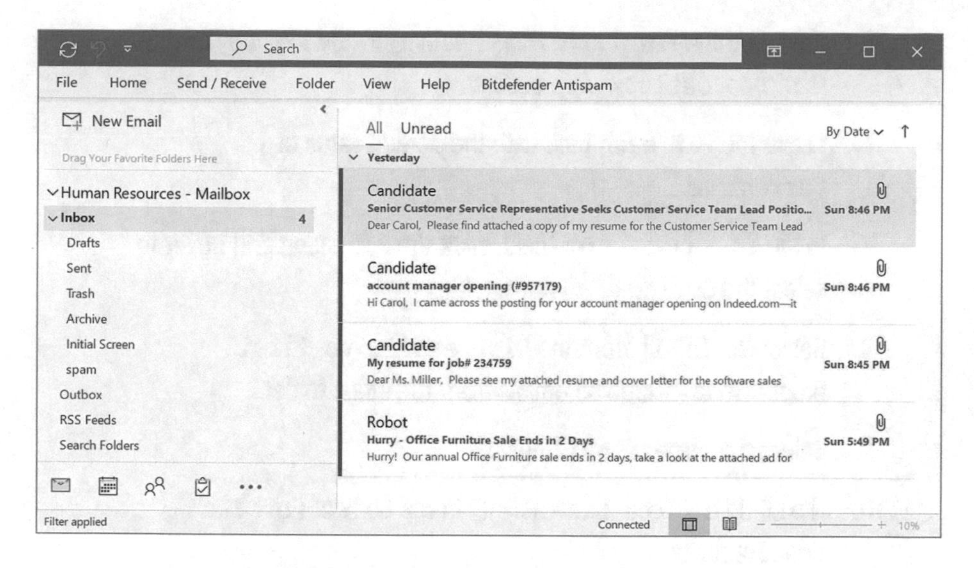

Figure 5-12. *The initial state of the Human Resources email inbox*

Source Code: Chapter_5-MailAutomationExercise

Setup: Here are step-by-step implementation instructions:

1. In `StudioX`, click the `Notebook` menu, and select `Configure Notebook`.
2. In the `Notebook file` field of the `Configure Notebook` screen, select the `C:\BookSamples\Chapter_05\Job_Posting_Details.xlsx` file. Click OK.
3. In the Designer panel, add the `Use Desktop Outlook App` activity to a blank process.
4. Select the appropriate email from the `Account` dropdown. For this exercise, we will specify the `Default Email Account`.

5. Next, add the For Each Email activity inside the body of Use Desktop Outlook App activity.
6. Leave the For each field with the default value of CurrentMail.
7. In the In emails from field, click the Plus icon and hover to select the Outlook ➤ Inbox folder.
8. Select No Limit from the List emails to first dropdown to include all emails from the inbox folder.
9. Select the Unread mail option.
10. Next, add a Write Line activity in the body of For Each Email activity.
11. In the Text field, click the Plus icon, and select CurrentMail ➤ Subject option.

Once you have completed the exercise, the final configuration of the **For Each Email** activity should resemble Figure 5-13. Figure 5-14 shows the list of received emails in the Output panel.

***Figure 5-13.** Final configuration of the For Each Email activity exercise*

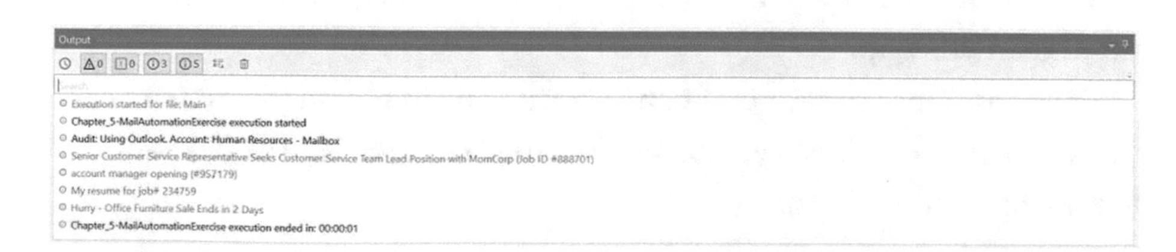

***Figure 5-14.** The output of the For Each Email activity exercise*

Mark Email As Read/Unread

The **Mark Email As Read/Unread** activity allows you to mark a specified email message as read or unread.

Configuration

This section provides instructions on how to configure a **Mark Email As Read/Unread** activity, shown in Figure 5-15.

Figure 5-15. *Activity card for Mark Email As Read/Unread*

Email: This is a required configuration available on the activity card. This configuration specifies the email whose state needs to be updated.

In the case of the Desktop Outlook app, you can also specify the email that is currently selected by choosing the `Outlook` ➤ `Selected Email` option. This will act on the selected email at execution time.

If used in a `For Each Email` activity, you can specify the `CurrentMail` message in the loop.

Mark as: This is a required configuration available on the activity card. This configuration allows you to set the email state as `Read` or `Unread`. By default, the value is `Read`.

EXERCISE

Goal: Building on our previous exercise, use the `Mark Email As Read/Unread` activity to mark all emails in the inbox as read. Figure 5-12 shows the current state of the inbox.

Source Code: Chapter_5-MailAutomationExercise

Setup: Here are step-by-step implementation instructions:

1. In `StudioX`, add a `Mark Email As Read/Unread` activity within the `For Each Email` activity after the `Write Line` activity. No additional configuration is needed. By default, the `CurrentMail` is selected, and the email state is set as `Read`.

Once you have completed the exercise, the final configuration of the **Mark Email As Read/Unread** activity should resemble Figure 5-16. Once the automation runs, all emails in the inbox are marked as read, as shown in Figure 5-17.

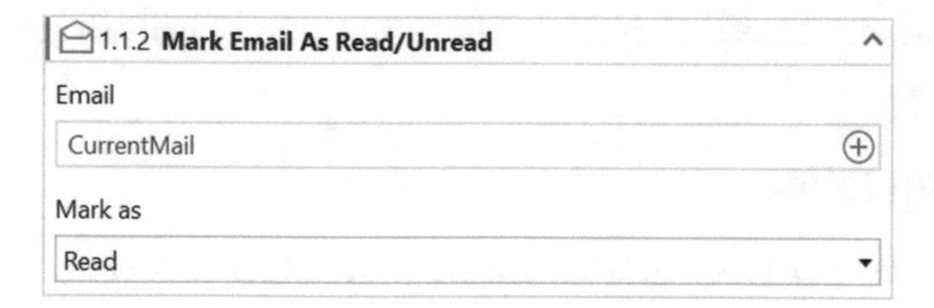

Figure 5-16. *Final configuration for Mark Email As Read/Unread exercise*

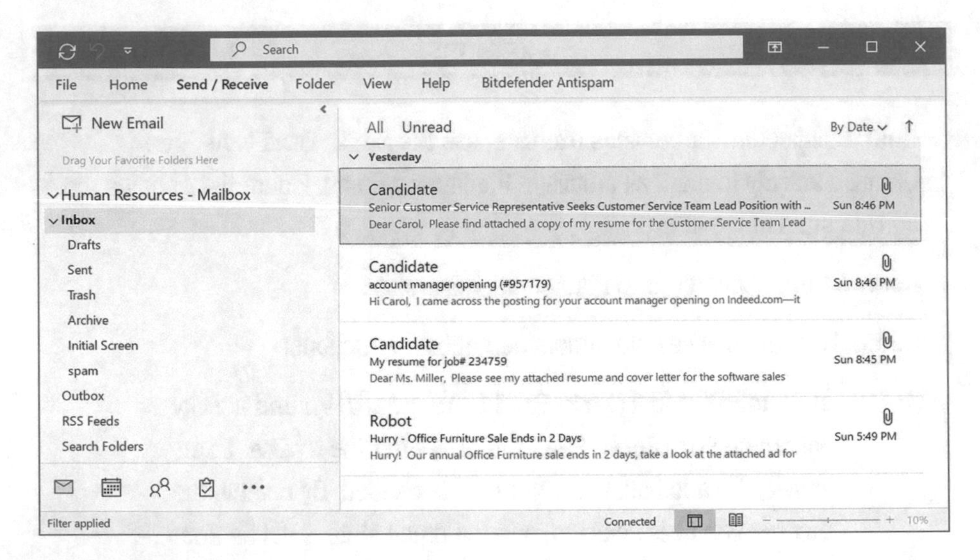

Figure 5-17. *The state of the Human Resources email inbox after the Mark Email As Read/Unread exercise*

Forward Email

The **Forward Email** activity allows you to forward an email to a specific recipient.

Configuration

This section provides instructions on how to configure a **Forward Email** activity, shown in Figure 5-18.

***Figure 5-18.** Activity card for Forward Email*

Email: This is a required configuration available on the activity card. This configuration allows you to specify the email that will be forwarded.

In the case of the Desktop Outlook app, you can also specify the email that is currently selected by choosing the `Outlook` ➤ `Selected Email` option. This will act on the selected email at execution time.

If used in a `For Each Email` activity, you can specify the `CurrentMail` message in the loop.

Add 'To' recipients: This is a required configuration available on the activity card. This configuration specifies the email recipient(s) of the forwarded email.

Add 'Cc' recipients: This is an optional configuration available on the activity card. This configuration specifies the secondary email recipient(s) of the forwarded message.

Add 'Bcc' recipients: This is an optional configuration available on the `Properties` panel. This configuration specifies the hidden email recipient(s) of the forwarded message.

New subject: This is an optional configuration available on the activity card. If specified, the subject of the forwarded email will be changed. Otherwise, the original subject will be used.

Body: This is an optional configuration available on the activity card. You can select the `Text` option to type in the body text or select the `Use Word Document` option to specify the file path of a Word document that will be used as an email body.

Save as draft: This is an optional configuration available on the activity card. This configuration is check-marked by default, indicating email will be saved as a draft. This is encouraged to be set during development time.

Attachments: This is an optional configuration available on the activity card. You can type or browse for a file or folder to attach to the forwarded email.

Max body document size: This is an optional configuration available on the `Properties` panel. You can specify the maximum size of the Word document that is being used in body text. By default, the size is set to `2 MB`.

Sent on behalf of: This is an optional configuration available on the `Properties` panel. You can specify an email address that will show up as on behalf of in the `From` field of the forwarded email.

EXERCISE

Goal: Building on our previous exercise, use the `Forward Email` activity to forward the email message to the hiring manager if it is related to Job Req `888701`. Hiring manager information will be read from cell `D2` of the `Job_Posting_Details` notebook. Figure 5-17 shows the current state of the inbox.

Source Code: Chapter_5-MailAutomationExercise

Setup: Here are step-by-step implementation instructions:

1. In StudioX, add an If activity within the For Each Email activity after the Mark Email As Read/Unread activity. This will be used to filter for emails where the subject contains the Job Req id 888701.
2. In the Condition field of the If activity, click the Plus icon, and open the Condition Builder.
3. On the left side of the condition, click the Plus icon, and hover over CurrentMail to select Subject.
4. In the condition dropdown, select contains.
5. On the right side of the condition, click the Plus icon, hover over Notebook, and select Indicate in Excel. Select cell B2 and click Confirm. Figure 5-19 shows the complete condition. This condition is checking if the subject of the current email contains the job req id mentioned in cell B2 of Job Postings sheet of the Job_Posting_Details.xlsx worksheet.

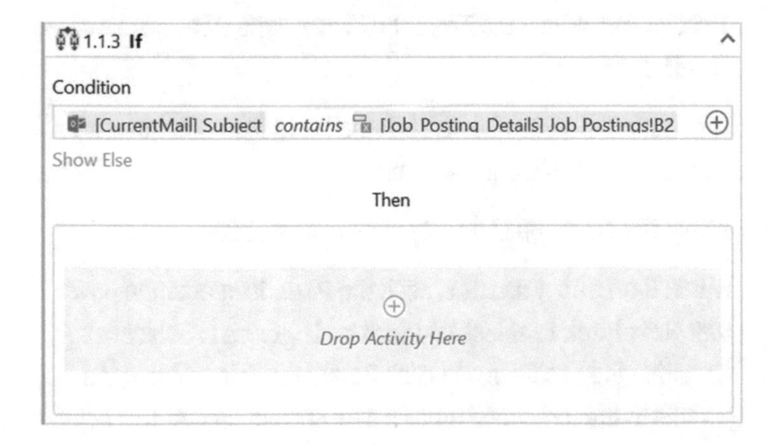

Figure 5-19. *Condition to check if the email subject contains a specific job req id*

6. Next, add the `Forward Email` activity in the Then block of `If` activity.
7. Leave the `Email` field with the default value, that is, `CurrentMail`.
8. In the `Add 'To' recipients` field, click the `Plus` icon and hover over `Notebook` to select `Indicate in Excel` option.
9. Within the Excel file, select cell D2 to specify the relevant hiring manager's email and click `Confirm`, as shown in Figure 5-20.

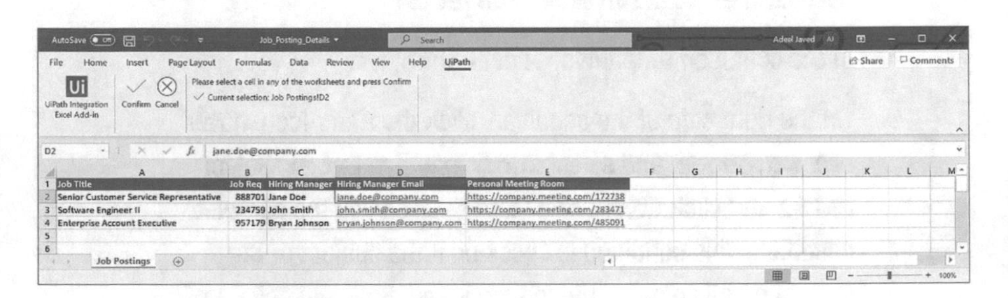

***Figure 5-20.** Selecting the hiring manager to forward the email to*

10. Select the `Plus` icon of the `New subject` field and select the `Text` option. Within the `Text Builder`, type `Please review candidate`.
11. Within the `Text Builder`, click the `Plus` icon and hover over `CurrentMail` and choose `From`.
12. Within the `Text Builder`, type `for position`.
13. Within the `Text Builder`, click the `Plus` icon again to hover over `Notebook` to select `Indicate in Excel`. Select cell A2 within the Excel file and click `Confirm`. This will add the job title to the forwarded email's new subject line. At this point, `Text Builder` should resemble Figure 5-21. Click `Save`.

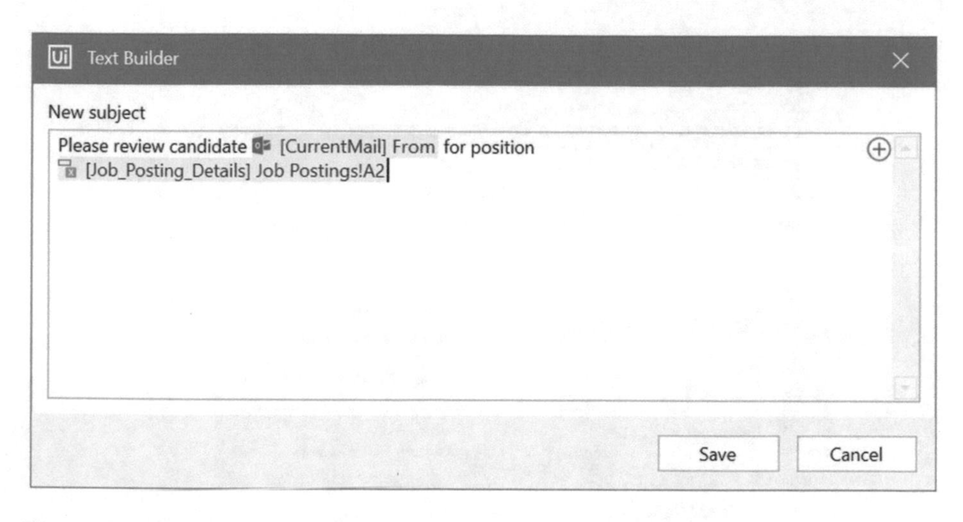

***Figure 5-21.** Configuration of new subject of forwarded email*

14. Next, in the Body field, click the `Plus` icon, select the `Text` option, and type `Here are the forwarded details of the candidate`. Click `Save`.

15. Leave the `Save as draft` option as checked.

Once you have completed the exercise, the final configuration of the **Forward Email** activity should resemble Figure 5-22. Figure 5-23 shows the email in the Drafts folder that was created by this activity.

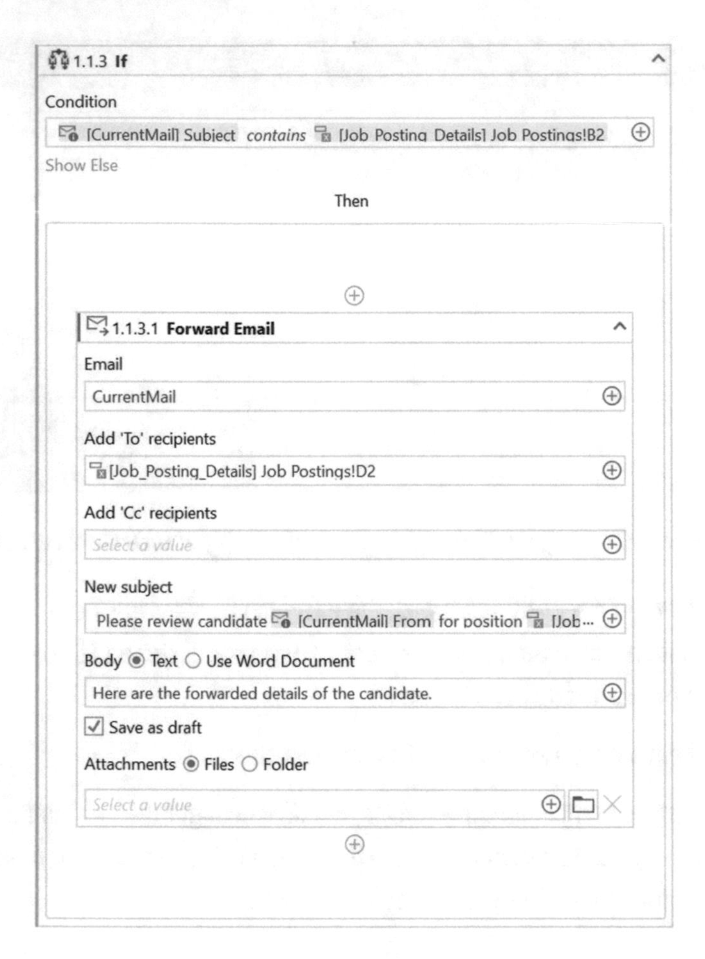

Figure 5-22. *Final configuration for Forward Email activity exercise*

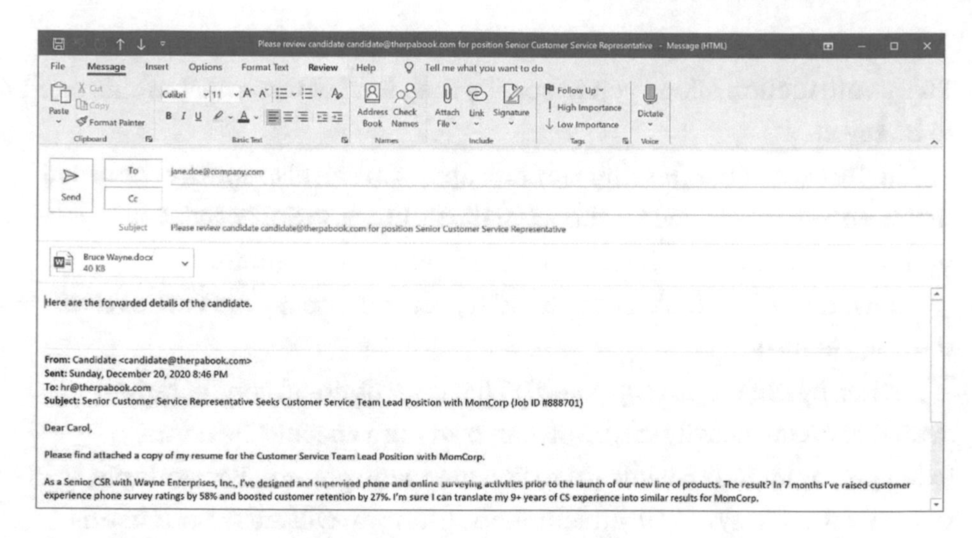

Figure 5-23. *Forwarded email in the Drafts folder*

Save Email Attachments

The **Save Email Attachments** activity allows you to save email attachments to a specific folder.

Configuration

This section provides instructions on how to configure a **Save Email Attachments** activity, shown in Figure 5-24.

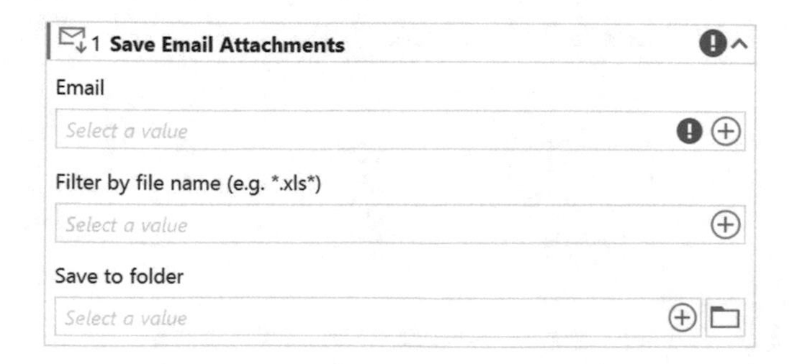

Figure 5-24. *Activity card for Save Email Attachments*

Email: This is a required configuration available on the activity card. This configuration allows you to specify the email whose attachments need to be saved.

In the case of the Desktop Outlook app, you can also specify the email that is currently selected by choosing the `Outlook` ➤ `Selected Email` option. This will act on the selected email at execution time.

If used in a `For Each Email` activity, you can specify the `CurrentMail` message in the loop.

Filter by file name (e.g. *.xls*): This is an optional configuration available on the activity card. You can provide a specific file name, extension or include a wildcard to fit a file name pattern. For example, if you only want to save PDF attachments, then you will enter *.pdf in this field.

Save to folder: This is an optional configuration available on the activity card. This configuration allows you to specify a folder where attachments will be saved. If left blank, it will save to the project folder. If the folder does not exist, it will create one.

Tip In common business processes, this folder is typically a shared network folder, but local for exercise purposes.

Attachments: This is an optional configuration available on the `Properties` panel. This configuration allows you to save a reference to the downloaded attachments for later use.

EXERCISE

Goal: Building on our previous exercise, use the Save Email Attachments activity to save resumes included in emails related to Job Req 888701. Attachments will be saved in C:\BookSamples\Chapter_05\Downloads\Resumes folder. Figure 5-17 shows the current state of the inbox.

Source Code: Chapter_5-MailAutomationExercise

Setup: Here are step-by-step implementation instructions:

1. In StudioX, add the Save Email Attachments activity in the Then block of If activity after the Forward Email activity from the previous exercise.
2. Leave the Email field value as CurrentMail.
3. Leave the Filter by file name field blank as we want all file names and extensions saved.
4. In the Save to folder field, click the Folder icon, and select the C:\BookSamples\Chapter_05\Downloads\Resumes folder.

Once you have completed the exercise, the final configuration of the **Save Email Attachments** activity should resemble Figure 5-25. Figure 5-26 shows the target folder that contains attachments saved after this activity was run.

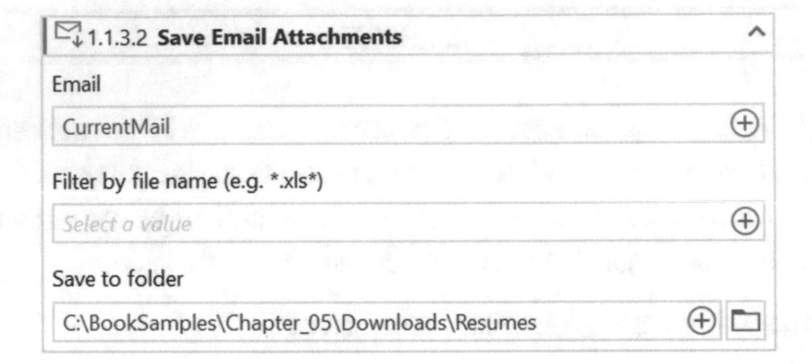

***Figure 5-25.** Final configuration of Save Email Attachments activity exercise*

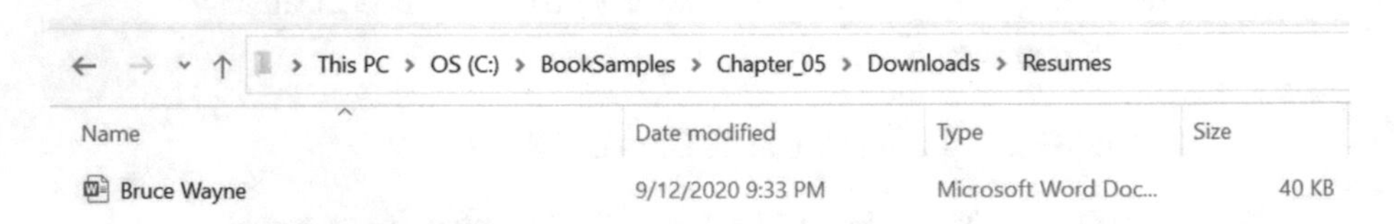

***Figure 5-26.** Download folder containing saved attachments*

Save Email

The **Save Email** activity allows you to save an email message as a .msg or a .eml file in a specified folder.

Configuration

This section provides instructions on how to configure a **Save Email** activity, shown in Figure 5-27.

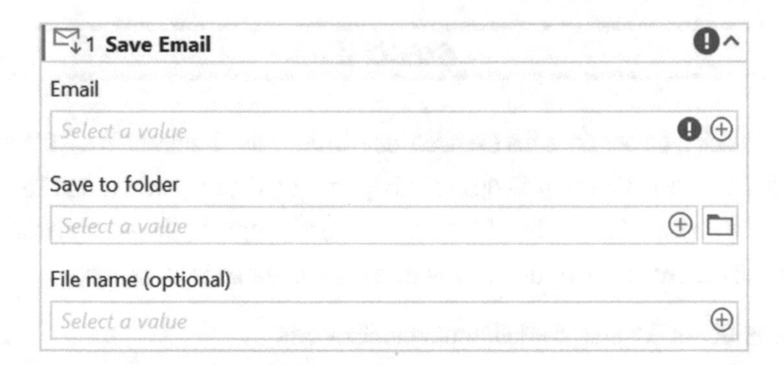

***Figure 5-27.** Activity card for Save Email*

Email: This is a required configuration available on the activity card. This configuration allows you to specify the email message that will be saved as a .msg file (in case of Outlook) or as a .eml file (in case of Gmail) in the specified folder.

In the case of the Desktop Outlook app, you can also specify the email that is currently selected by choosing the `Outlook` ➤ `Selected Email` option. This will act on the selected email at execution time.

If used in a `For Each Email` activity, you can specify the `CurrentMail` message in the loop.

Save to folder: This is an optional configuration available on the activity card. This configuration allows you to specify a folder where the email message will be saved. If left blank, it will save to the project folder. If the folder does not exist, it will create one.

Tip In common business processes, this folder is typically a shared network folder but local for exercise purposes.

File name (optional): This is an optional configuration available on the activity card. This configuration allows you to specify the name of the saved file. If left blank, the subject of the email is used as the file name.

EXERCISE

Goal: Building on our previous exercise, use `Save Email` activity to save the email message in `C:\BookSamples\Chapter_05\Downloads\Emails` folder if it is related to Job Req `888701`. Set the file name as the sender, followed by email received date. Figure 5-17 shows the current state of the inbox.

Source Code: Chapter_5-MailAutomationExercise

Setup: Here are step-by-step implementation instructions:

1. In `StudioX`, add the `Save Email` activity in the Then block of `If` activity after the `Save Email Attachments` activity from the previous exercise.
2. Leave the `Email` field value as `CurrentMail`.
3. In the `Save to folder` field, select the `Folder` icon, and browse to specify the `C:\BookSamples\Chapter_05\Downloads\Emails` folder.
4. From the `File` name field, click the `Plus` icon and select the `Text` option. In the `Text Builder`, click the `Plus` icon, hover over `CurrentMail`, and select `From`.
5. Next, within the `Text Builder`, click the `Plus` icon again, hover over `CurrentMail`, and select `Date (as text)`. Click `Save`. This will set the sender, and email received date as the saved message file name.

Once you have completed the exercise, the final configuration of the **Save Email** activity should resemble Figure 5-28. Figure 5-29 shows the target folder that contains attachments saved after this activity was run.

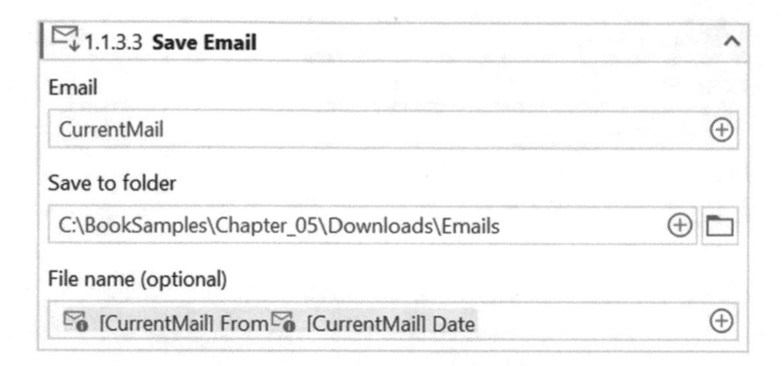

***Figure 5-28.** Final configuration of Save Email exercise*

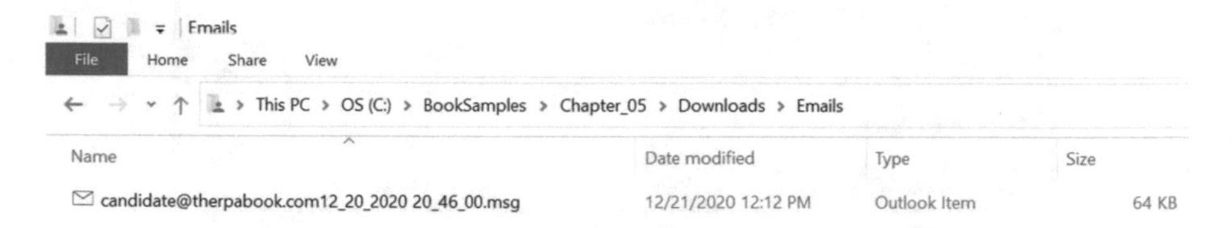

***Figure 5-29.** Download folder containing saved email messages*

Send Email

The **Send Email** activity allows you to send an email message.

Configuration

This section provides instructions on how to configure a **Send Email** activity, shown in Figure 5-30.

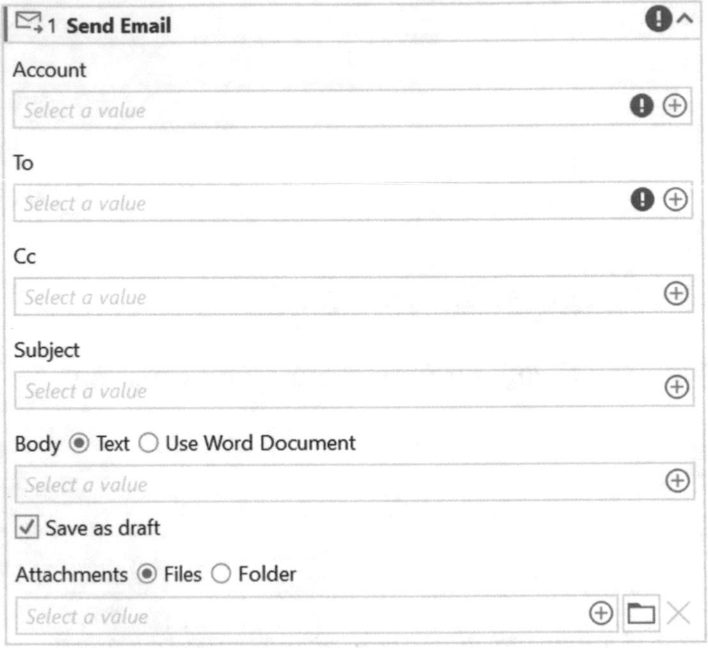

Figure 5-30. Activity card for Send Email

Account: This is a required configuration available on the activity card. This configuration allows you to specify the account email address to send the email from.

To: This is a required configuration available on the activity card. This configuration specifies the recipient(s) of the email.

Cc: This is an optional configuration available on the activity card. This configuration specifies the secondary recipient(s) of the email.

Bcc: This is an optional configuration available on the `Properties` panel. This configuration specifies the hidden recipient(s) of the email.

Subject: This is an optional configuration available on the activity card. This configuration specifies the subject of the email.

Body: This is an optional configuration available on the activity card. You can select the `Text` option to type in the body text or select the `Use Word Document` option to specify the file path of a Word document that will be used as an email body.

Max body document size: This is an optional configuration available on the Properties panel. You can specify the maximum size of the Word document that is being used in the body text. By default, the size is set to 2 MB.

Save as draft: This is an optional configuration available on the activity card. This configuration is check-marked by default, indicating email will be saved as a draft. This is encouraged to be set during development time.

Attachments: This is an optional configuration available on the activity card. You can type or browse for a file or folder to attach to the email.

EXERCISE

Goal: Building on our previous exercise, use the Send Email activity to send the interviewee preparation material if the application is related to Job Req 888701. Figure 5-17 shows the current state of the inbox.

Source Code: Chapter_5-MailAutomationExercise

Setup: Here are step-by-step implementation instructions:

1. In StudioX, add the Send Email activity in the Then block of If activity after the Save Email activity from the previous exercise.
2. Leave the Account field as is with the default value of Outlook.
3. In the To field, click the Plus icon, and hover over CurrentMail to select From. This will set the interviewee as the email recipient.
4. Next, in the Subject field, click the Plus icon, select Text option, and type Interview Prep Material. Click Save.

5. Next, in the `Body` field, click the `Plus` icon, select `Text` option, and type `Please review the attached MomCorp Interview Preparation Guide containing helpful links, videos, practice problems`. Click `Save`.

6. Leave the `Save draft` checked.

7. In the `Attachments` field, select the `Files` option, click the `Folder` icon, and select `C:\BookSamples\Chapter_05\PreparationGuide\MomCorp.docx` file.

Once you have completed the exercise, the final configuration of the **Send Email** activity should resemble Figure 5-31. Figure 5-32 shows the email in the Drafts folder that was created by this activity.

***Figure 5-31.** Final configuration of Send Email exercise*

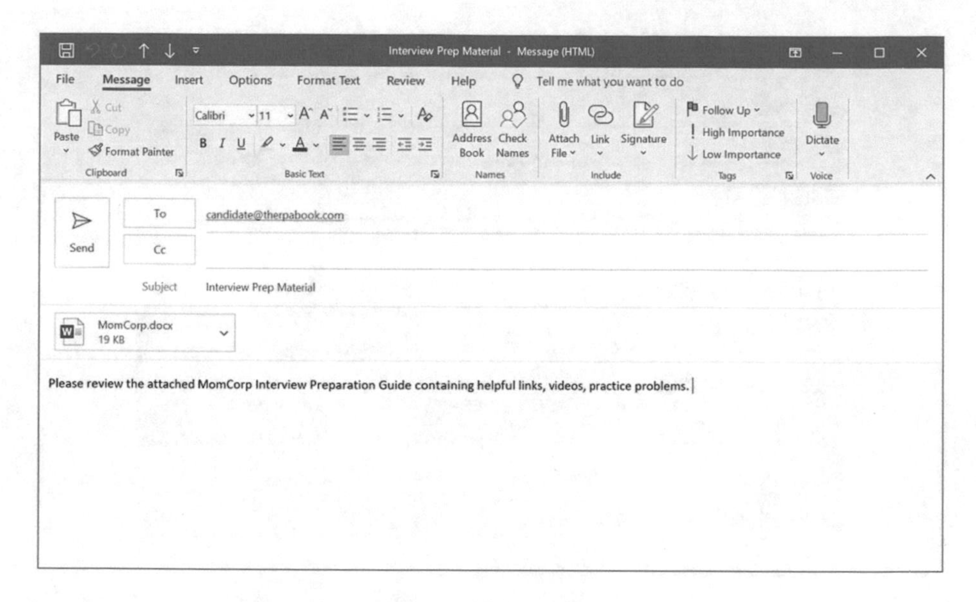

Figure 5-32. *New email in the Drafts folder*

Send Calendar Invite

The **Send Calendar Invite** activity allows you to create and send a calendar invite.

Configuration

This section provides instructions on how to configure a **Send Calendar Invite** activity, shown in Figure 5-33.

Figure 5-33. *Activity card for Send Calendar Invite*

Account: This is a required configuration available on the activity card. This configuration allows you to specify the email address account to send the calendar invite from.

Title/Subject: This is an optional configuration available on the activity card. This configuration specifies the title/subject of the calendar invite.

Required attendees: This is an optional configuration available on the activity card. This configuration specifies the participants required to attend the meeting.

Optional attendees: This is an optional configuration available on the activity card. This configuration specifies the participants not required to attend the meeting.

Start date: This is a required configuration available on the activity card. This configuration allows you to specify the start date of the meeting invite.

Start time: This is a required configuration available on the activity card. This configuration allows you to specify the start time of the meeting invite.

Duration: This is a required configuration available on the activity card. This configuration allows you to specify the duration of the meeting. You must provide duration either in hours/minutes/seconds or check the `All day event` option. If the `All day event` is checked, it sets the meeting to the full day.

Location: This is an optional configuration available on the activity card. This configuration allows you to specify the location of the meeting.

Description: This is an optional configuration available on the activity card. This configuration allows you to specify the description or body of the meeting invite in `Text` format.

Reminder: This is a required configuration available on the activity card. This configuration allows you to specify when to show the meeting reminder. By default, the value is set to `15 minute(s)`.

Show as: This is a required configuration available on the activity card. This configuration allows you to specify how the meeting is marked on participants' calendars. You can select one of these options `Free`, `Tentative`, `Busy`, or `Out of Office`. By default, this is set to `Busy`.

Recurrence: This is an optional configuration available on the activity card. This configuration allows you to set the meeting to repeat on a defined frequency. Figure 5-34 shows the configuration to define recurrence.

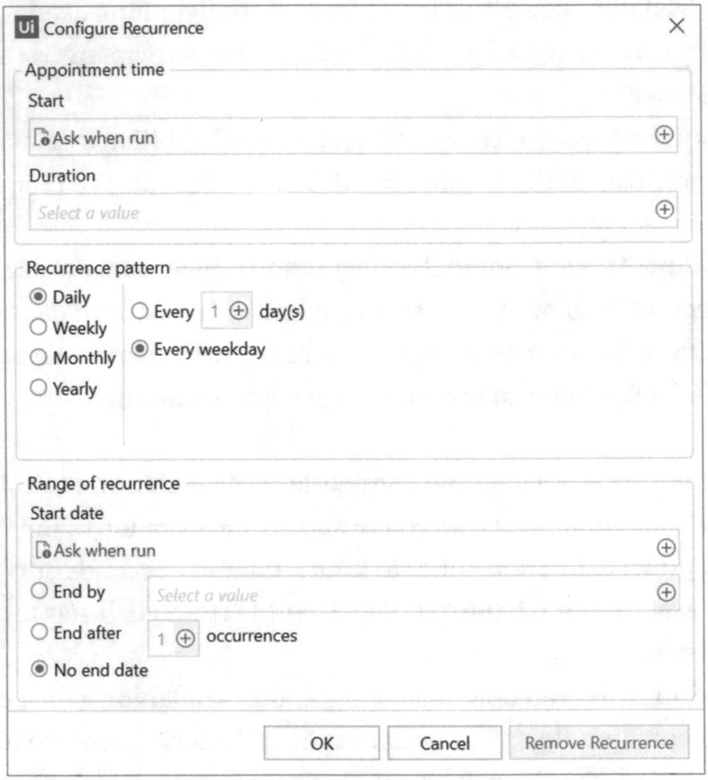

Figure 5-34. *Recurrence configurations*

Save without sending: This is an optional configuration available on the activity card. This configuration is check-marked by default, indicating calendar invites will be saved and not sent. This is recommended to be set during development time.

EXERCISE

Goal: Building on our previous exercise, use Send Calendar Invite activity to send the hiring manager and the candidate an initial interview meeting invite if the application is related to Job Req 888701. Here are some additional configurations for the invite:

- Send an invite to the hiring manager and interviewee.
- Prompt for the meeting invite date and time.
- Set meeting reminder to an hour before the event.
- Specify the hiring manager's personal meeting room link as the location of the invite. This is found in JobPostingDetails.xlsx Excel file under the Personal Meeting Room column.

Figure 5-17 shows the current state of the inbox.

Source Code: Chapter_5-MailAutomationExercise

Setup: Here are step-by-step implementation instructions:

1. In StudioX, add the Send Calendar Invite activity in the Then block of If activity after the Send Email activity from the previous exercise.
2. In the Title/Subject field, click the Plus icon, select Text option, and type Initial interview with.
3. Next, from within the Text Builder, click the Plus icon and hover over CurrentMail, and choose From. Click Save. This specifies the interviewee in the title of the calendar invite.
4. In the Required attendees field, click the Plus icon, and select the Text option. Next, in the Text Builder, click the Plus icon and hover over CurrentMail and choose From.

5. Next, with the Text Builder still open, type a semicolon (;). The semicolon is used by Outlook to separate multiple recipients.
6. Next, from within the Text Builder, click the Plus icon and hover over Notebook to select Indicate in Excel. Select cell D2 and click the Confirm button. Click Save.
7. In both the Start date and Start time fields, click the Plus icon, and select Ask when run.
8. Extend the default duration of the meeting by clicking the Plus icon in the Duration field and specifying 1 in the Hours field.
9. Then, in the Location field, click the Plus icon and hover over Notebook to select Indicate in Excel. Select cell E2 to specify the hiring manager's personal meeting room link as the meeting location. Click Confirm.
10. After clicking the Plus icon in the Description field, select the Text option, and type the description shown in Figure 5-35. Indicate cells A2 and C2 for the Job Title and Hiring Manager Name.
11. Change the Reminder field to 1 hour(s).

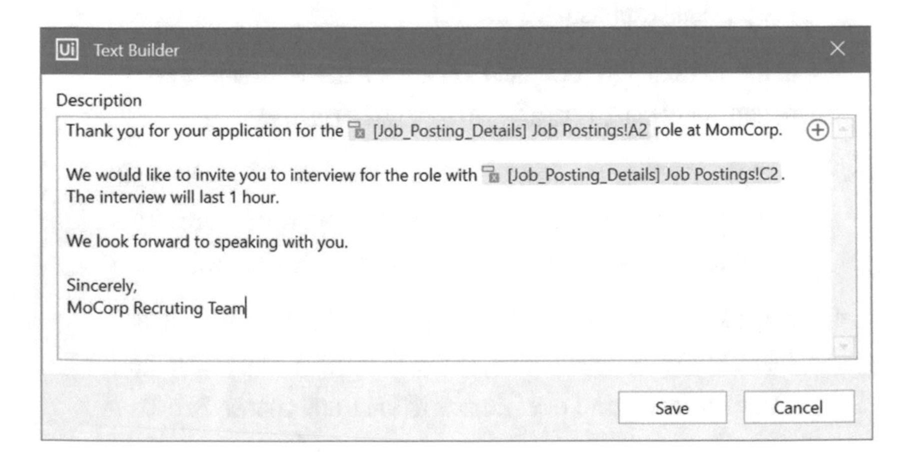

***Figure 5-35.** Detailed description of the Calendar Invite*

Once you have completed the exercise, the final configuration of the **Send Calendar Invite** activity should resemble Figure 5-36. Figure 5-37 shows the calendar invite that was created by this activity.

Figure 5-36. *Final configuration of Send Calendar Invite exercise*

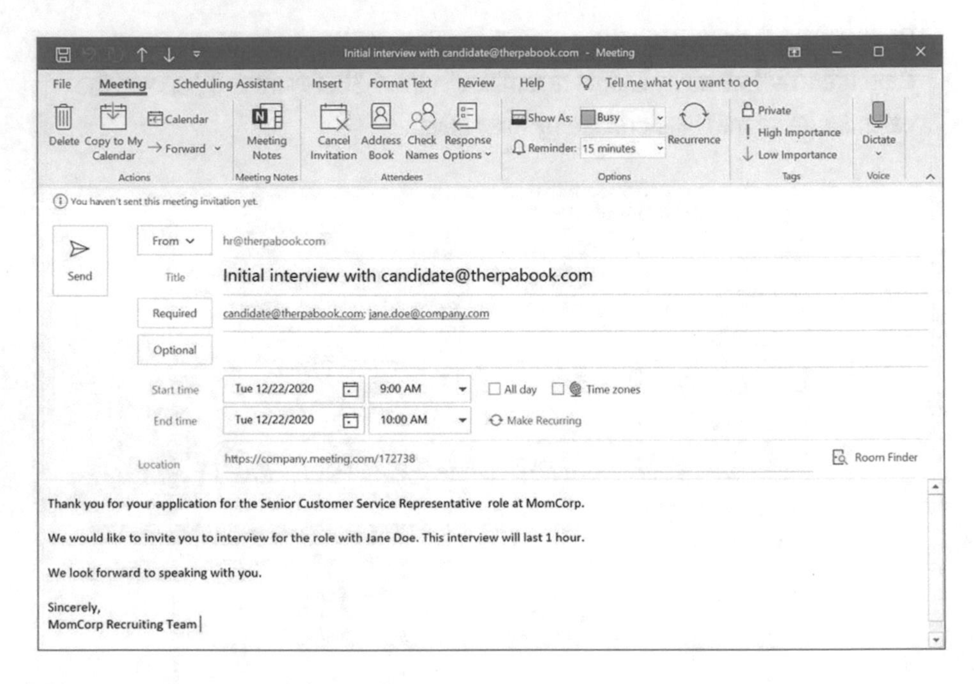

Figure 5-37. *Saved calendar invite*

Move Email

The **Move Email** activity allows you to move a specified email to a folder.

Configuration

This section provides instructions on how to configure a **Move Email** activity, shown in Figure 5-38.

1 Move Email

Email

Select a value

Move to

Select a value

Figure 5-38. *Activity card for Move Email*

Email: This is a required configuration available on the activity card. This configuration specifies the email that will be moved.

In the case of the Desktop Outlook app, you can also specify the email that is currently selected by choosing the `Outlook` ➤ `Selected Email` option. This will act on the selected email at execution time.

If used in a `For Each Email` activity, you can specify the `CurrentMail` message in the loop.

Move to: This is a required configuration available on the activity card. This configuration specifies the folder where the email will be moved.

EXERCISE

Goal: Building on our previous exercise, use `Move Email` activity to move the candidate email to `Inbox` ➤ `Initial Screen` folder if it is related to Job Req `888701`. Figure 5-17 shows the current state of the inbox.

Source Code: Chapter_5-MailAutomationExercise

Setup: Here are step-by-step implementation instructions:

1. In `StudioX`, add the `Move Email` activity in the Then block of `If` activity after the `Send Calendar Invite` activity from the previous exercise.

2. Select the `Plus` icon in `Move to` field to select the `Inbox\Initial Screen` Outlook folder.

Once you have completed the exercise, the final configuration of the **Move Email** activity should resemble Figure 5-39. Figure 5-40 shows the state of the Inbox after this activity has run, while Figure 5-41 shows the state of the Initial Screen folder where the email has been moved.

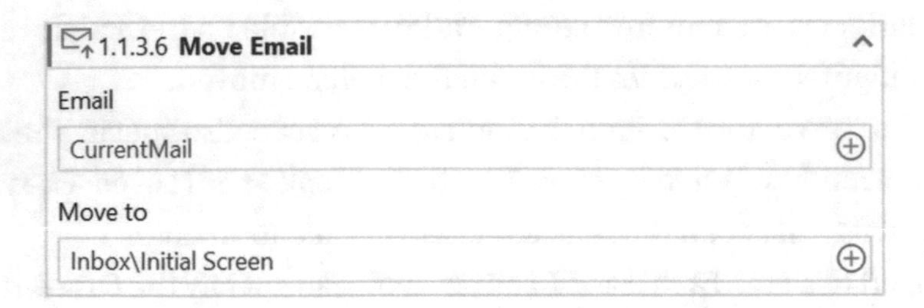

Figure 5-39. *Final configuration for Move Email exercise*

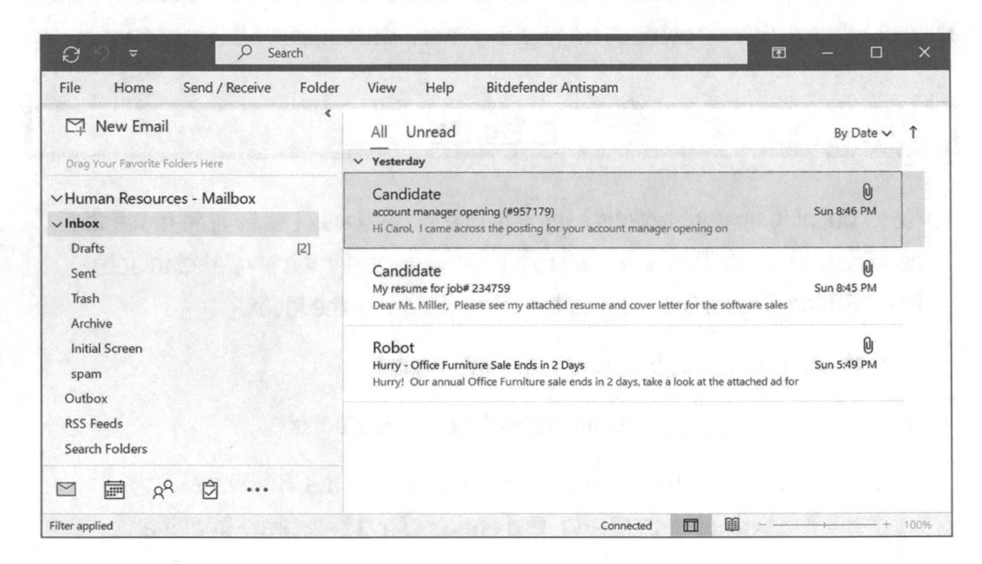

Figure 5-40. *The state of the Human Resources email inbox after the Move Email exercise*

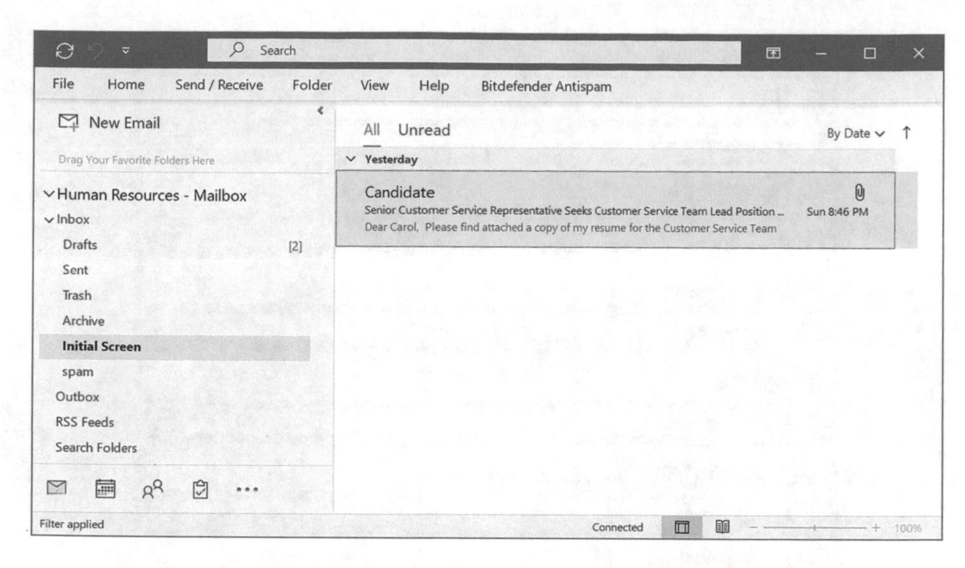

***Figure 5-41.** The state of the Initial Screen folder after the Move Email exercise*

Reply to Email

The **Reply to Email** activity allows you to reply to a specified email message.

Configuration

This section provides instructions on how to configure a **Reply To Email** activity, shown in Figure 5-42.

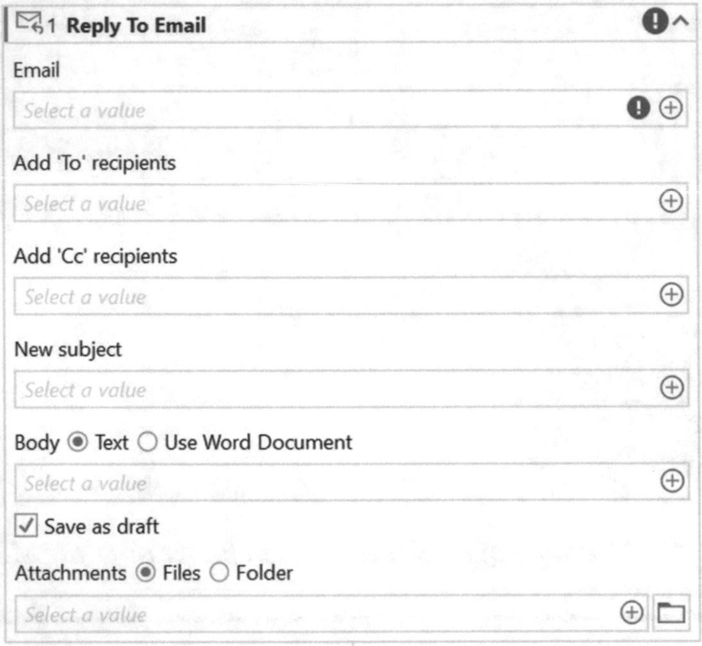

***Figure 5-42.** Activity card for Reply To Email*

Email: This is a required configuration available on the activity card. This configuration allows you to specify the email for which the reply will be sent.

In the case of the Desktop Outlook app, you can also specify the email that is currently selected by choosing the `Outlook` ➤ `Selected Email` option. This will act on the selected email at execution time.

If used in a `For Each Email` activity, you can specify the `CurrentMail` message in the loop.

Reply to all: This is an optional configuration available on the `Properties` panel. This configuration allows you to specify if the reply should go out to all original recipients of the email. By default, the option is unchecked.

Add 'To' recipients: This is a required configuration available on the activity card. This configuration specifies the recipient(s) of the reply email.

Add 'Cc' recipients: This is an optional configuration available on the activity card. This configuration specifies the secondary recipient(s) of the reply message.

Add 'Bcc' recipients: This is an optional configuration available on the activity card. This configuration specifies the hidden email recipient(s) of the reply message.

New subject: This is an optional configuration available on the activity card. If specified, the subject of the reply email will be changed. Otherwise, the original subject will be used.

Body: This is an optional configuration available on the activity card. You can select the `Text` option to type in the body text or select the `Use Word Document` option to specify the file path of a Word document that will be used as an email body.

Max body document size: This is an optional configuration available on the `Properties` panel. You can specify the maximum size of the Word document that is being used in the body text. By default, the size is set to `2 MB`.

Save as draft: This is an optional configuration available on the activity card. This configuration is check-marked by default, indicating email will be saved as a draft. This is encouraged to be set during development time.

Attachments: This is an optional configuration available on the activity card. You can type or browse for a file or folder to attach to the reply email.

EXERCISE

Goal: Building on our previous exercise, use Reply to Email activity to reply to the candidate emails related to Job Req ids 234759 or 957179. Figure 5-40 shows the current state of the inbox.

Source Code: Chapter_5-MailAutomationExercise

Setup: Here are step-by-step implementation instructions:

1. In StudioX, add an If activity in the Else block of If activity from the previous exercise. This is to filter for emails with subject lines that contain the specific Job Req ids referenced from the Notebook.
2. Click the Plus icon in the Condition field to open the Condition Builder.
3. On the left side of the condition, click the Plus icon and hover over the CurrentMail to select Subject.
4. In the condition dropdown, select contains.
5. On the right side of the condition, click the Plus icon and hover over the Notebook to select Indicate in Excel. Select cell B3 and click Confirm.
6. Click the + Add button within the Condition Builder to add a condition.
7. Select Any (OR) from the dropdown on the top left.
8. Repeat steps 3 and 4.
9. On the right side of the condition, click the Plus icon and hover over the Notebook to select Indicate in Excel. Select cell B4 and click Confirm.

10. Add a `Reply to Email` activity in the Then block of the newly created `If` activity.

11. Leave the `Email` field as is with the default value of `CurrentMail`.

12. In the To field, click the `Plus` icon and hover over `CurrentMail`, and select `From`. This will send the email response to the original sender.

13. Next, in the `New subject` field, click the `Plus` icon, and select the `Text` option to open the `Text Builder`. Type `Thank you for your interest` as the email subject line. Click `Save`.

14. In the Body field, select the `Use Word Document` option. Click the `Folder` icon to browse for the file and select `C:\BookSamples\Chapter_05\EmailTemplates\HiringPausedResponse.docx` file. Figure 5-43 shows the contents of the Word document used as an email response body.

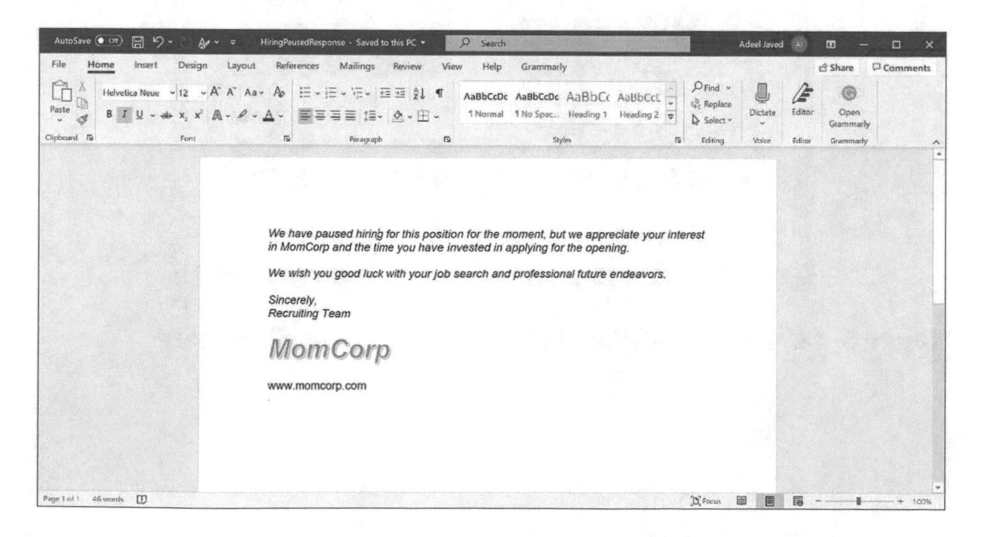

Figure 5-43. *Word document template for an email response*

Once you have completed the exercise, the final configuration of the **Reply to Email** activity should resemble Figure 5-44. Figure 5-45 shows the email in the Drafts folder generated by this activity.

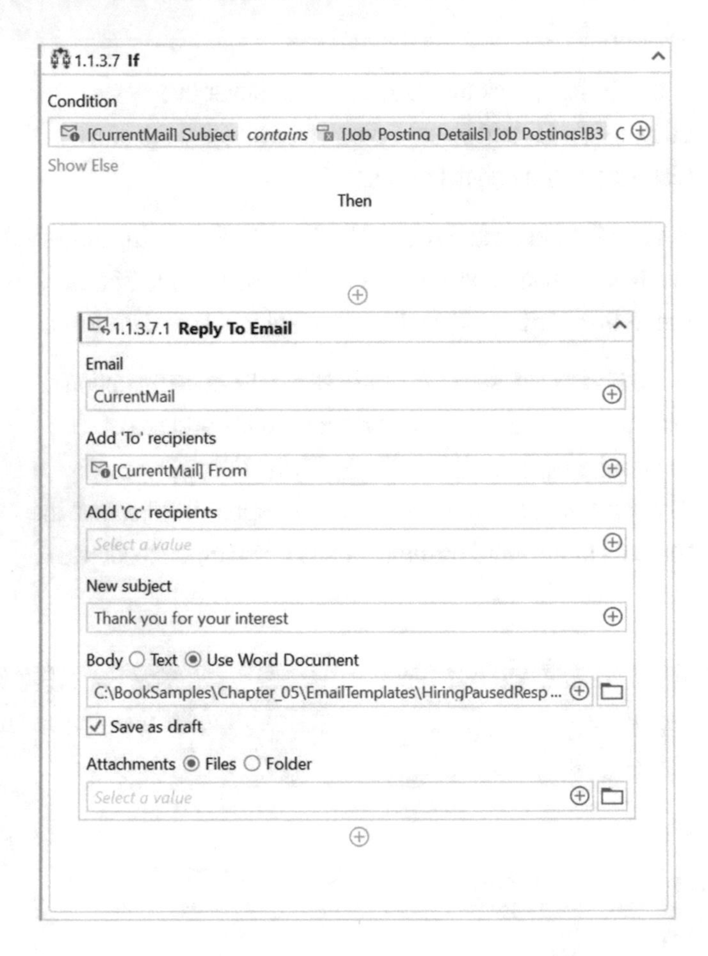

***Figure 5-44.** Final configuration for Reply to Email exercise*

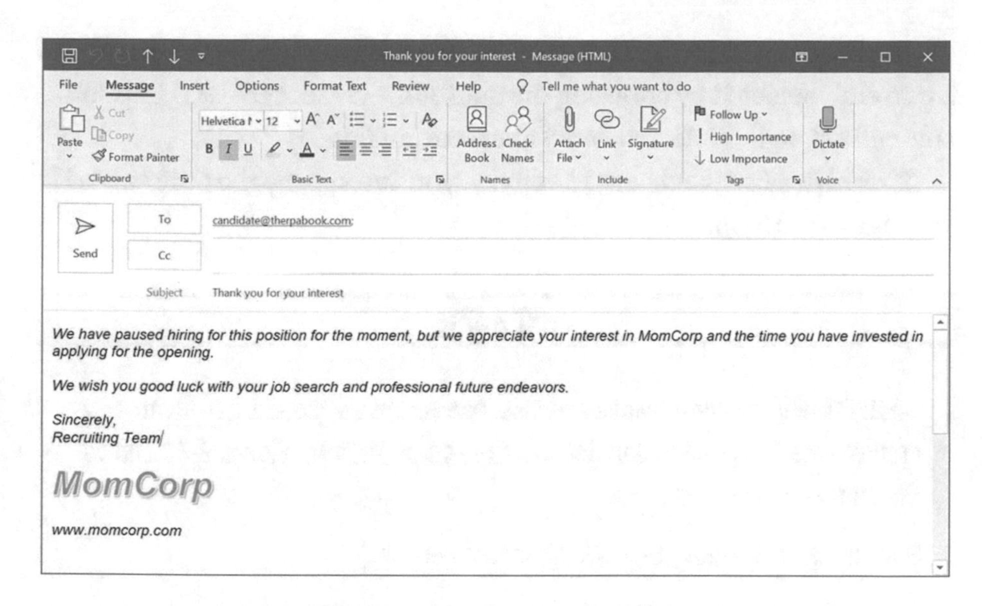

Figure 5-45. *Draft reply message generated by Reply to Email activity*

Archive Email

The **Archive Email** activity allows you to archive an email message.

Configuration

This section provides instructions on how to configure an **Archive Email** activity, shown in Figure 5-46.

Figure 5-46. *Activity card for Archive Email*

Email: This is a required configuration available on the activity card. This configuration specifies an email to archive.

In case of the Desktop Outlook app, you can also specify the email that is currently selected by choosing the `Outlook` ➤ `Selected Email` option. This will act on the selected email at execution time.

If used in a `For Each Email` activity, you can specify the `CurrentMail` message in the loop.

EXERCISE

Goal: Building on our previous exercise, use `Archive Email` activity to archive emails related to Job Req ids `234759` or `957179`. Figure 5-40 shows the current state of the inbox.

Source Code: Chapter_5-MailAutomationExercise

Setup: Here are step-by-step implementation instructions:

1. In `StudioX`, add an `Archive Email` activity in the `Then` block of `If` activity after the `Reply Email` activity.
2. Leave the `Email` field as is with the default value of `CurrentMail`.

Once you have completed the exercise, the final configuration of the **Archive Email** activity should resemble Figure 5-47. Figure 5-48 shows the Archive folder where the emails have been archived.

***Figure 5-47.** Final configuration of Archive Email exercise*

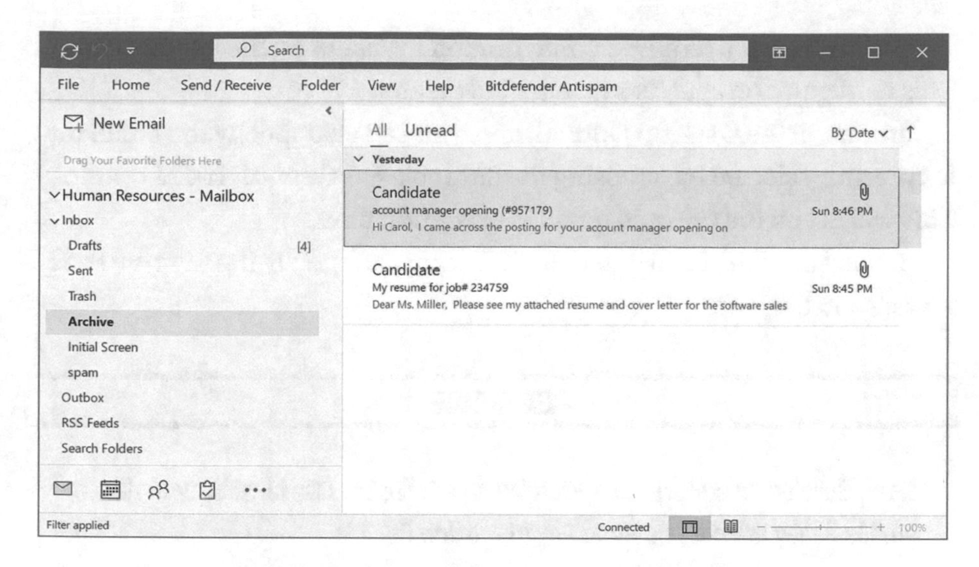

Figure 5-48. *Archive folder of the Human Resources mailbox*

Delete Email

The **Delete Email** activity allows you to delete a specified email message.

Configuration

This section provides instructions on how to configure a **Delete Email** activity, shown in Figure 5-49.

Figure 5-49. *Activity card for Delete Email*

Email: This is a required configuration available on the activity card. This configuration specifies an email to delete.

In case of the Desktop Outlook app, you can also specify the email that is currently selected by choosing the `Outlook` ➤ `Selected Email` option. This will act on the selected email at execution time.

If used in a `For Each Email` activity, you can specify the `CurrentMail` message in the loop.

EXERCISE

Goal: Building on our previous exercise, use `Delete Email` activity to delete emails if they are not related to any of the Job Req ids.

Source Code: Chapter_5-MailAutomationExercise

Setup: Here are step-by-step implementation instructions:

1. Click the `Show Else` option within the `If` activity. This step is in continuation of the previous exercise.
2. Add a `Delete Email` activity in the `Else` block of the `If` activity. This will delete emails where the subject does not contain the Job Req ids within the `Notebook`.

Once you have completed the exercise, the final configuration of the **Delete Email** activity should resemble Figure 5-50. Figure 5-51 shows the Trash folder with deleted emails.

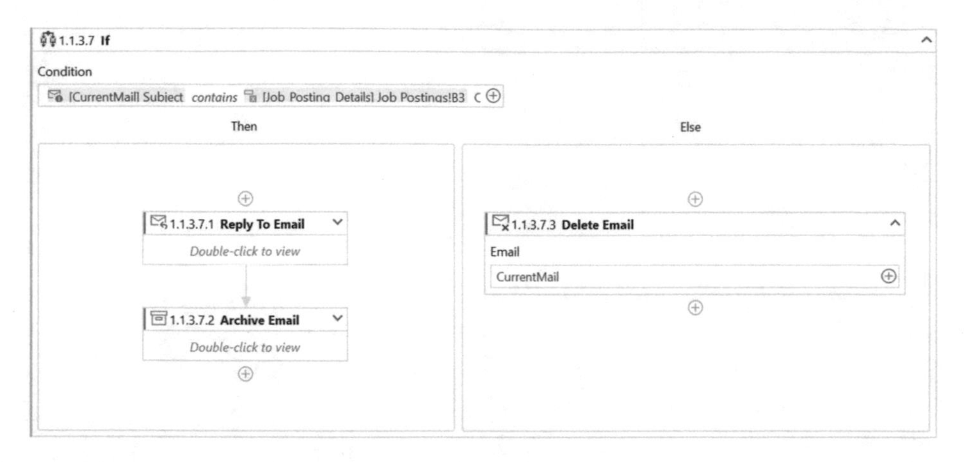

Figure 5-50. *Final configuration of Delete Email exercise*

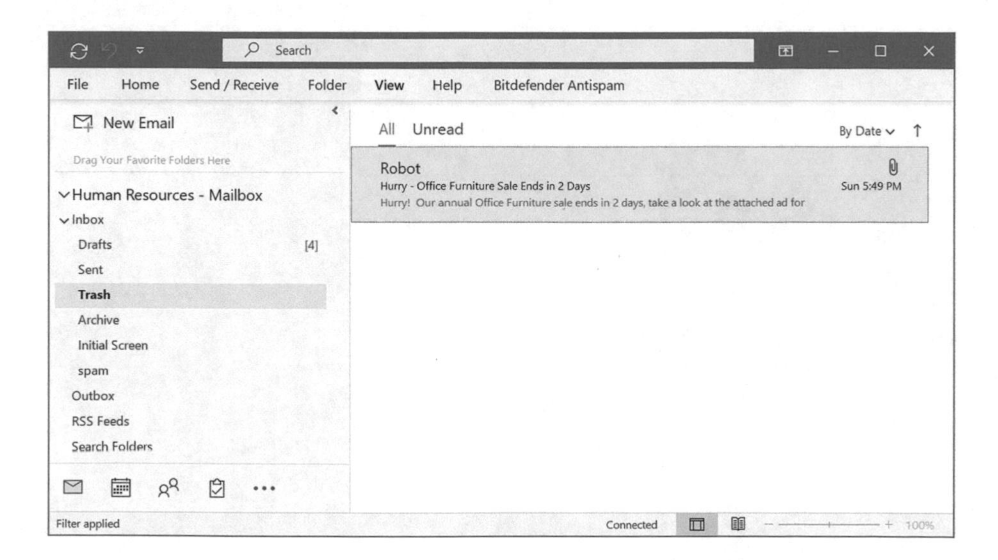

Figure 5-51. *Trash folder of the Human Resources mailbox*

CHAPTER 6

Word Automation

Microsoft Word is the most popular and universal word processing program used in organizations across the world. Word is often utilized for creating legal documents, financial files, and procurement documents and customizing templates and requirements documents. UiPath StudioX Word automation provides citizen developers with automation capabilities for critical tasks like reading content, writing text, adding images, inserting tables, exporting as PDF, and creating HTML-rich email bodies, for example, for ones that include tables, signatures, or logos.

Learning Objectives

At the end of this chapter, you will learn how to

- Define a Word scope to automate
- Save document as a different file
- Read text from a document
- Replace all occurrences of text within a document
- Set the text of a document bookmark
- Replace a picture based on Alt Text
- Insert a data table relative to text or a bookmark

A. Javed et al., *Robotic Process Automation using UiPath StudioX*,
https://doi.org/10.1007/978-1-4842-6794-3_6

- Add a picture at the end of a Word document
- Append text within a Word document
- Export a Word document as a PDF

Sample Overview

Throughout this chapter, we will be using a simplified version of invoice generation process to showcase the usage of all Word automation activities. Concepts discussed in this chapter can be applied to filling information from different data sources to complete a templated Word document such as master service agreements, statements of work, request for proposals, prescriptions, and service desk tickets.

This section will familiarize you with the prerequisites for all exercises in this chapter.

Word Setup

This section provides information on how to enable bookmarks and add alternate text to images in Word.

Bookmarks

First, ensure bookmarks are shown by navigating to the Advanced Options from `File` ➤ `Options` ➤ `Advanced`. Within the `Show document content` section, confirm `Show bookmarks` is checked. Figure 6-1 shows the setting configured.

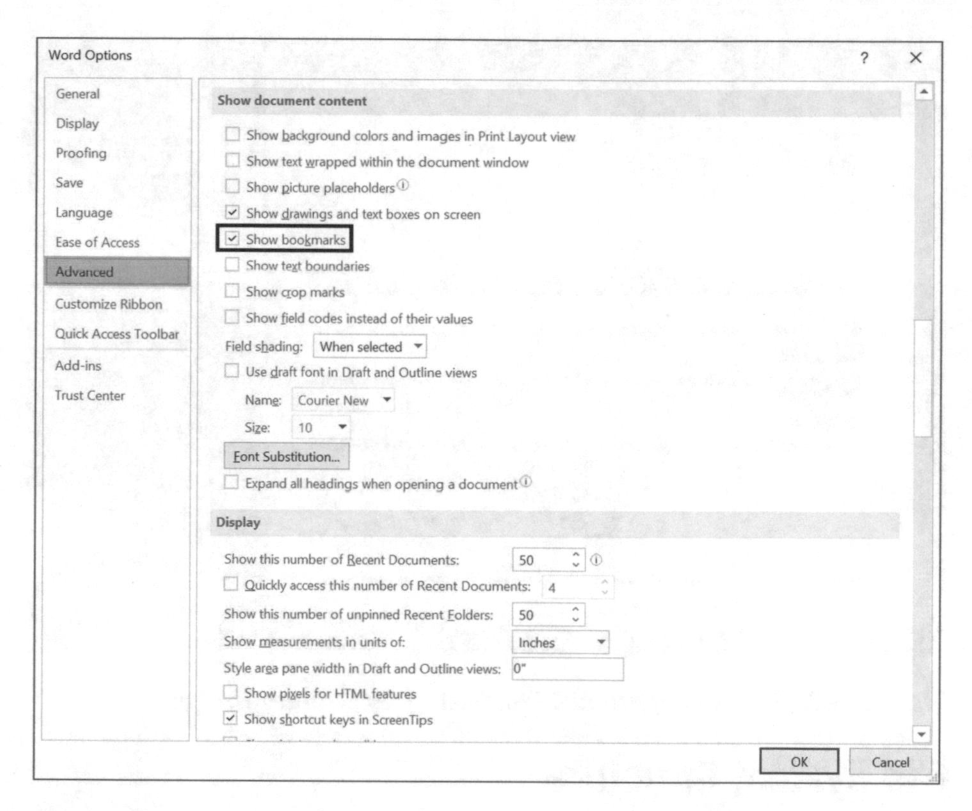

Figure 6-1. *Show bookmarks from Advanced Word Options*

Alt Text

StudioX uses alternate text (Alt Text) of images in Word to find and replace pictures. To view Alt Text of a picture, select the image within the Word document, then right-click, and choose Edit Alt Text. Figure 6-2 shows the Alt Text for a placeholder picture within the invoice template.

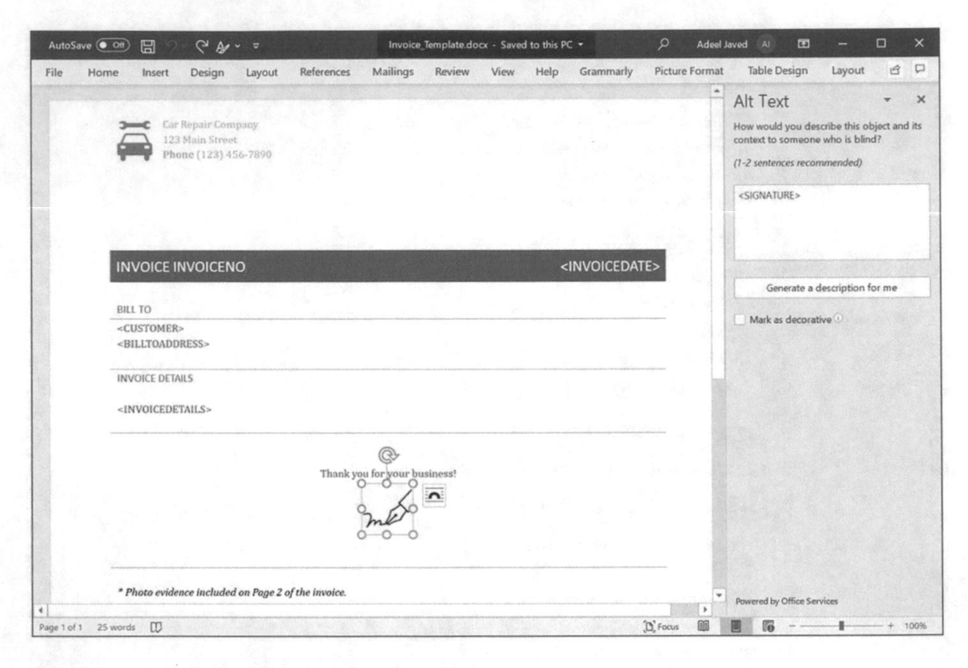

Figure 6-2. *Alt Text of invoice template placeholder image*

File System Structure

Next, download the source code from the book's site, and make sure you move the entire BookSamples folder to your C:\ drive. All exercises in this chapter assume the folder paths will be **C:\BookSamples\Chapter_06**. Figure 6-3 shows the physical folder structure required for this sample. This folder structure comes with the source code.

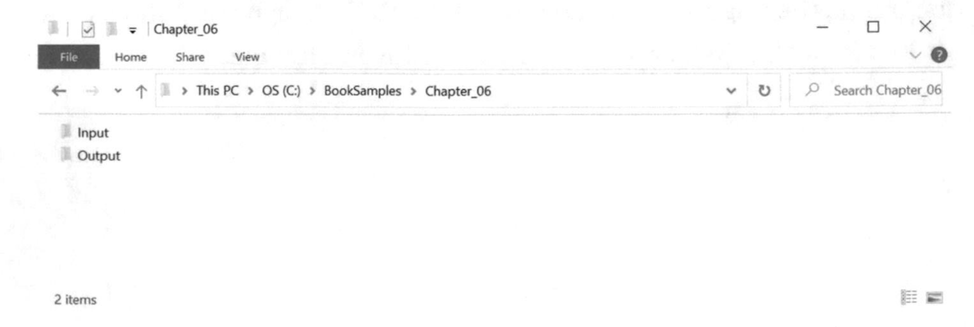

Figure 6-3. *Folder structure used for the Invoice Generation exercise*

Input: This folder contains the invoice template along with images and an Excel file as input data. Figure 6-4 shows the contents of the Input folder.

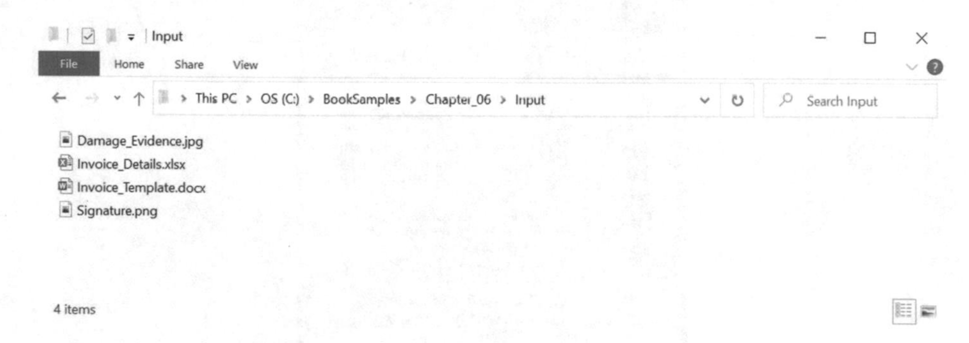

Figure 6-4. *Contents of the Input folder*

Output: This empty folder is to store the completed invoice template and will be where the final PDF is created and saved. Figure 6-5 shows the contents of the Output folder.

Figure 6-5. *Contents of the Output folder*

Activities Reference

As shown in Figure 6-6, all Word automation activities can be found under the Word category. The following sections will provide instructions on how to configure and use each activity.

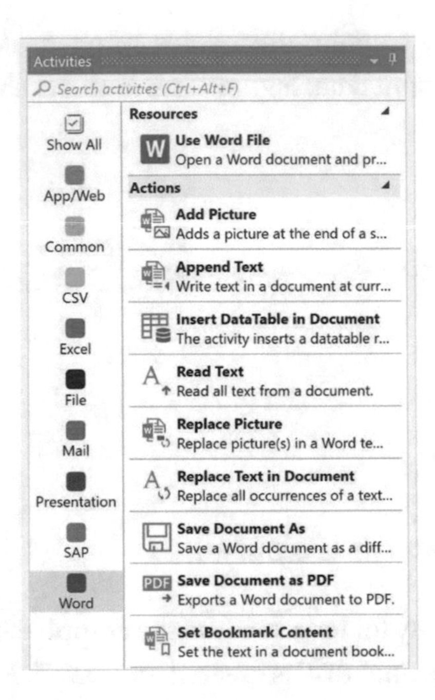

Figure 6-6. *Activities for Word automation*

Use Word File

The **Use Word File** activity allows you to define a Word scope for your automation. When the process is executed, the Word document is opened and then closed on completion.

This activity will contain all the actions that you want to take on the Word document. For example, if you want to export a document as a PDF, the **Save Document as PDF** activity will be nested in the body of **Use Word File** activity.

Configuration

This section provides instructions on how to configure a **Use Word File** activity, shown in Figure 6-7.

Note To configure this activity, make sure your Microsoft Word desktop application is signed in.

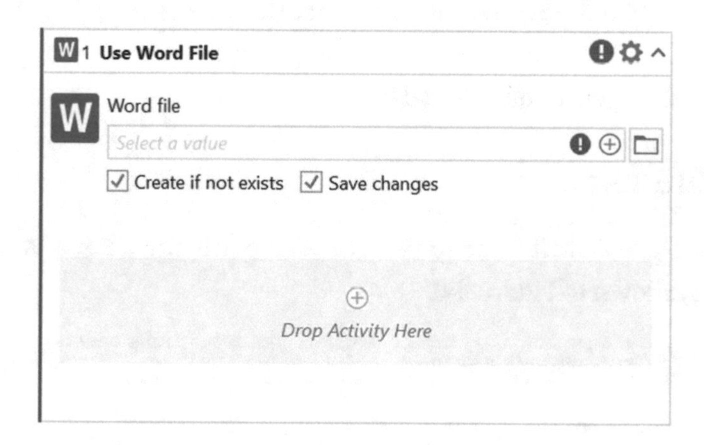

Figure 6-7. *Activity card for Use Word File*

Word File: This is a required configuration available on the activity card. This configuration allows you to specify the Word document you plan to automate.

Create if not exists: This is an optional configuration available on the activity card. When checked, if the document does not exist, one will be created. By default, this option is checked.

Auto save: This is an optional configuration available on the activity card. When checked, changes will be saved when the process is run. By default, this option is checked.

Save Document As

The **Save Document As** activity allows you to save the document as a new file or replaces an existing. Any activities after the **Save Document As** activity will affect the Word document that has been saved. This activity is recommended to use at the very beginning of an automation to prevent accidental changes to the original file.

Configuration

This section provides instructions on how to configure a **Save Document As** activity, shown in Figure 6-8.

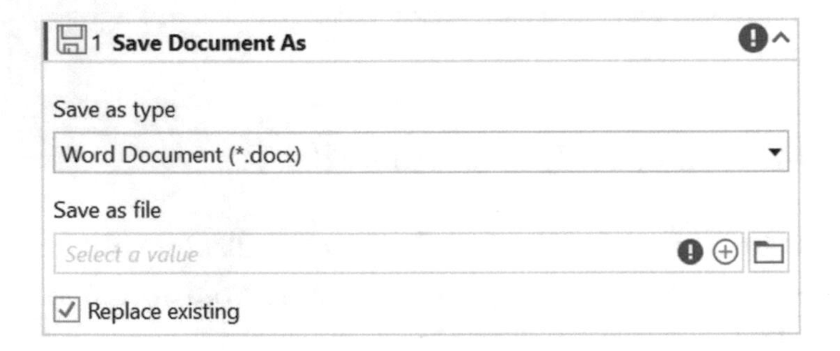

Figure 6-8. *Activity card for Save Document As*

Save as type: This is a required configuration available on the activity card. This configuration allows you to specify the file extension to save the file as. As shown in Figure 6-9, you can see the list of extensions available.

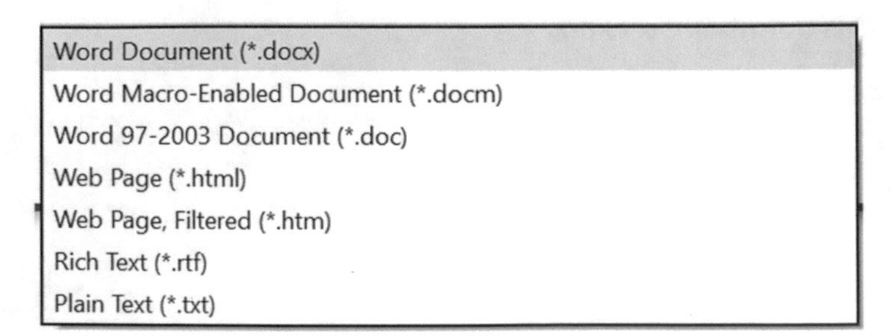

Figure 6-9. *Extensions of the Save Document As activity*

Save as file: This is a required configuration available on the activity card. You can type or browse for a file to save the document as.

Replace existing: This is an optional configuration available on the activity card. When checked, the file will be replaced if it already exists. If left unchecked and if the file already exists, an error will notify the file already exists. By default, this option is checked.

EXERCISE

Goal: Use the `Use Word File` and `Save Document As` activities to create a copy of the `C:\BookSamples\Chapter_06\Input\Invoice_Template.docx` file. The new file will be created in the `C:\BookSamples\Chapter_06\Output` folder with `INV_CustomerName.docx` name. This will be used so the original invoice template is not altered when the automation is executed. Figure 6-5 shows the current state of the Output folder.

Source Code: Chapter_6-WordAutomationExercise

Setup: Here are step-by-step implementation instructions:

1. In `StudioX`, add the `Use Word File` activity to a blank process.
2. In the `Word file` field, click the `Folder` icon to select `C:\BookSamples\Chapter_06\Input\Invoice_Template.docx` file.
3. Next, add the `Save Document As` activity inside the body of `Use Word File` activity.
4. Leave the `Save as type` field with the default value of `Word Document (*.docx)`.
5. Click the `Plus` icon of the `Save as file` field and select the `Text` option. Within the `Text Builder`, type `C:\BookSamples\Chapter_06\Output\INV`.

6. Within the `Text Builder`, click the `Plus` icon, and hover over `Notebook` to select `Indicate in Excel`. Select cell D2 and click `Confirm`.

7. Next, within `Text Builder`, type `.docx`. Click `Save`.

8. Leave the `Replace existing` field checked.

Once you have completed the exercise, the final configuration of the **Save Document As** activity should resemble Figure 6-10. Figure 6-11 shows the copy of invoice template saved in the Output folder.

Figure 6-10. *Final configuration of the Save Document As activity exercise*

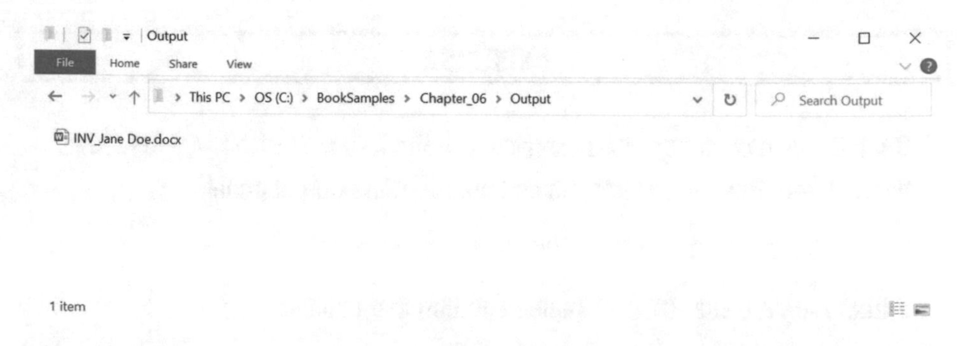

***Figure 6-11.** The output of the Save Document As activity exercise*

Read Text

The **Read Text** activity allows you to read all text from a document to save to the Project Notebook or clipboard to reference within an automation.

Configuration

This section provides instructions on how to configure a **Read Text** activity, shown in Figure 6-12.

***Figure 6-12.** Activity card for Read Text*

Save to: This is a required configuration available on the activity card. This configuration specifies where the document text will be saved.

EXERCISE

Goal: Building on our previous exercise, use the `Read Text` activity to read all the text from the invoice template and print it in the Output panel.

Source Code: Chapter_6-WordAutomationExercise

Setup: Here are step-by-step implementation instructions:

1. In `StudioX`, add the `Read Text` activity within the `Use Word File` activity after the `Save Document As` activity.
2. Then, in the `Save to` field, click the `Plus` icon and select `Copy to clipboard`.
3. Next, add the `Write Line` activity to the body of `Use Word File` activity after the `Read Text` activity.
4. In the `Text` field, click the `Plus` icon, and select `Paste from clipboard`.

Once you have completed the exercise, the final configuration of the **Read Text** activity should resemble Figure 6-13. Once the automation runs, all text from the invoice template document is printed in the Output panel, as shown in Figure 6-14.

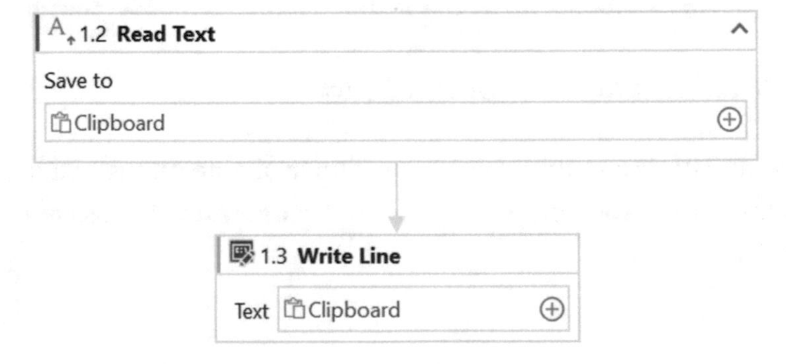

***Figure 6-13.** Final configuration for Read Text exercise*

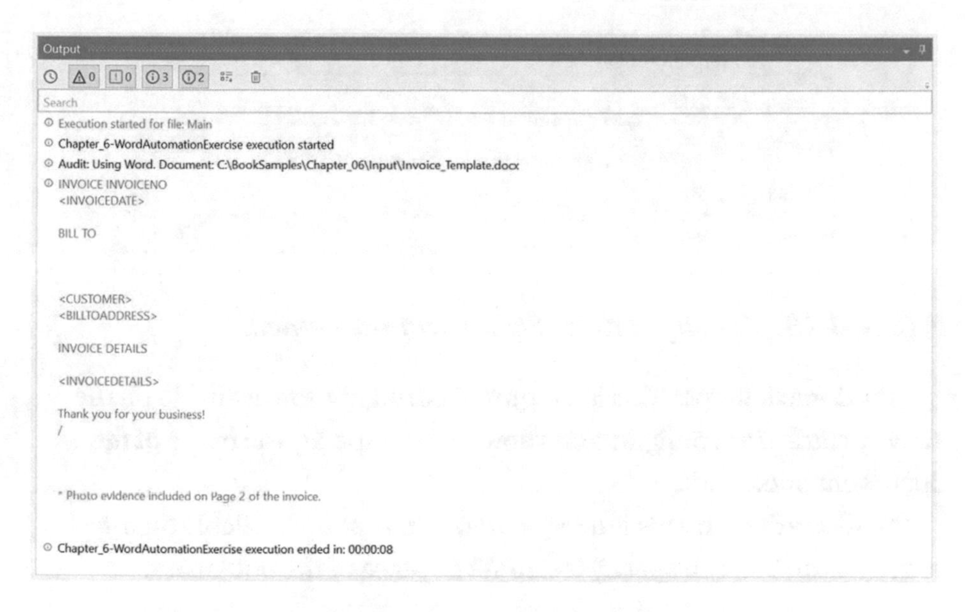

***Figure 6-14.** The output from the Read Text exercise*

Set Bookmark Content

The **Set Bookmark Content** activity allows you to set the text of a document bookmark. This is especially useful when editing documents based on a template as it allows you to remove whole sections of a document, for example, the payment section of a contract for an unpaid internship.

Configuration

This section provides instructions on how to configure a **Set Bookmark Content** activity, shown in Figure 6-15.

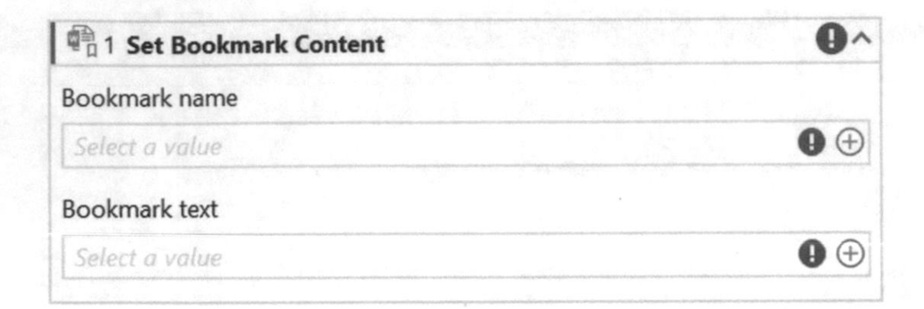

***Figure 6-15.** Activity card for Set Bookmark Content*

Bookmark name: This is a required configuration available on the activity card. This configuration allows you to specify the name of the document bookmark.

Bookmark text: This is a required configuration available on the activity card. You can specify text to set in place of the bookmark.

Tip The "Word Setup" section demonstrates how to show document bookmarks.

EXERCISE

Goal: Building on our previous exercise, use the `Set Bookmark Content` activity to replace the document bookmark, `INVOICENO`, with the invoice number value from the `Notebook`. Figure 6-2 shows the current state of the invoice template.

Source Code: Chapter_6-WordAutomationExercise

Setup: Here are step-by-step implementation instructions:

1. In `StudioX`, add the `Set Bookmark Content` activity within the `Use Word File` activity after the `Write Line` activity. This will be used to replace the bookmark text with the invoice number.

2. Click the Plus icon of the Bookmark name field and select the Text option. Within the Text Builder, type INVOICENO. Click Save.

3. In the Bookmark text field, click the Plus icon and hover over Notebook to select Indicate in Excel option.

4. Within the Excel file, select cell A2 of the Invoice Details sheet to specify the invoice number and click Confirm.

Once you have completed the exercise, the final configuration of the **Set Bookmark Content** activity should resemble Figure 6-16. Figure 6-17 shows the generated invoice after this activity was run.

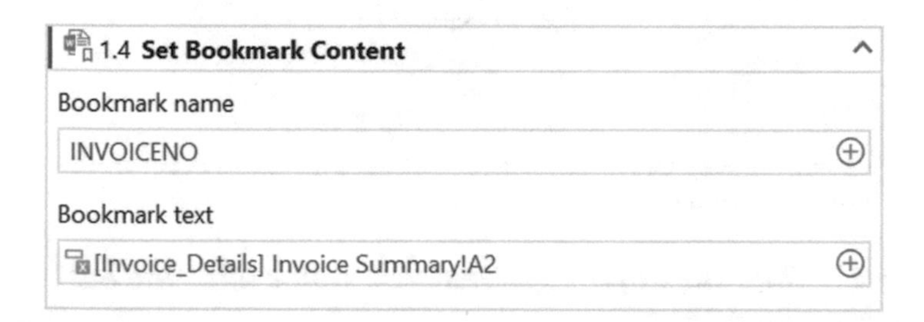

Figure 6-16. *Final configuration of Set Bookmark Content activity exercise*

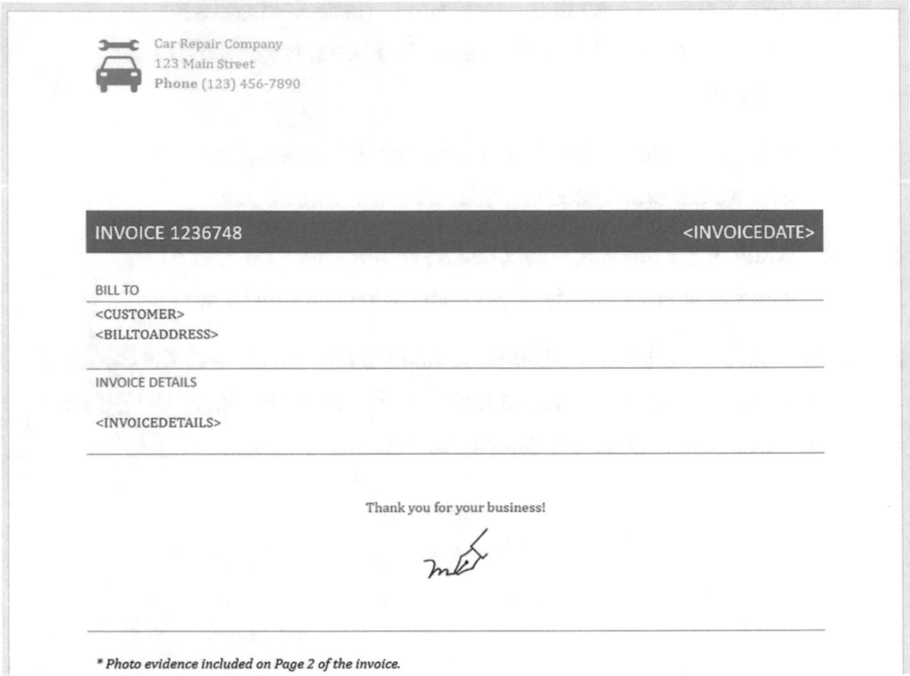

Car Repair Company
123 Main Street
Phone (123) 456-7890

INVOICE 1236748 <INVOICEDATE>

BILL TO
<CUSTOMER>
<BILLTOADDRESS>

INVOICE DETAILS
<INVOICEDETAILS>

Thank you for your business!

* Photo evidence included on Page 2 of the invoice.

***Figure 6-17.** The state of the generated invoice after the Set Bookmark Content exercise*

Replace Text in Document

The **Replace Text in Document** activity allows you to replace all occurrences of text within a document with other text. This can be used when editing documents based on a template, for example, the Name, Surname, and Address on an insurance form.

Tip To ensure only text intended is replaced, it is common to have parentheses or brackets around the text to have it uniquely identified, for example, specifying <Company Name> or {Invoice_No}.

Configuration

This section provides instructions on how to configure a **Replace Text in Document** activity, shown in Figure 6-18.

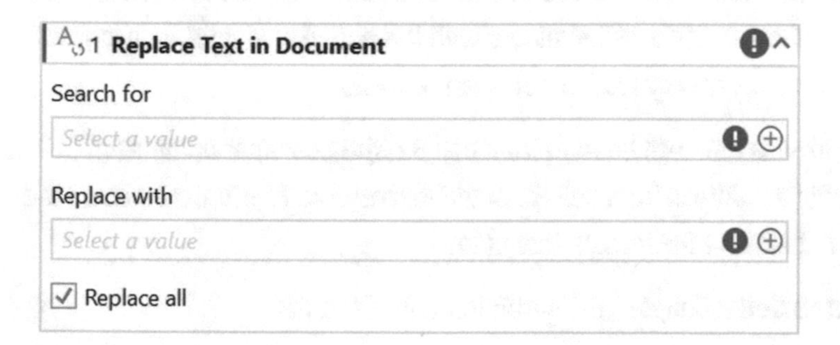

***Figure 6-18.** Activity card for Replace Text in Document*

Search for: This is a required configuration available on the activity card. This configuration allows you to specify the text that will be searched for to replace the text. As a note, the search is not case-sensitive.

Replace with: This is a required configuration available on the activity card. This configuration specifies the text that will be put in place of the searched text.

Replace all: This is an optional configuration available on the activity card. When checked, all occurrences of the text searched for will be replaced. If left unchecked, only the first occurrence of the text searched will be replaced. By default, this option is checked.

Found: This is an optional configuration available on the `Properties` panel. This configuration returns true or false depending on if the `Search for` text was found.

EXERCISE

Goal: Building on our previous exercise, use the Replace Text in Document activity to replace the <INVOICEDATE>, <CUSTOMER>, and <BILLTOADDRESS> placeholders with the actual customer details from Invoice Summary worksheet of Notebook.

The invoice date will be read from cell B2, the customer name from cell D2, and bill to address from cell E2 of the Notebook. Figure 6-17 shows the current state of the invoice template.

Source Code: Chapter_6-WordAutomationExercise

Setup: Here are step-by-step implementation instructions:

1. In StudioX, add a Replace Text in Document activity within the Use Word File activity after the Set Bookmark Content activity. This will be used to replace the invoice date value.
2. Click the Plus icon in the Search for field and select the Text option. Within the Text Builder, type <INVOICEDATE>. Click Save.
3. In the Replace with field, click the Plus icon and hover over Notebook to select the Indicate in Excel option. Within the Excel file, select cell B2 to specify the invoice date and click Confirm.
4. Add the Replace Text in Document activity within the Use Word File activity after the Replace Text in Document activity. This will be used to replace the customer name.
5. Click the Plus icon in the Search for field and select the Text option. Within the Text Builder, type <CUSTOMER>. Click Save.

6. In the Replace with field, click the Plus icon and hover over Notebook to select the Indicate in Excel option. Within the Excel file, select cell D2 to specify the customer name and click Confirm.

7. Add the Replace Text in Document activity within the Use Word File activity after the Replace Text in Document activity. This will be used to replace the bill to address.

8. Click the Plus icon in the Search for field and select the Text option. Within the Text Builder, type <BILLTOADDRESS>. Click Save.

9. In the Replace with field, click the Plus icon and hover over Notebook to select the Indicate in Excel option. Within the Excel file, select cell E2 to specify the bill to address and click Confirm.

Once you have completed the exercise, the final configuration of the **Replace Text in Document** activity should resemble Figure 6-19. Figure 6-20 shows the generated invoice with the placeholder texts replaced by this activity.

1.5 **Replace Text in Document**

Search for

<INVOICEDATE>

Replace with

[Invoice_Details] Invoice Summary!B2

☑ Replace all

1.6 **Replace Text in Document**

Search for

<CUSTOMER>

Replace with

[Invoice_Details] Invoice Summary!D2

☑ Replace all

1.7 **Replace Text in Document**

Search for

<BILLTOADDRESS>

Replace with

[Invoice_Details] Invoice Summary!E2

☑ Replace all

Figure 6-19. *Final configuration for Replace Text in Document activity exercise*

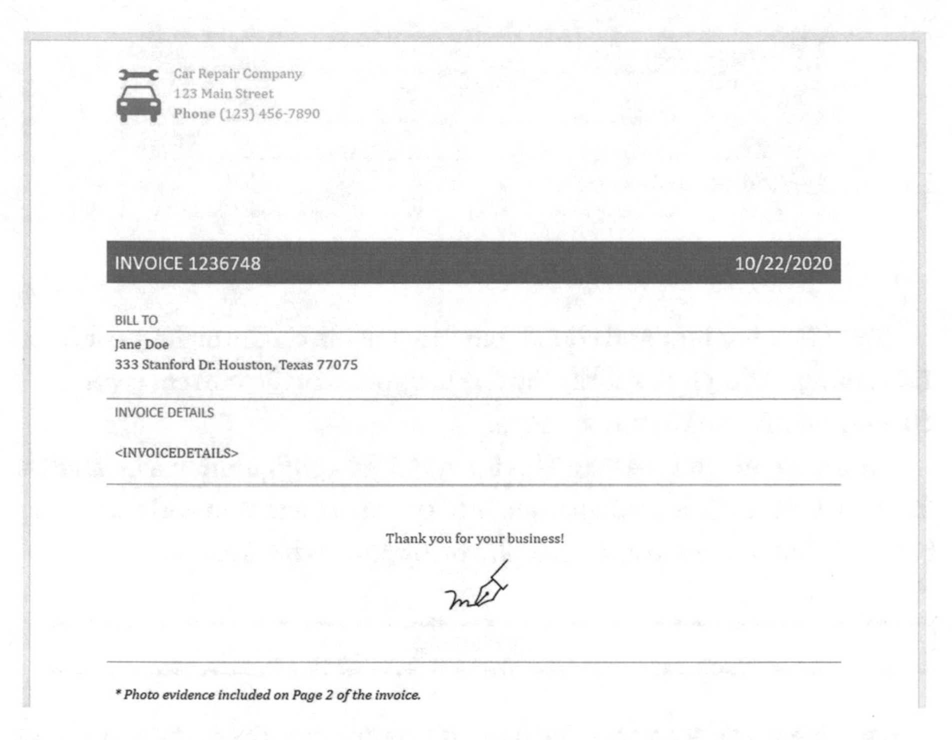
Car Repair Company
123 Main Street
Phone (123) 456-7890

INVOICE 1236748 10/22/2020

BILL TO
Jane Doe
333 Stanford Dr. Houston, Texas 77075

INVOICE DETAILS

<INVOICEDETAILS>

Thank you for your business!

* Photo evidence included on Page 2 of the invoice.

Figure 6-20. *The state of the generated invoice after the Replace Text in Document exercise*

Append Text

The **Append Text** activity allows you to write text in a document at the current caret position. This can be used to append text to the end of a document.

Configuration

This section provides instructions on how to configure an **Append Text** activity, shown in Figure 6-21.

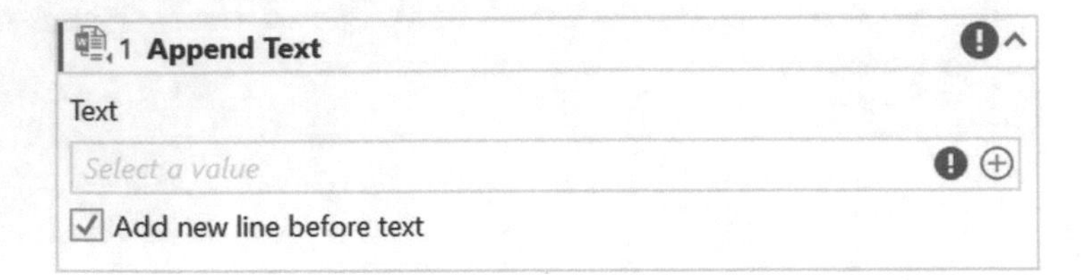

***Figure 6-21.** Activity card for Append Text*

Text: This is a required configuration available on the activity card. This configuration is to specify the text to append at the current caret position within the Word document.

Add new line before text: This is a required configuration available on the activity card. This configuration is to specify whether to add a new line before the appended text. By default, the option is checked.

EXERCISE

Goal: Building on our previous exercise, use the `Append Text` activity to append text related to payment terms. The payment date will be pulled from cell `C2` from `Invoice Summary` worksheet of the `Notebook`. Figure 6-20 shows the current state of the generated invoice.

Source Code: Chapter_6-WordAutomationExercise

Setup: Here are step-by-step implementation instructions:

1. In `StudioX`, add the `Append Text` activity in the `Use Word File` activity after the `Replace Text in Document` activity from the previous exercise.

2. Select the `Plus` icon of the `Text` field and select the `Text` option. Within the `Text Builder`, type `Terms: Please pay invoice by`.

3. Next, select the `Plus` icon in the `Text Builder` and hover over and select `Notebook` ➤ `Indicate in Excel`.

4. Within the Excel file, select cell `C2` of the `Invoice Details` sheet to specify the payment due date and click `Confirm`. Click Save.

Once you have completed the exercise, the final configuration of the **Append Text** activity should resemble Figure 6-22. Figure 6-23 shows the state of the generated invoice after this activity has run.

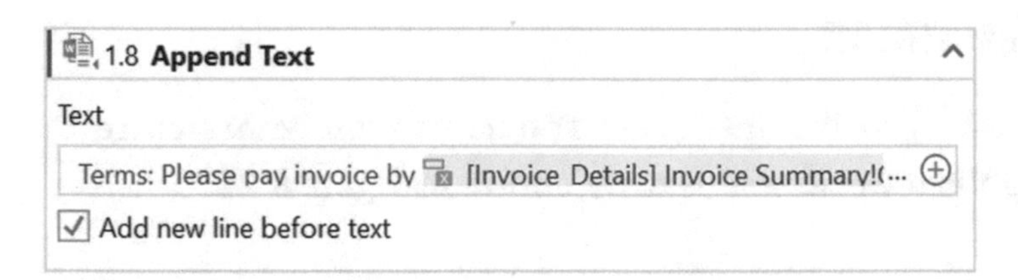

Figure 6-22. *Final configuration for Append Text exercise*

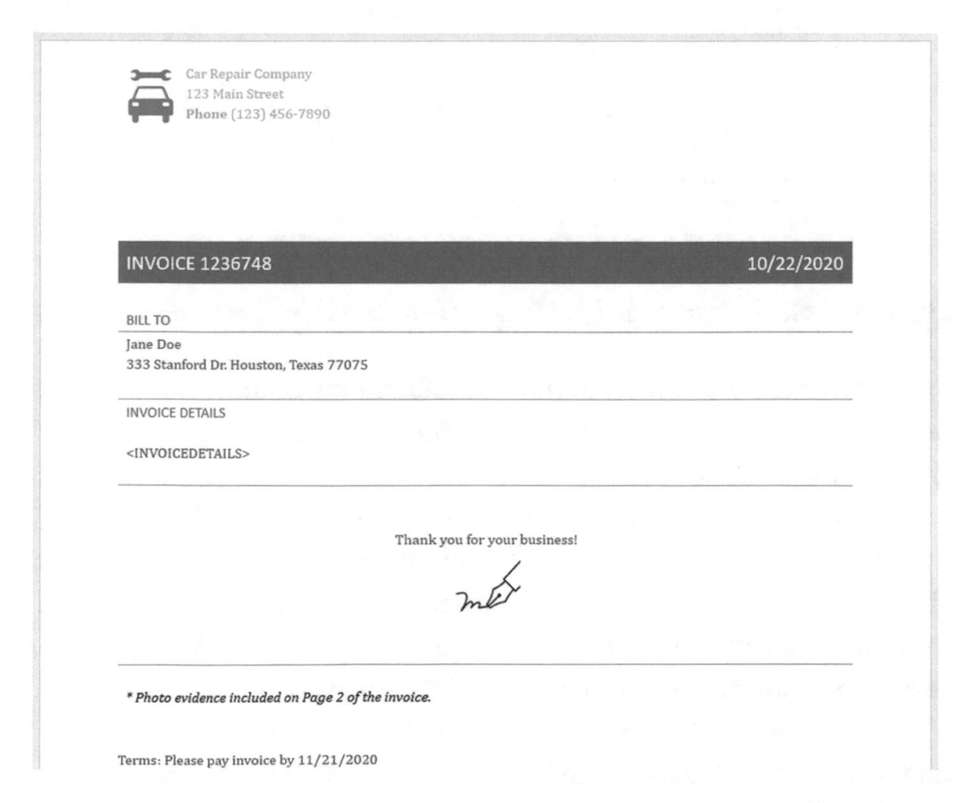
Car Repair Company
123 Main Street
Phone (123) 456-7890

INVOICE 1236748 10/22/2020

BILL TO

Jane Doe
333 Stanford Dr. Houston, Texas 77075

INVOICE DETAILS

<INVOICEDETAILS>

Thank you for your business!

** Photo evidence included on Page 2 of the invoice.*

Terms: Please pay invoice by 11/21/2020

Figure 6-23. *The state of the generated invoice after the Append Text exercise*

Insert DataTable in Document

The **Insert DataTable in Document** activity allows you to add a table at specific points within a document, for example, a table from Excel or a web page.

Configuration

This section provides instructions on how to configure an **Insert DataTable in Document** activity, shown in Figure 6-24.

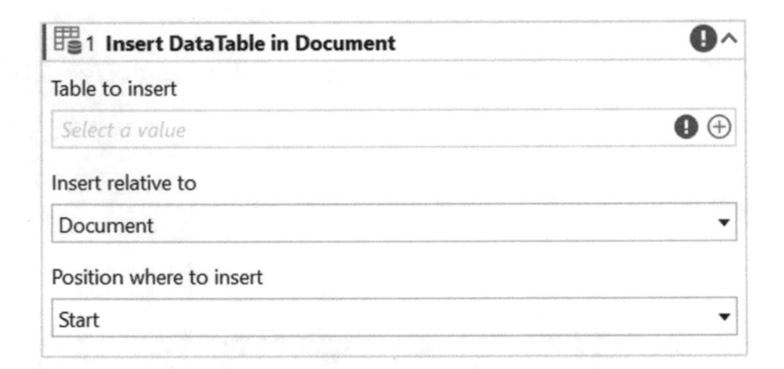

Figure 6-24. *Activity card for Insert DataTable in Document*

Table to insert: This is a required configuration available on the activity card. This configuration allows you to specify the data table to insert in the document.

Insert relative to: This is a required configuration available on the activity card and is set to `Document` by default. This configuration specifies the relative location of where the table will be inserted. You can choose to insert a datable relative to the start or end of the document, a bookmark, or text. Depending on this selection, the configuration fields will change.

Position where to insert: This is a required configuration available on the activity card and is set to `Start` by default. This configuration specifies where to insert the table based on the "insert using position of" setting within Word. You have the option to specify where to insert a data table in place of text or a bookmark or at the start or end of a document. Depending on the selection of the `insert relative to`, the configuration fields will change.

EXERCISE

Goal: Building on our previous exercise, use `Insert DataTable in Document` activity to replace all occurrences of the placeholder text `<INVOICEDETAILS>` with the data table from `Invoice Details` worksheet of `Notebook`. Figure 6-23 shows the current state of the invoice template.

Source Code: Chapter_6-WordAutomationExercise

Setup: Here are step-by-step implementation instructions:

1. In `StudioX`, add the `Insert DataTable in Document` activity in the `Use Word File` activity after the `Append Text` activity from the previous exercise.
2. Select the `Plus` icon of the `Table to insert` field and hover over and select `Notebook` ➤ `Invoice Table [Sheet]` ➤ `Table1 [Table]`.
3. In the `Insert relative to` field, select `Text` from the dropdown.
4. Select the `Plus` icon of the `Text to search for` field and select the `Text` option. Within the `Text Builder`, type `<table>`. Click `Save`.
5. Leave the `Text occurrence` field as is with the default value of `All`.
6. In the `Position where to insert` field, select `Replace` from the dropdown.

Once you have completed the exercise, the final configuration of the **Insert DataTable in Document** activity should resemble Figure 6-25. Figure 6-26 shows the state of the invoice template after this activity has run and the invoice details data table has been added.

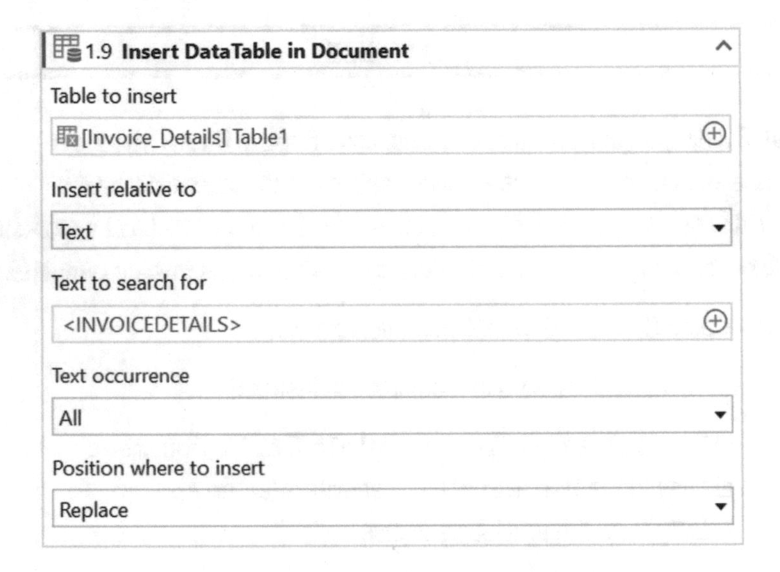

***Figure 6-25.** Final configuration of Insert DataTable in Document exercise*

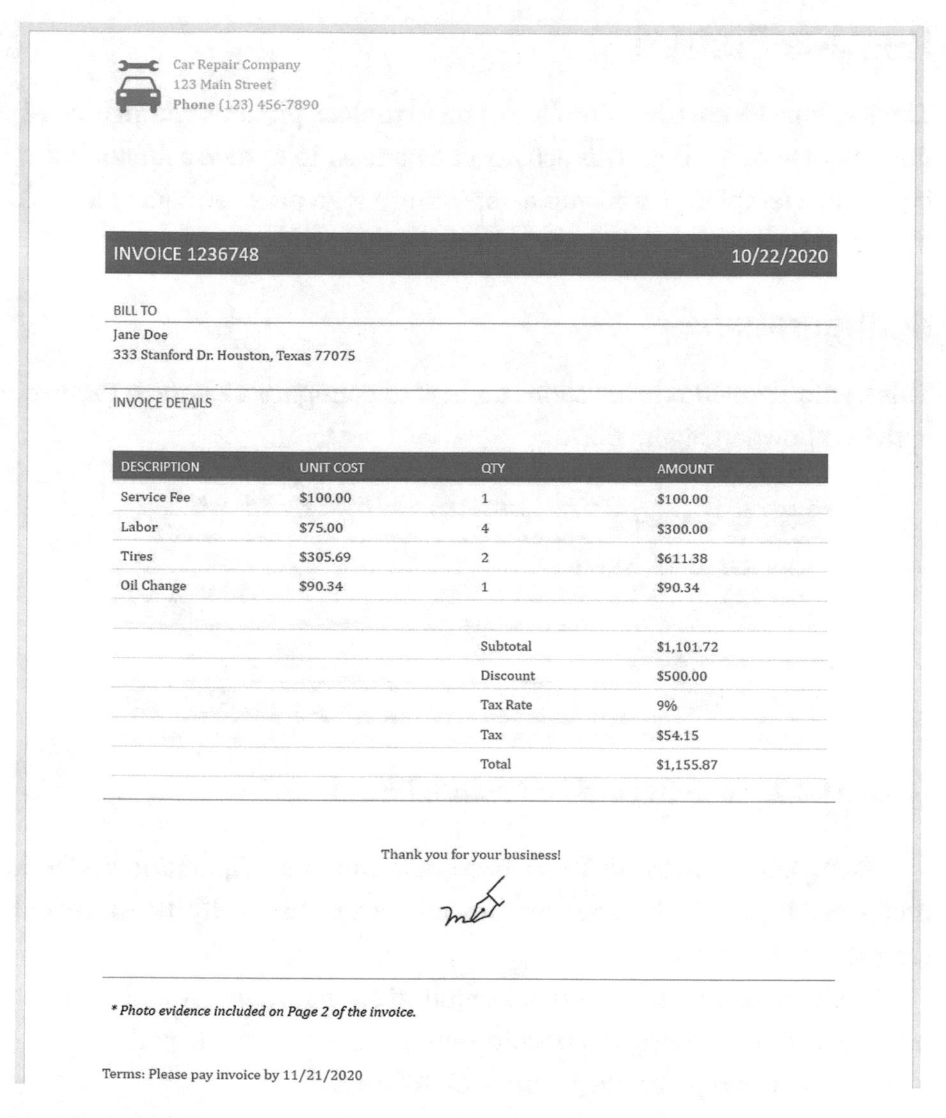

Car Repair Company
123 Main Street
Phone (123) 456-7890

INVOICE 1236748 10/22/2020

BILL TO

Jane Doe
333 Stanford Dr. Houston, Texas 77075

INVOICE DETAILS

DESCRIPTION	UNIT COST	QTY	AMOUNT
Service Fee	$100.00	1	$100.00
Labor	$75.00	4	$300.00
Tires	$305.69	2	$611.38
Oil Change	$90.34	1	$90.34
		Subtotal	$1,101.72
		Discount	$500.00
		Tax Rate	9%
		Tax	$54.15
		Total	$1,155.87

Thank you for your business!

* Photo evidence included on Page 2 of the invoice.

Terms: Please pay invoice by 11/21/2020

***Figure 6-26.** The state of the generated invoice after the Insert DataTable in Document exercise*

Replace Picture

The **Replace Picture** activity allows you to replace pictures within a Word document with Alt Text. This activity can be used to update a document based on a template, for example, replacing a logo on a contract or a photo of an employee on an onboarding form.

Configuration

This section provides instructions on how to configure a **Replace Picture** activity, shown in Figure 6-27.

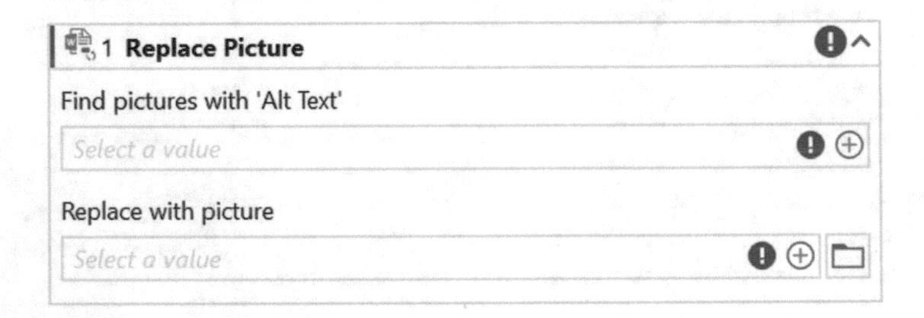

Figure 6-27. *Activity card for Replace Picture*

Find pictures with 'Alt Text': This is a required configuration available on the activity card. This configuration allows you to specify the Alt Text of the picture to replace.

Replace with picture: This is a required configuration available on the activity card. This configuration allows you to specify the file path of the picture that will replace the picture with Alt Text.

Tip The "Word Setup" section shows how to view the Alt Text of the placeholder picture.

EXERCISE

Goal: Building on our previous exercise, use the `Replace Picture` activity to replace the placeholder image with the Alt Text, `<SIGNATURE>`, with the actual signature image located at `C:\BookSamples\Chapter_06\Input\Signature.png`. Figure 6-26 shows the current state of the invoice template before this exercise.

Source Code: Chapter_6-WordAutomationExercise

Setup: Here are step-by-step implementation instructions:

1. In `StudioX`, add the `Replace Picture` activity within the `Use Word File` activity after the `Insert DataTable in Document` activity from the previous exercise.
2. Select the `Plus` icon of the `Find pictures with 'Alt Text'` field and select the `Text` option. Within the `Text Builder`, type `<SIGNATURE>`. Click `Save`.
3. Select the `Plus` icon of the `Replace with picture` field and select the `Folder` icon to browse for the file and select `C:\BookSamples\Chapter_06\Input\ Signature.png` file.

Once you have completed the exercise, the final configuration of the **Replace Picture** activity should resemble Figure 6-28. Figure 6-29 shows the placeholder image with 'Alt Text' replaced with the actual signature after this activity was run.

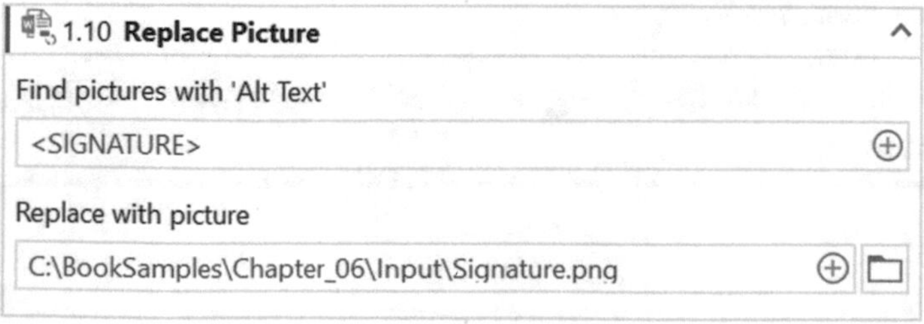

1.10 **Replace Picture**

Find pictures with 'Alt Text'

<SIGNATURE>

Replace with picture

C:\BookSamples\Chapter_06\Input\Signature.png

***Figure 6-28.** Final configuration of Replace Picture exercise*

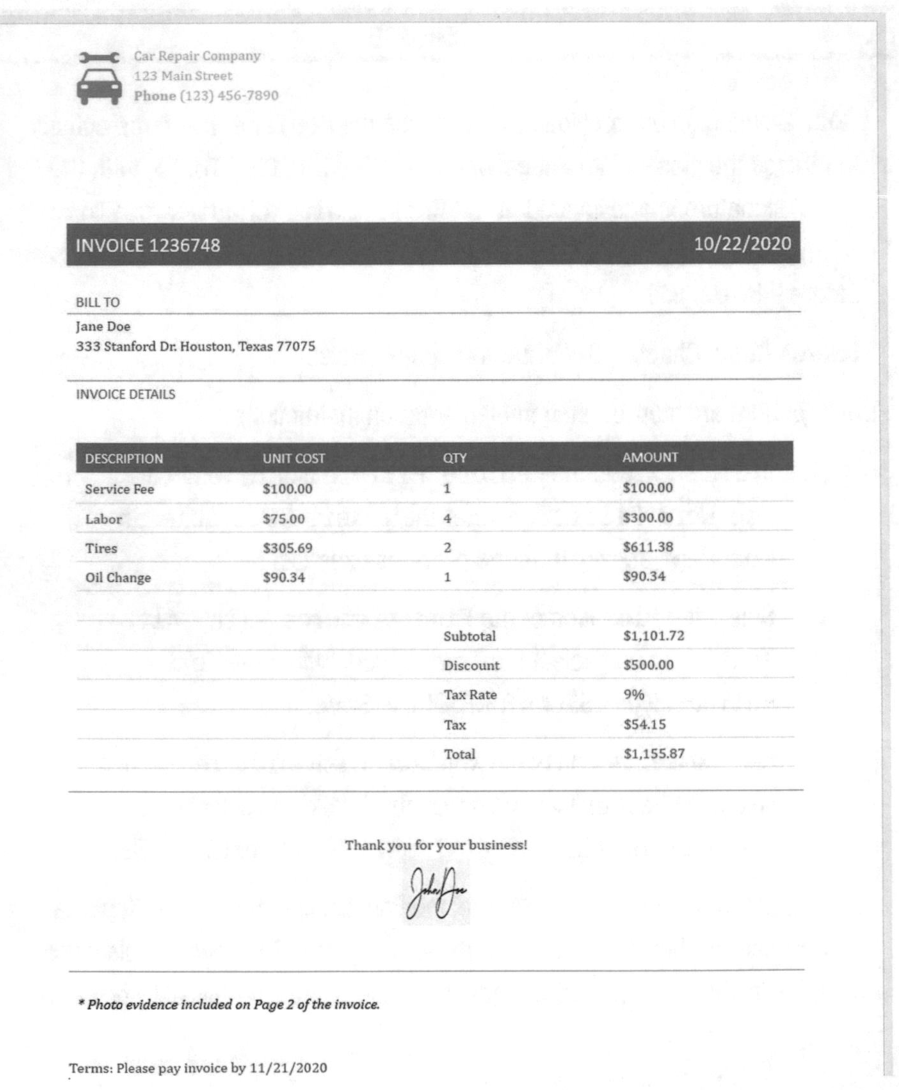

Car Repair Company
123 Main Street
Phone (123) 456-7890

INVOICE 1236748 10/22/2020

BILL TO

Jane Doe
333 Stanford Dr. Houston, Texas 77075

INVOICE DETAILS

DESCRIPTION	UNIT COST	QTY	AMOUNT
Service Fee	$100.00	1	$100.00
Labor	$75.00	4	$300.00
Tires	$305.69	2	$611.38
Oil Change	$90.34	1	$90.34
		Subtotal	$1,101.72
		Discount	$500.00
		Tax Rate	9%
		Tax	$54.15
		Total	$1,155.87

Thank you for your business!

** Photo evidence included on Page 2 of the invoice.*

Terms: Please pay invoice by 11/21/2020

Figure 6-29. *The state of the generated invoice after the Replace Picture exercise*

Add Picture

The **Add Picture** activity allows you to add a picture at the end of your Word document.

Tip This activity can be used to append an email signature to the body of an email using the **Send Outlook Email** activity.

Configuration

This section provides instructions on how to configure an **Add Picture** activity, shown in Figure 6-30.

Figure 6-30. *Activity card for Add Picture*

Picture to insert: This is a required configuration available on the activity card. This configuration allows you to specify the picture to insert at the end of a Word document.

EXERCISE

Goal: Building on our previous exercise, use `Add Picture` activity to insert photo evidence of damage located at `C:\BookSamples\Chapter_06\Input\Damage_Evidence.jpg`. Figure 6-29 shows the current state of the invoice template.

Source Code: Chapter_6-WordAutomationExercise

Setup: Here are step-by-step implementation instructions:

1. In `StudioX`, add the `Add Picture` activity in the `Use Word File` activity after the `Replace Picture` activity from the previous exercise.

2. Select the `Plus` icon of the `Picture to insert` field and select the `Folder` icon to browse for the file and select `C:\BookSamples\Chapter_06\Input\ Damage_Evidence.jpg` file. Figure 6-33 shows the signature to insert at the end of the invoice template.

Once you have completed the exercise, the final configuration of the **Add Picture** activity should resemble Figure 6-31. Figure 6-32 shows the generated invoice with a new picture added after this activity was run.

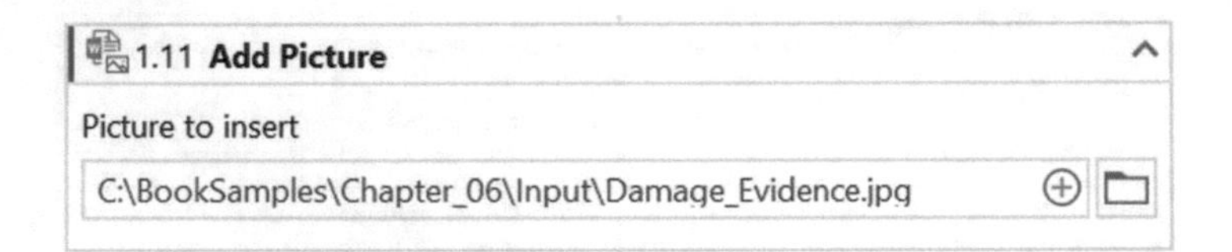

Figure 6-31. *Final configuration of Add Picture exercise*

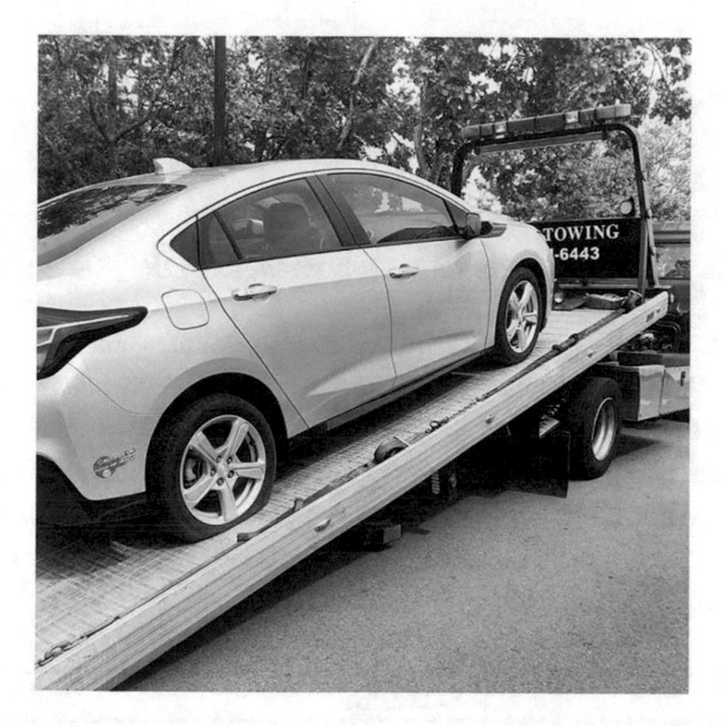

Figure 6-32. *The state of the generated invoice after the Add Picture exercise*

Save Document as PDF

The **Save Document as PDF** activity allows you to export the Word document as a PDF file. This is commonly used when needing to email a document as a PDF to prevent further editing.

Configuration

This section provides instructions on how to configure a **Save Document as PDF** activity, shown in Figure 6-33.

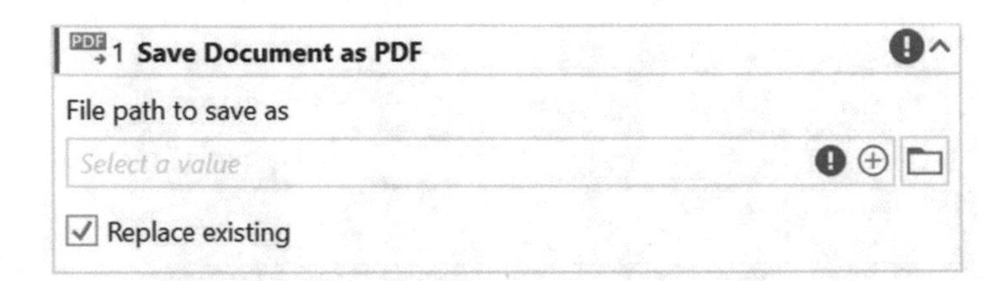

Figure 6-33. *Activity card for Save Document as PDF*

File path to save as: This is a required configuration available on the activity card. This configuration allows you to specify the file path of where the exported PDF file is saved.

Replace existing: This is an optional configuration available on the activity card. When checked, the file will be replaced if it already exists. If left unchecked and if the file already exists, an error will notify the file already exists. By default, this option is checked.

EXERCISE

Goal: Building on our previous exercise, use `Save Document as PDF` activity to export the generated invoice to a PDF version.

Source Code: Chapter_6-WordAutomationExercise

Setup: Here are step-by-step implementation instructions:

1. In `StudioX`, add the `Save Document as PDF` activity in the `Use Word File` activity after the `Add Picture` activity from the previous exercise.

2. Click the `Plus` icon of the `File path to save as` field and select the `Text` option. Within the `Text Builder`, type `C:\BookSamples\Chapter_06\Output\INV`.

3. Within the `Text Builder`, click the `Plus` icon, and hover over `Notebook` to select `Indicate in Excel`. Select cell `D2` and click `Confirm`.

4. Next, within `Text Builder`, type `.pdf`. Click `Save`.

Once you have completed the exercise, the final configuration of the **Save Document as PDF** activity should resemble Figure 6-34. Figure 6-35 shows the final state of the Output folder.

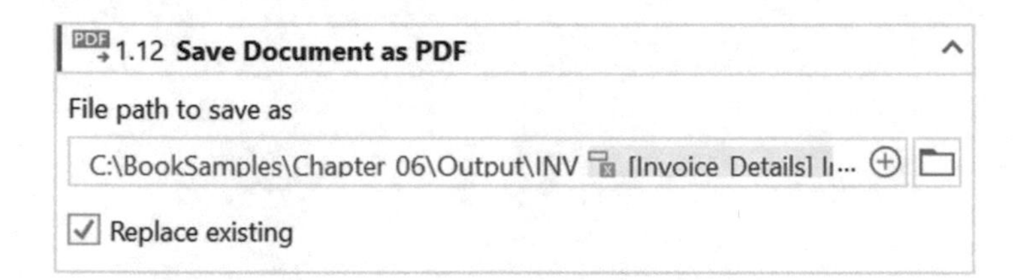

Figure 6-34. *Final configuration of the Save Document as PDF activity*

Figure 6-35. *Contents of the Output folder*

CHAPTER 7

Excel Automation

While Excel is quite powerful and widely used in today's working environment, automating in Excel has historically required skills and training. UiPath's StudioX integrates intuitively with Microsoft Excel simplifying citizen developers' experience when automating tasks such as extracting, filtering, manipulating, writing to, reading, and transferring data.

Learning Objectives

At the end of this chapter, you will learn how to

- Define an Excel scope to automate
- Insert, delete, duplicate, rename, and iterate through worksheets
- Insert, delete, and split columns
- Insert, delete, and iterate through Excel rows
- Write values and formulas into cells
- Read values and formulas from cells
- Create, change, and refresh a pivot table
- Copy, append, fill, write to, auto-fill, clear, and sort a range

A. Javed et al., *Robotic Process Automation using UiPath StudioX*,
https://doi.org/10.1007/978-1-4842-6794-3_7

- Save an Excel file in different formats, including PDF
- Export data to CSV format
- Apply filters and VLookups and run macros

Sample Overview

The sample scenario utilized in this chapter is Employee Onboarding. The exercise consists of three Excel workbooks used by the Human Resources (HR) team at a small firm that has recently hired 20 new employees.

The purpose of these Excel files are to provide the HR team with a centralized location to manage the new hired employee data and report on their onboarding status as they acclimate to the organization.

EmployeeOnboarding.xlsx: This workbook is used to manage the list of employees with details such as new hire start dates, salaries, and required training. This will be the main file that most of the exercises in the chapter will use in order to organize sheets, add input data, fill in formulas, and create tables for analysis.

EmployeeOnboardingInput.xlsx: This file provides five new employees' data that will need to be appended to the EmployeeOnboarding Excel file containing the original 20 employees.

EmployeeOnboarding Final.xlsm: This is the final version of the New Hires List sheet that needs to be formatted and have a Macro executed so that it is ready to be delivered.

Figures 7-1, 7-2, and 7-3 display the spreadsheets before executing any Excel automation exercises detailed in this chapter.

	A	B	C	D	E
1	ID	Full Name	Start Date	Department	Status
2	1	Brown, Sylvia	1/6/2020	Operations	
3	2	Carter, Samantha	4/10/2020	Accounting	
4	3	Clark, Jacob	7/20/2020	Human Resources	
5	4	Davis, Samuel	1/6/2020	Operations	
6	5	Davis, Sasha	7/20/2020	Information Technology	
7	6	Hill, Karen	1/6/2020	Operations	
8	7	Johnson, Elijah	7/20/2020	Information Technology	
9	8	Johnson, Adam	4/10/2020	Human Resources	
10	9	Jones, Daniel	7/20/2020	Information Technology	
11	10	Khan, Zain	1/6/2020	Operations	
12	11	Lane, Tamara	7/20/2020	Human Resources	
13	12	Lopez, Maya	4/10/2020	Accounting	
14	13	Miller, Raymond	7/20/2020	Information Technology	
15	14	Patel, Priya	1/6/2020	Accounting	
16	15	Sanchez, Gabriella	7/20/2020	Information Technology	
17	16	Singh, Aditya	4/10/2020	Human Resources	
18	17	Smith, Carolyn	1/6/2020	Accounting	
19	18	Smith, John	7/20/2020	Human Resources	
20	19	Williams, Jane	4/10/2020	Information Technology	
21	20	Young, Alyssa	1/6/2020	Accounting	
22					

New Hires List | New Hire Checklist | New Hire Salaries

***Figure 7-1.** Sample Employee Onboarding Excel file – New Hires List sheet*

	A	B	C	D	E	F	G	H	I
1	ID	Last Name	First Name	Orientatior	Employee Hanbool	Policy Trainin	Benefits Packag	Direct Deposit Setup	Technology Setu
2	1	Brown	Sylvia	Complete	Complete	Complete	Complete	Complete	Complete
3	2	Carter	Samantha	Not Started	Not Started	In Progress	Not Started	Not Started	Not Started
4	3	Clark	Jacob	Not Started	Not Started	In Progress	In Progress	Not Started	In Progress
5	4	Davis	Samuel	Complete	Complete	Complete	Complete	Complete	Complete
6	5	Davis	Sasha	Complete	Complete	Complete	Complete	Complete	Complete
7	6	Hill	Karen	Complete	Complete	Complete	Complete	Complete	Complete
8	7	Johnson	Elijah	Not Started	In Progress	In Progress	Not Started	In Progress	Not Started
9	8	Johnson	Adam	Complete	Complete	Complete	Complete	Complete	Complete
10	9	Jones	Daniel	Not Started	Not Started	In Progress	Not Started	Not Started	Not Started
11	10	Khan	Zain	Complete	Complete	Complete	Complete	Complete	Complete
12	11	Lane	Tamara	Not Started	Not Started	In Progress	Not Started	Not Started	Not Started
13	12	Lopez	Maya	Complete	Complete	Complete	Complete	Complete	Complete
14	13	Miller	Raymond	Not Started	In Progress	Not Started	In Progress	Not Started	Not Started
15	14	Patel	Priya	Complete	Complete	Complete	Complete	Complete	Complete
16	15	Sanchez	Gabriella	Not Started	In Progress	In Progress	Not Started	In Progress	Not Started
17	16	Singh	Aditya	Complete	Complete	Complete	Complete	Complete	Complete
18	17	Smith	Carolyn	Complete	Complete	Complete	Complete	Complete	Complete
19	18	Smith	John	Not Started	Not Started	Not Started	In Progress	Not Started	Not Started
20	19	Williams	Jane	Complete	Complete	Complete	Complete	Complete	Complete
21	20	Young	Alyssa	Not Started	Not Started	Not Started	In Progress	Not Started	Not Started
22									

New Hires List | New Hire Checklist | New Hire Salaries

Figure 7-2. *Sample Employee Onboarding Excel file – New Hire Checklist sheet*

	A	B	C	D
1	ID	Salary	Estimated Bonus	
2	1	$45,000		
3	2	$100,000		
4	3	$75,000		
5	4	$50,000		
6	5	$65,000		
7	6	$125,000		
8	7	$80,000		
9	8	$30,000		
10	9	$85,000		
11	10	$40,000		
12	11	$105,000		
13	12	$55,000		
14	13	$60,000		
15	14	$45,000		
16	15	$70,000		
17	16	$110,000		
18	17	$80,000		
19	18	$30,000		
20	19	$85,000		
21	20	$60,000		
22				

New Hires List | New Hire Checklist | New Hire Salaries

Figure 7-3. *Sample Employee Onboarding Excel file – New Hire Salaries sheet*

Activities Reference

As shown in Figure 7-4, all Excel automation activities can be found under the Excel category/tile. The following sections will provide instructions on how to configure and use each activity.

Figure 7-4. *Activities for Excel automation*

Use Excel File

The **Use Excel File** activity found under Resources allows you to create or open an Excel file to utilize the file's data in your automation.

This is the foundation for the Excel automation and serves as the parent activity integrating StudioX with Excel. For example, copying and pasting, sorting, filtering, inserting, and deleting data activities can be added as child activities within the **Use Excel File** activity.

Configuration

This section provides instructions on how to configure a **Use Excel File** activity, shown in Figure 7-5.

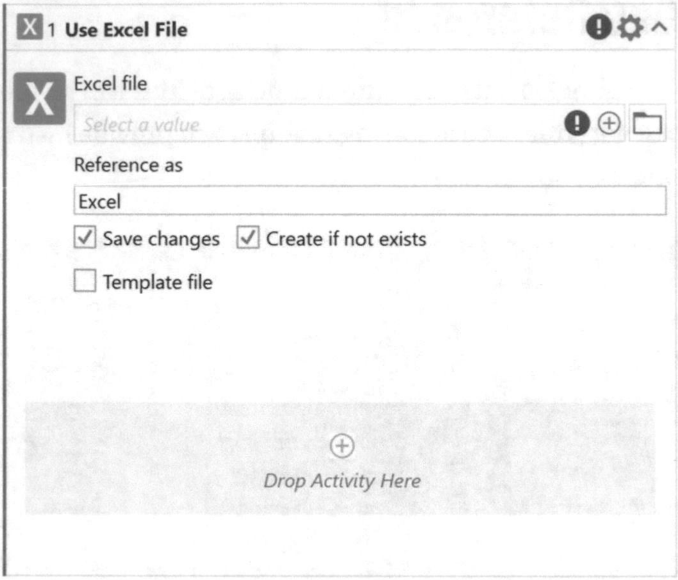

Figure 7-5. *Use Excel File activity card*

Excel file: This is a required configuration available on the activity card. This configuration gives StudioX the full file path for the workbook data that the Excel automation will use.

Reference as: This is a required configuration available on the activity card. It is recommended that this field is populated with a unique and descriptive name to reference the Excel file for the automation project, for example, InvoiceData or EmployeeOnboarding. By default, the field is populated with `Excel`.

Save changes: This is an optional configuration available on the activity card. This configuration ensures that the workbook is saved after each action is taken on an activity. By default, this option is checked. If this option is unchecked, then the message `Save off` is displayed next to the Excel file in the Data Manager panel.

Create if not exists: This is an optional configuration available on the activity card. This configuration ensures that a blank file is created if it does not exist in the target location. By default, this option is checked.

Edit password: This is an optional configuration available on the `Properties` panel. This field is used for editing a password-protected Excel file. Enter the password in this field if necessary.

Password: This is an optional configuration available on the `Properties` panel. This field is used for opening a password-protected Excel file. Enter the password in this field if necessary.

ReadOnly: This is an optional configuration available on the `Properties` panel. If checked, the workbook will open in read-only mode for automation. This option will allow the automation to extract data from a workbook even if it is password protected. By default, this option is not checked.

Note The following Exercise for Use Excel File activity must be done prior to completing the exercises in any of the other Excel activities sections.

EXERCISE

Goal: Use the Use Excel File activity to open the EmployeeOnboarding.xlsx Excel workbook located in the file path C:\BookSamples\Chapter_07\EmployeeOnboarding.xlsx.

Setup: Here are step-by-step implementation instructions:

1. In `StudioX`, add the `Use Excel File` activity to a blank process.
2. Next, click the `Folder` icon in the `Excel file` field and select `C:\BookSamples\Chapter_07\EmployeeOnboarding.xlsx` file.
3. Next, enter a name for this file in the `Reference as` field. For this example, enter the name `EmployeeOnboarding`.
4. Leave the default configurations as is for this example.

Once you have completed the exercise, the final configuration of the **Use Excel File** activity should resemble Figure 7-6. The output of this example will read the EmployeeOnboarding.xlsx from the selected file path and add it as an Excel file resource for this StudioX project.

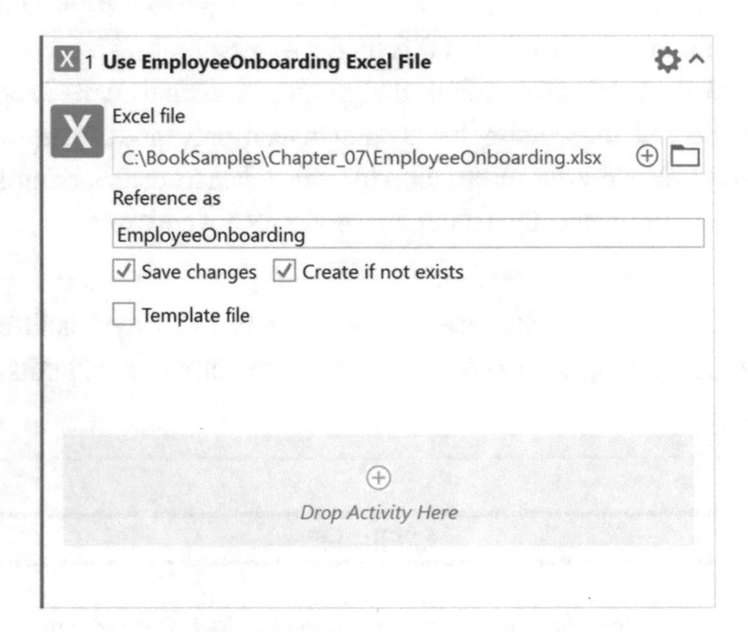

***Figure 7-6.** EmployeeOnboarding .xlsx populated in Use Excel File activity*

Insert Sheet

The **Insert Sheet** activity allows you to add an empty sheet to a workbook.

Configuration

This section provides instructions on how to configure an **Insert Sheet** activity, displayed in Figure 7-7.

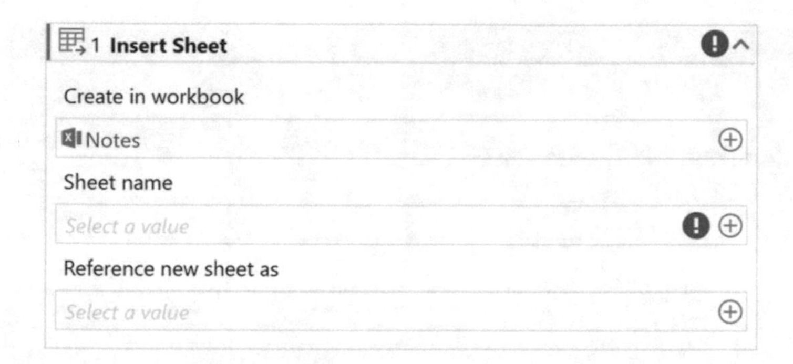

***Figure 7-7.** Insert Sheet activity card*

Create in workbook: This is a required configuration available on the activity card. This configuration provides StudioX with the workbook to insert as the new sheet. The `Create in workbook` field is pre-populated with the workbook specified in the parent `Use Excel File` activity. You can click the `Plus` icon to reference a different workbook.

Sheet name: This is a required configuration available on the activity card. This configuration determines the name of the new sheet.

Reference new sheet as: This is an optional configuration available on the activity card. This configuration allows you to specify the name you want to use to reference the sheet in further activities. This is commonly configured through the `Text Builder` option allowing you to give the reference name.

EXERCISE

Goal: Building on our previous exercise, use the `Insert Sheet` activity to add a new sheet to the `EmployeeOnboarding` Excel file. Figure 7-8 shows the initial state of `EmployeeOnboarding.xlsx`.

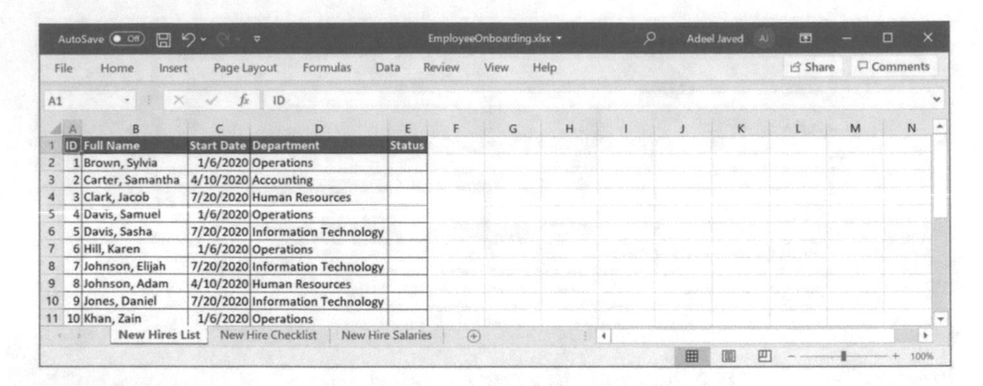

ID	Full Name	Start Date	Department	Status
1	Brown, Sylvia	1/6/2020	Operations	
2	Carter, Samantha	4/10/2020	Accounting	
3	Clark, Jacob	7/20/2020	Human Resources	
4	Davis, Samuel	1/6/2020	Operations	
5	Davis, Sasha	7/20/2020	Information Technology	
6	Hill, Karen	1/6/2020	Operations	
7	Johnson, Elijah	7/20/2020	Information Technology	
8	Johnson, Adam	4/10/2020	Human Resources	
9	Jones, Daniel	7/20/2020	Information Technology	
10	Khan, Zain	1/6/2020	Operations	

Figure 7-8. *Displays the EmployeeOnboarding Excel workbook before the automation is executed*

Source Code: Chapter_7_ExcelSheetActivitiesExercise

Setup: Here are step-by-step implementation instructions:

1. In `StudioX`, add the `Insert Sheet` activity to the body of the parent `Use Excel File` activity in the `Designer` panel.
2. The `EmployeeOnboarding` file will be auto-populated in the `Create in workbook` field. Leave it as is: `[EmployeeOnboarding]`.
3. Next, click the `Plus` icon in the `Sheet name` field, select `Text Builder`, and type `New Data`.
4. Leave `Reference new sheet as` blank.

Once you have completed the exercise, the final configuration of the **Insert Sheet** activity should resemble Figure 7-9.

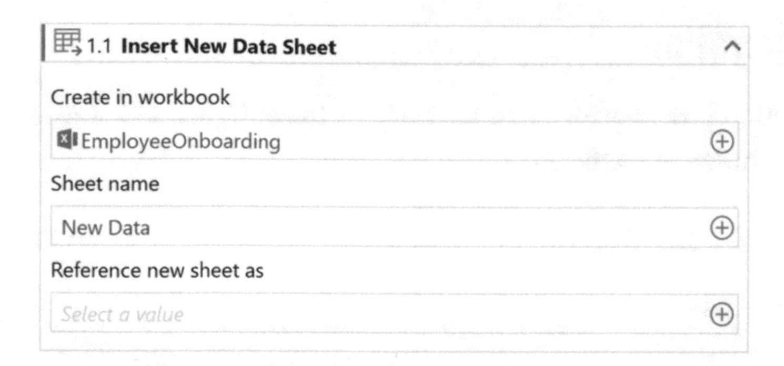

***Figure 7-9.** Displays the final configuration for the Insert Sheet activity*

Figure 7-10 shows the output of the Insert Sheet activity that has added a New Data sheet to the EmployeeOnboarding Excel file.

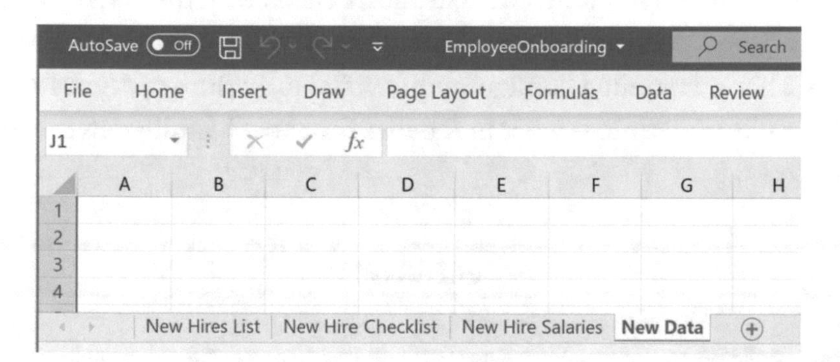

***Figure 7-10.** A new sheet is added to the EmployeeOnboarding Excel file*

Rename Sheet

The **Rename Sheet** activity changes the name of an existing sheet in a workbook.

Configuration

This section provides instructions on how to configure a **Rename Sheet** activity, shown in Figure 7-11.

Figure 7-11. *Rename Sheet activity card*

From: This is a required configuration available on the activity card. This configuration identifies the existing sheet that needs to be renamed.

To: This is a required configuration available on the activity card. This configuration determines what to rename the sheet identified in the `From` field.

EXERCISE

Goal: Building on our previous exercise, use the `Rename Sheet` activity to update the name of the `New Data` sheet to `New Hire Statistics`. Figure 7-10 shows the sheets in the `EmployeeOnboarding.xlsx` file before the automation.

Source Code: Chapter_7_ExcelSheetActivitiesExercise

Setup: Here are step-by-step implementation instructions:

1. In `StudioX`, add the `Rename Sheet` activity to the body of the parent `Use Excel File` activity after the `Insert Sheet` activity from previous exercise.

2. Next, click the `Plus` icon in the `From` field, and hover over `EmployeeOnboarding` workbook to select the `New Data [Sheet]`.

3. Next, click the `Plus` icon in the `To` field, select the `Text` option, and using `Text Builder`, type in `New Hire Statistics`.

Once you have completed the exercise, the configuration of the **Rename Sheet** activity should resemble Figure 7-12.

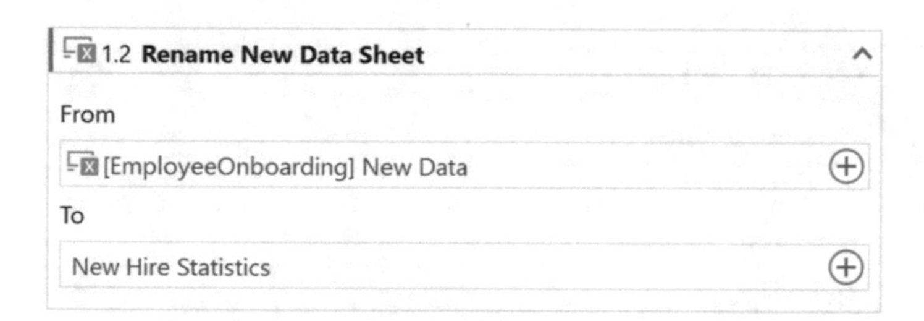

Figure 7-12. *Displays the final configuration for the Rename Sheet activity*

Figure 7-13 shows the output of the **Rename Sheet** activity that has updated the name of the New Data sheet to New Hires Statistics.

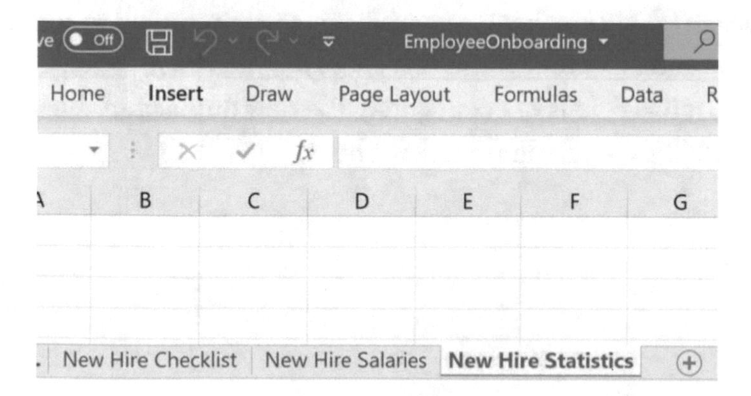

Figure 7-13. *New Data sheet renamed to New Hire Statistics*

Duplicate Sheet

The **Duplicate Sheet** activity creates a copy of a chosen sheet within the same workbook.

Configuration

This section provides instructions on how to configure a **Duplicate Sheet** activity, shown in Figure 7-14.

Figure 7-14. *Duplicate Sheet activity card*

Sheet to duplicate: This is a required configuration available on the activity card. This configuration supplies the name of the existing sheet that will be copied.

Rename to: Although not required, it is recommended to configure this field to enter a name for the new sheet that will be meaningful for the workbook. If this field is not configured, the new duplicated sheet will be created with the default name of `New Sheet`.

EXERCISE

Goal: Building on our previous exercise, use the `Duplicate Sheet` activity to create a duplicate of the `New Hire Salaries` sheet in the `EmployeeOnboarding` workbook. Figure 7-13 displays the sheets in the `EmployeeOnboarding.xlsx` prior to the exercise.

Source Code: Chapter_7_ExcelSheetActivitiesExercise

Setup: Here are step-by-step implementation instructions:

1. In `StudioX`, add the `Duplicate Sheet` activity to the body of the parent `Use Excel File` activity in the `Designer` panel.
2. Next, click the `Plus` icon in the `Sheet to duplicate` field, and hover over the `EmployeeOnboarding` workbook to select the `New Hire Salaries [Sheet]`.
3. Next, click the `Plus` icon in the `Rename to` field, select the `Text Builder` option, and type in `Salaries Copy`.

Once you have completed the exercise, the final configuration of the **Duplicate Sheet** activity should resemble Figure 7-15.

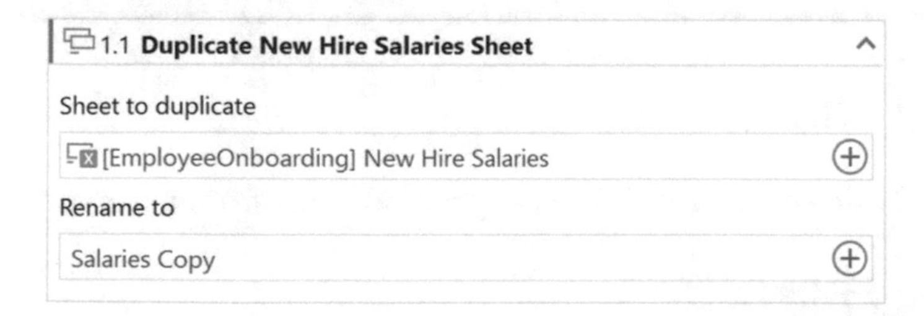

Figure 7-15. *Displays the final configuration for the Duplicate Sheet activity*

The output of this example duplicates the New Hire Salaries sheet with the name Salaries Copy as displayed in Figure 7-16.

ID	Salary	Estimated Bonus
1	$45,000	
2	$100,000	
3	$75,000	
4	$50,000	
5	$65,000	
6	$125,000	
7	$80,000	
8	$30,000	
9	$85,000	
10	$40,000	
11	$105,000	
12	$55,000	
13	$60,000	
14	$45,000	
15	$70,000	
16	$110,000	
17	$80,000	
18	$30,000	
19	$85,000	
20	$60,000	

... New Hire Checklist | New Hire Salaries | **Salaries Copy** | New Hire Statistics

***Figure 7-16.** Displays the duplicated Salaries Copy sheet*

Delete Sheet

The **Delete Sheet** activity deletes a specified sheet from a workbook.

Configuration

This section provides instructions on how to configure a **Delete Sheet** activity, shown in Figure 7-17.

Figure 7-17. *Delete Sheet activity card*

Select sheet: This is a required configuration available on the activity card. This configuration identifies the exact sheet to be deleted from the workbook.

EXERCISE

Goal: Building on our previous exercise, use the `Delete Sheet` activity to delete the `Salaries Copy` sheet from the `EmployeeOnboarding` Excel file. Figure 7-16 displays the `EmployeeOnboarding.xlsx` sheet before the exercise.

Source Code: Chapter_7_ExcelSheetActivitiesExercise

Setup: Here are step-by-step implementation instructions:

1. In StudioX, add the `Delete Sheet` activity under the parent `Use Excel File` activity in the `Designer` panel.
2. Next, click the `Plus` icon in the `Select sheet` field, and from `EmployeeOnboarding` workbook, select `Salaries Copy [Sheet]`.

Once you have completed the exercise, the final configuration of the **Delete Sheet** activity should resemble Figure 7-18.

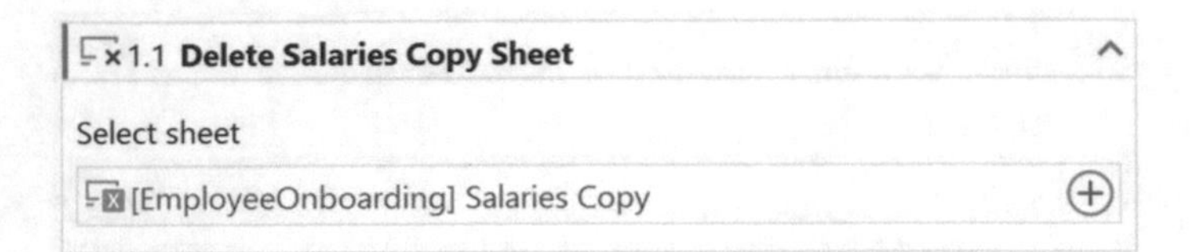

***Figure 7-18.** Displays the final configuration for the Delete Sheet activity*

The output of this example deletes the sheet with the name Salaries Copy in the EmployeeOnboarding.xlsx workbook, as shown in Figure 7-19.

	A	B	C	D	E
1	ID	Salary	Estimated Bonus		
2	1	$45,000			
3	2	$100,000			
4	3	$75,000			
5	4	$50,000			
6	5	$65,000			
7	6	$125,000			
8	7	$80,000			
9	8	$30,000			
10	9	$85,000			
11	10	$40,000			
12	11	$105,000			
13	12	$55,000			
14	13	$60,000			
15	14	$45,000			
16	15	$70,000			
17	16	$110,000			
18	17	$80,000			
19	18	$30,000			
20	19	$85,000			
21	20	$60,000			

... New Hire Checklist | **New Hire Salaries** | New Hire Statistics

***Figure 7-19.** Displays that the Salaries Copy sheet is now deleted*

For Each Excel Sheet

The **For Each Excel Sheet** activity allows you to execute one or more activities against each sheet within the same workbook.

Note You must configure at least one child activity within the For Each Excel Sheet activity card to iterate through multiple actions on a set of sheets.

Configuration

This section provides instructions on how to configure a **For Each Excel Sheet** activity, as shown in Figure 7-20.

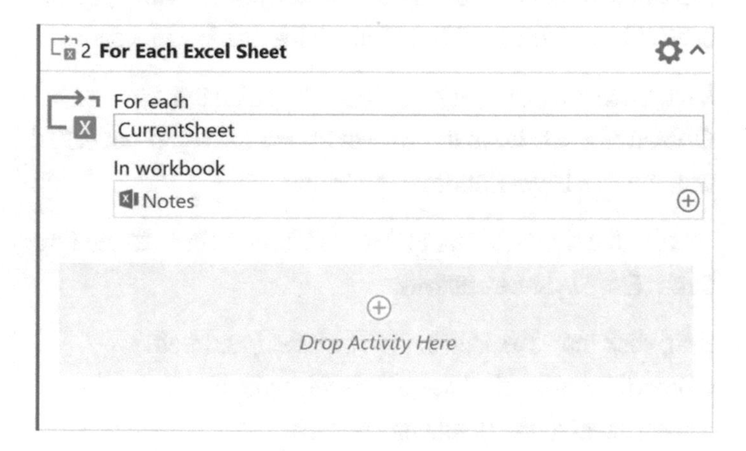

Figure 7-20. *For Each Excel Sheet activity card*

For each: This is a required configuration available on the activity card. This configuration allows you to provide a name to reference the current sheet which is being iterated through; this is especially important for referencing in the child activities. The default text is entered as `CurrentSheet` and can be changed based on what data you are iterating through. You can simply rename this field or leave the default.

In workbook: This is a required configuration available on the activity card. This configuration identifies the workbook containing the sheets that need to be iterated through. If this activity is inside of a parent Use Excel File activity, then the field will be auto-populated with the parent Excel file.

EXERCISE

Goal: Building on our previous exercise, use the `For Each Excel Sheet` activity to iterate through the sheets in the `EmployeeOnboarding` Excel file and print their names in the Output panel.

Source Code: Chapter_7_ExcelSheetActivitiesExercise

Setup: Here are step-by-step implementation instructions:

1. In `StudioX`, add the `For Each Excel Sheet` activity in the body of `Use Excel File` activity in the `Designer` panel.
2. Next, leave the `For each` field with the default text of `CurrentSheet`; leave the `In workbook` field auto-populated with the `EmployeeOnboarding` Excel.
3. Then, add a `Write line` activity as a child activity to the `For Each Excel Sheet` activity.
4. Next, click the `Plus` icon in the `Text` field, and from `CurrentSheet`, select Name. This will write the name of each current sheet to the Output panel.

Once you have completed the exercise, the final configuration of the **For Each Excel Sheet** activity should resemble Figure 7-21.

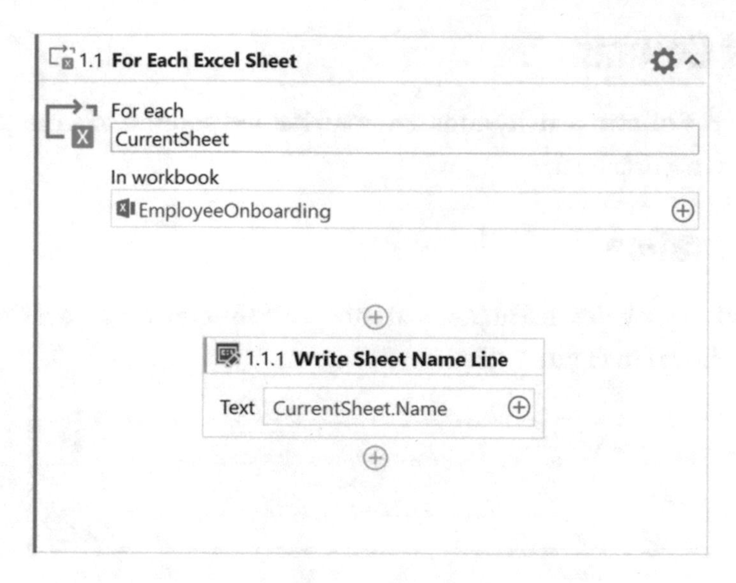

Figure 7-21. *Displays the final configuration for the For Each Excel Sheet activity*

The output of this exercise writes the name of each sheet in the EmployeeOnboarding.xlsx workbook to the Output panel, as shown in Figure 7-22.

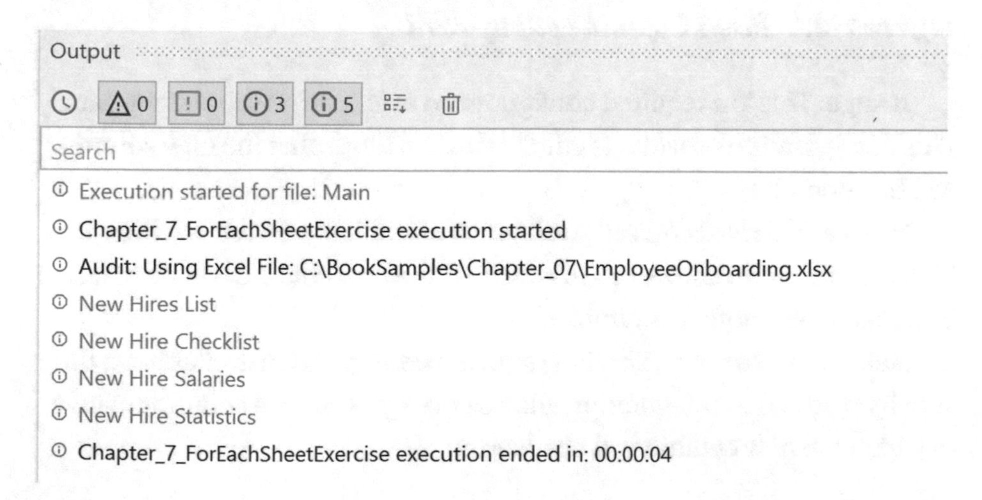

Figure 7-22. *Displays the names of all four worksheets written to the Output panel*

Insert Column

The **Insert Column** activity adds a newly defined column to a range, sheet, or table in a worksheet.

Configuration

This section provides instructions on how to configure an **Insert Column** activity, shown in Figure 7-23.

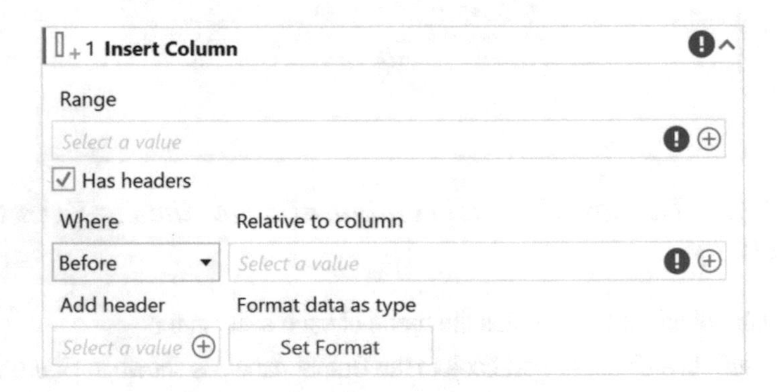

Figure 7-23. *Insert Column activity card*

Range: This is a required configuration available on the activity card. This configuration provides StudioX with the range that the new column will be added to.

Where: This is a required configuration available on the activity card. This configuration allows you to specify if the new column will be inserted before or after another column.

Relative to column: This is a required configuration available on the activity card. This configuration allows you to specify the column relative to which the new column will be inserted.

Add header: This is an optional configuration available on the activity card. This configuration provides the new column with a header name while executing the automation. It is recommended that this field is configured to ensure that the new column contains a header name.

Format data as type: This is an optional configuration available on the activity card. This configuration allows you to set the format for the cells being added in the new column. The configuration for this field is selected through the `Category` dropdown menu populated with the available formats that can be chosen for the cells. The `Category` options include `General`, `Number`, `Date`, `Time`, `Percentage`, `Currency`, `Text`, or `Custom`.

EXERCISE

Goal: Use the `Use Excel File` and `Insert Column` activities to add a `Last Name` and a `First Name` column to the `New Hires List` sheet after the `Full Name` column. Figure 7-24 shows the `New Hires List` range before the exercise.

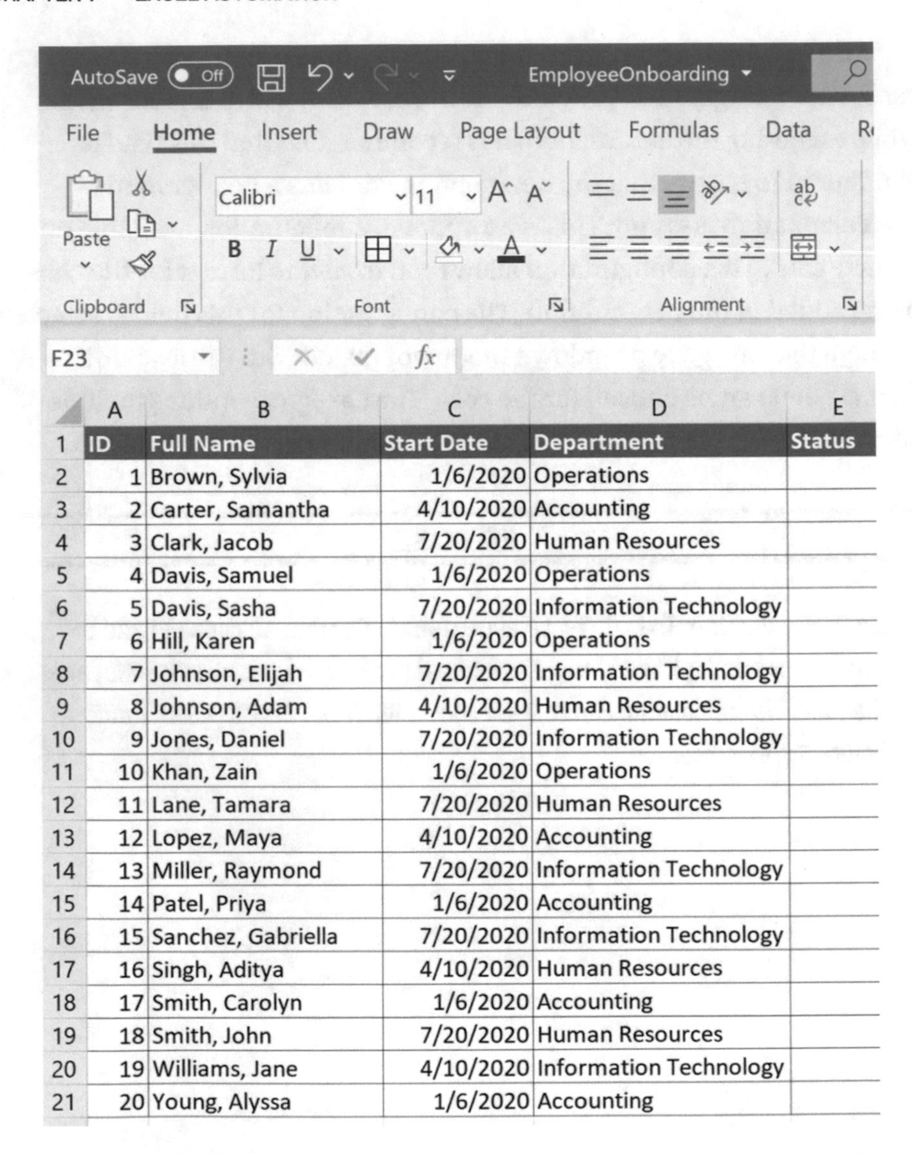

	A	B	C	D	E
1	ID	Full Name	Start Date	Department	Status
2	1	Brown, Sylvia	1/6/2020	Operations	
3	2	Carter, Samantha	4/10/2020	Accounting	
4	3	Clark, Jacob	7/20/2020	Human Resources	
5	4	Davis, Samuel	1/6/2020	Operations	
6	5	Davis, Sasha	7/20/2020	Information Technology	
7	6	Hill, Karen	1/6/2020	Operations	
8	7	Johnson, Elijah	7/20/2020	Information Technology	
9	8	Johnson, Adam	4/10/2020	Human Resources	
10	9	Jones, Daniel	7/20/2020	Information Technology	
11	10	Khan, Zain	1/6/2020	Operations	
12	11	Lane, Tamara	7/20/2020	Human Resources	
13	12	Lopez, Maya	4/10/2020	Accounting	
14	13	Miller, Raymond	7/20/2020	Information Technology	
15	14	Patel, Priya	1/6/2020	Accounting	
16	15	Sanchez, Gabriella	7/20/2020	Information Technology	
17	16	Singh, Aditya	4/10/2020	Human Resources	
18	17	Smith, Carolyn	1/6/2020	Accounting	
19	18	Smith, John	7/20/2020	Human Resources	
20	19	Williams, Jane	4/10/2020	Information Technology	
21	20	Young, Alyssa	1/6/2020	Accounting	

Figure 7-24. *Displays the New Hires List before the automation is executed*

Source Code: Chapter_7_ExcelColumnActivitiesExercise

Setup: Here are step-by-step implementation instructions:

1. In StudioX, add the Use Excel File activity and configure with the C:\BookSamples\Chapter_07\EmployeeOnboarding.xlsx file as demonstrated in the first exercise.
2. Then, drag the Insert Column activity in the body of Use Excel File activity in the Designer panel.
3. Next, click the Plus icon in the Range field, and select the Indicate in Excel option for the EmployeeOnboarding workbook. Once Excel is open, select range A1:E21 in the New Hires List sheet.
4. Next, click Where dropdown and select After.
5. Next, click the Plus icon in the Relative to column field, and hover over Range to select the Full Name column header.
6. Next, click the Plus icon in the Add header field, select the Text option, and type in Last Name.
7. Leave the default options of Has headers and Format data as type as is.
8. Add a second Insert Column activity under the first Insert Column activity.
9. The Range configuration will be the same as step 3.
10. Next, click the Plus icon in the Relative to column field, and navigate through the Range option to select the Start Date column header.
11. Leave the default options of Has headers, Where, and Format data as type as is.

12. Next, click the `Plus` icon in the `Add header` field, select the `Text` option, and type in `First Name`.

Note Before you can integrate with Microsoft Excel to use the Indicate in Excel functionality, you will need to install the relevant extension from Home ➤ `Tools` ➤ `Excel Add-in`.

Once you have completed the exercise, the final configuration of the **Insert Column** activity should resemble Figure 7-25.

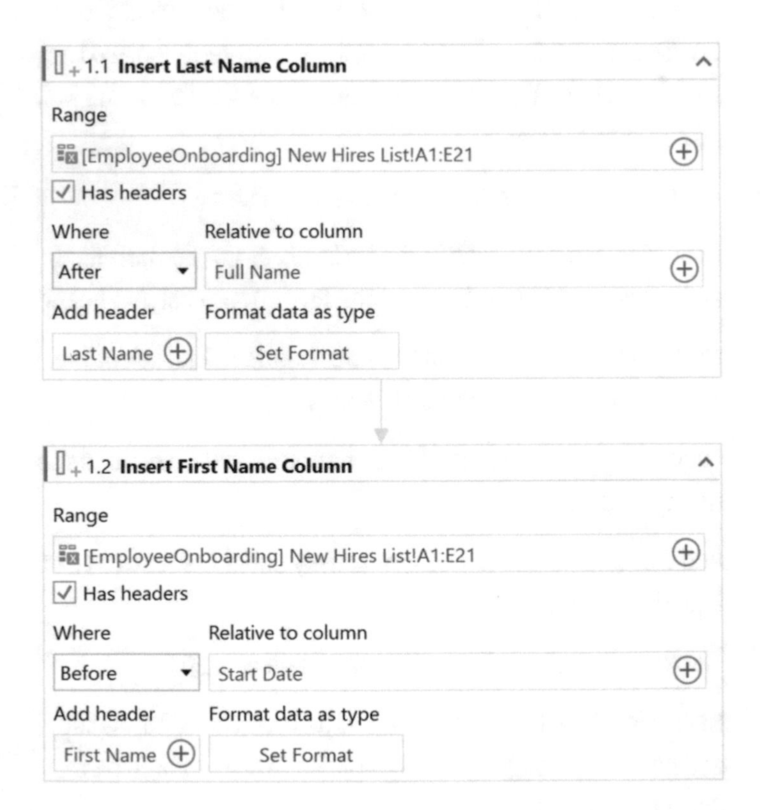

***Figure 7-25.** Displays the final configuration for the Insert Column activity*

Figure 7-26 shows the output of the Insert Column activity that has added a Last Name and First Name column after Full Name in the New Hires List sheet.

	A	B	C	D	E	F	G
1	ID	Full Name	Last Name	First Name	Start Date	Department	Status
2	1	Brown, Sylvia			1/6/2020	Operations	
3	2	Carter, Samantha			4/10/2020	Accounting	
4	3	Clark, Jacob			7/20/2020	Human Resources	
5	4	Davis, Samuel			1/6/2020	Operations	
6	5	Davis, Sasha			7/20/2020	Information Technology	
7	6	Hill, Karen			1/6/2020	Operations	
8	7	Johnson, Elijah			7/20/2020	Information Technology	
9	8	Johnson, Adam			4/10/2020	Human Resources	
10	9	Jones, Daniel			7/20/2020	Information Technology	
11	10	Khan, Zain			1/6/2020	Operations	
12	11	Lane, Tamara			7/20/2020	Human Resources	
13	12	Lopez, Maya			4/10/2020	Accounting	
14	13	Miller, Raymond			7/20/2020	Information Technology	
15	14	Patel, Priya			1/6/2020	Accounting	
16	15	Sanchez, Gabriella			7/20/2020	Information Technology	
17	16	Singh, Aditya			4/10/2020	Human Resources	
18	17	Smith, Carolyn			1/6/2020	Accounting	
19	18	Smith, John			7/20/2020	Human Resources	
20	19	Williams, Jane			4/10/2020	Information Technology	
21	20	Young, Alyssa			1/6/2020	Accounting	

***Figure 7-26.** The Last Name and First Name columns added after Insert Column activities are executed*

Text To Columns

The **Text To Columns** activity allows you to split the text in a given cell, sheet, or range to separate columns.

Common examples for this activity include splitting first and last names from a single row to separate columns or separating a cell with a list of supplies into multiple columns.

Configuration

This section provides instructions on how to configure a **Text To Columns** activity, shown in Figure 7-27.

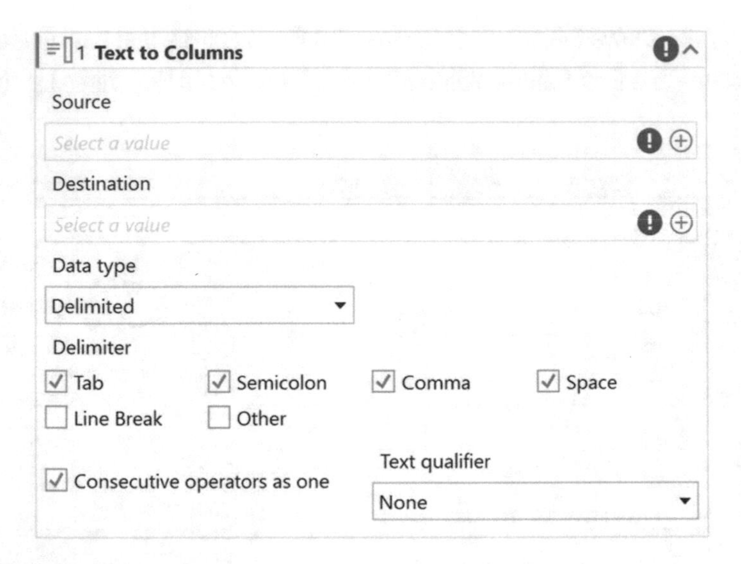

***Figure 7-27.** Text to Columns activity card*

Source: This is a required configuration available on the activity card. This configuration identifies which cell, range, or table contains the data that needs to be split into columns.

Destination: This is a required configuration available on the activity card. This configuration provides the location of the columns where the split data should be added.

Note The Destination for the Text To Columns activity must be in the same worksheet as the Source.

Data type: This is a required configuration available on the activity card. This configuration indicates what type of data the source range is separated by in order to split. The default selection for this field is `Delimited`, meaning that the data in the source is separated by a certain character. Data type can be changed to `Fixed width` if the source data needs to be separated based on a certain number of characters.

Delimiter: This is a required configuration available on the activity card. This configuration is only available when the `Data type` is `Delimited`. This configuration identifies what character to use as the separator for the source range data. By default, the `Tab`, `Semicolon`, `Comma`, and `Space` options are checked. `Line break` and `Other` are left unchecked and, if checked, will open a text field to specify the delimiter character.

Consecutive operators as one: This is a required configuration available on the activity card. This configuration is only available when the `Data type` is `Delimited`. This configuration allows you to specify if multiple characters are separating the data. By default, this option is checked.

For example, if the source cell has `notebooks;, clipboards` text, then the `;,` are going to function as a single delimiter. In this case, `notebooks` will be added to the first column and `clipboards` to the second.

Text qualifier: This is a required configuration available on the activity card. This configuration is only available when the `Data type` is `Delimited`. By default, this option is set to `none`, meaning that enclosed text qualifiers do not define the data being split. This option can be changed to a `single quote` or `double quote`.

For example, if the `Delimiter` is a comma and the `Text qualifier` is double quotes, the text `notebooks, pens", clipboards` will be separated as `notebooks, pens` in the first and `clipboards` in the second column.

Number of characters per column: This is a required configuration available on the activity card. This configuration is required only when the `Data type` is `Fixed width`. This field provides the length of how many characters to group for splitting into columns.

EXERCISE

Goal: Building on our previous exercise, use the Text To Columns activity to split the First and Last Names in the Full Name column on the New Hires List sheet. Figure 7-26 displays the New Hires List sheet before the exercise.

Source Code: Chapter_7_ExcelColumnActivitiesExercise

Setup: Here are step-by-step implementation instructions:

1. In StudioX, add the Text To Columns activity in the body of Use Excel File activity after the Insert Column activity from the previous exercise.
2. Next, click the Plus icon in the Source field, and select the Indicate in Excel option for the EmployeeOnboarding workbook. Once Excel is open, select range B2:B21 in the New Hires List sheet.
3. Next, click the Plus icon in the Destination field, and select the Indicate in Excel option for the EmployeeOnboarding workbook. Once Excel is open, select range C2:D21 in the New Hires List sheet.
4. Then, leave all of the default configurations as is for the Data type, Delimiter, Consecutive operations as one, and Text qualifier fields.

Once you have completed the exercise, the final configuration of the **Text To Columns** activity should resemble Figure 7-28.

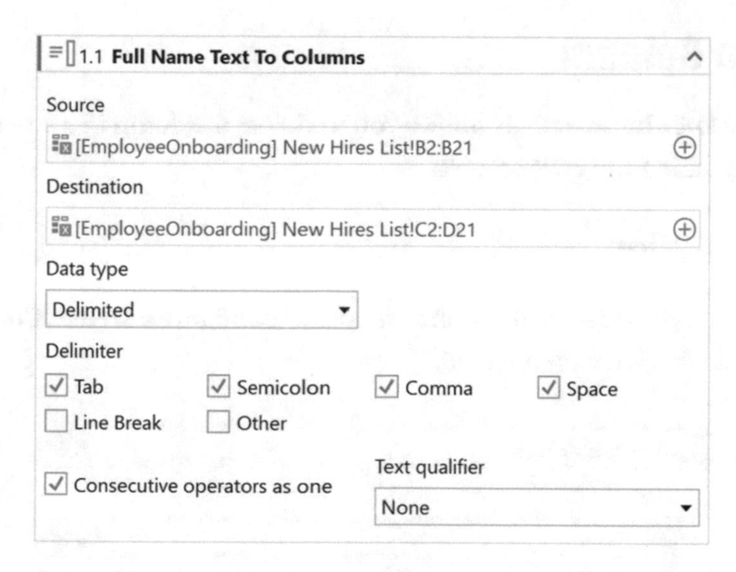

***Figure 7-28.** Displays the final configuration for the Text To Columns activity*

Figure 7-29 shows the output of the Text To Columns activity that has separated the Full Names into First and Last Name columns.

	A	B	C	D	E	F	G
1	ID	Full Name	Last Name	First Name	Start Date	Department	Status
2	1	Brown, Sylvia	Brown	Sylvia	1/6/2020	Operations	
3	2	Carter, Samantha	Carter	Samantha	4/10/2020	Accounting	
4	3	Clark, Jacob	Clark	Jacob	7/20/2020	Human Resources	
5	4	Davis, Samuel	Davis	Samuel	1/6/2020	Operations	
6	5	Davis, Sasha	Davis	Sasha	7/20/2020	Information Technology	
7	6	Hill, Karen	Hill	Karen	1/6/2020	Operations	
8	7	Johnson, Elijah	Johnson	Elijah	7/20/2020	Information Technology	
9	8	Johnson, Adam	Johnson	Adam	4/10/2020	Human Resources	
10	9	Jones, Daniel	Jones	Daniel	7/20/2020	Information Technology	
11	10	Khan, Zain	Khan	Zain	1/6/2020	Operations	
12	11	Lane, Tamara	Lane	Tamara	7/20/2020	Human Resources	
13	12	Lopez, Maya	Lopez	Maya	4/10/2020	Accounting	
14	13	Miller, Raymond	Miller	Raymond	7/20/2020	Information Technology	
15	14	Patel, Priya	Patel	Priya	1/6/2020	Accounting	
16	15	Sanchez, Gabriella	Sanchez	Gabriella	7/20/2020	Information Technology	
17	16	Singh, Aditya	Singh	Aditya	4/10/2020	Human Resources	
18	17	Smith, Carolyn	Smith	Carolyn	1/6/2020	Accounting	
19	18	Smith, John	Smith	John	7/20/2020	Human Resources	
20	19	Williams, Jane	Williams	Jane	4/10/2020	Information Technology	
21	20	Young, Alyssa	Young	Alyssa	1/6/2020	Accounting	

***Figure 7-29.** Displays the output of the Text To Columns activity*

Delete Column

The **Delete Column** activity allows you to delete a column in a sheet, range, or table from a workbook.

Configuration

This section provides instructions on how to configure a **Delete Column** activity, as shown in Figure 7-30.

Figure 7-30. *Delete Column activity card*

Source: This is a required configuration available on the activity card. This configuration provides StudioX with the range, sheet, or table that a column needs to be deleted from.

Column name: This is a required configuration available on the activity card. This configuration provides the column(s) that need to be deleted from the specified range, sheet, or table configured in the `Source` field. The most common ways to configure the Column name for this activity are by selecting a column header from the `Range` selector or using `Indicate in Excel` to select the desired column header. Alternatively, if the goal is to delete multiple columns, then you can use the `Text Builder` to identify the range or comma-delimited list of columns to delete.

Has headers: This is an optional configuration available on the activity card. This configuration is selected as a default so that StudioX identifies the column being deleted as having a header and identifies the column(s) based on the name in the first row rather than Excel column identifiers like A, B, C, and so on.

EXERCISE

Goal: Building on our previous exercise, use the `Delete Column` activity to delete the `Full Name` column from the `New Hires List` sheet in the `EmployeeOnboarding.xlsx` file. Figure 7-29 displays the `New Hires List` before the automation has been executed.

Source Code: Chapter_7_ExcelColumnActivitiesExercise

Setup: Here are step-by-step implementation instructions:

1. In `StudioX`, add the `Delete Column` activity in the body of `Use Excel File` activity after `Text to Columns` activity.
2. Next, click the `Plus` icon in the `Source` field, and navigate to the `EmployeeOnboarding` and select the `New Hires List` sheet.
3. Next, click the `Plus` icon in the `Column name` field and select `Range` ➤ `Full Name` option.
4. Leave the default configurations for `Has headers` as is for this example.

Once you have completed the exercise, the final configuration of the **Delete Column** activity should resemble Figure 7-31.

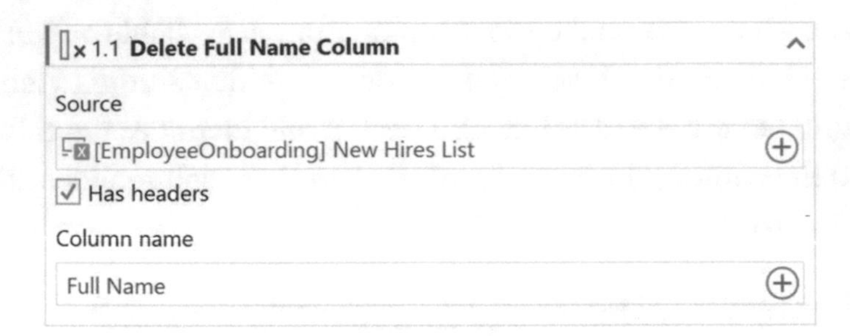

***Figure 7-31.** Displays the final configuration for the Delete Column activity*

The output of this example deletes the Full Name column from the New Hires List sheet as displayed in Figure 7-32.

	A	B	C	D	E	F
1	ID	Last Name	First Name	Start Date	Department	Status
2	1	Brown	Sylvia	1/6/2020	Operations	
3	2	Carter	Samantha	4/10/2020	Accounting	
4	3	Clark	Jacob	7/20/2020	Human Resources	
5	4	Davis	Samuel	1/6/2020	Operations	
6	5	Davis	Sasha	7/20/2020	Information Technology	
7	6	Hill	Karen	1/6/2020	Operations	
8	7	Johnson	Elijah	7/20/2020	Information Technology	
9	8	Johnson	Adam	4/10/2020	Human Resources	
10	9	Jones	Daniel	7/20/2020	Information Technology	
11	10	Khan	Zain	1/6/2020	Operations	
12	11	Lane	Tamara	7/20/2020	Human Resources	
13	12	Lopez	Maya	4/10/2020	Accounting	
14	13	Miller	Raymond	7/20/2020	Information Technology	
15	14	Patel	Priya	1/6/2020	Accounting	
16	15	Sanchez	Gabriella	7/20/2020	Information Technology	
17	16	Singh	Aditya	4/10/2020	Human Resources	
18	17	Smith	Carolyn	1/6/2020	Accounting	
19	18	Smith	John	7/20/2020	Human Resources	
20	19	Williams	Jane	4/10/2020	Information Technology	
21	20	Young	Alyssa	1/6/2020	Accounting	

***Figure 7-32.** Full Name column deleted after the Delete Column activity is executed*

Insert Rows

The **Insert Rows** activity allows you to add one or multiple rows to a defined range, sheet, or table.

Configuration

This section provides instructions on how to configure an **Insert Rows** activity, as shown in Figure 7-33.

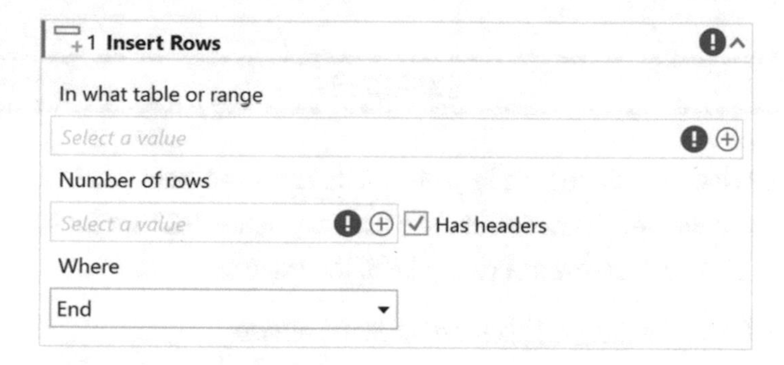

Figure 7-33. *Insert Rows activity card*

In what table or range: This is a required configuration available on the activity card. This configuration provides StudioX with the range that the new row(s) will be added to.

Number of rows: This is a required configuration available on the activity card. This configuration determines how many rows should be added to the selected range.

Where: This is a required configuration available on the activity card. This configuration is default selected as `End`, meaning that newly inserted rows will be added to the end of the range selected in the `In what table or range` field. This field configuration can be changed to `Start` or `Specific Index`.

Row number: This is a required configuration available on the activity card. This configuration is required only when `Specific Index` is selected in the `Where` dropdown. This configuration allows you to specify an exact row number where the new row(s) should be inserted.

Has headers: This is an optional configuration available on the activity card. This configuration is selected by default, that is, the first row in the range will be counted as a header. For example, if you are inserting three rows at the start of a table and Has headers is selected, then the three rows will be inserted after the header row in the table.

EXERCISE

Goal: Use the `Use Excel File` and the `Insert Rows` activities to add two rows in the `New Hires List` sheet at cell. Figure 7-32 shows the `New Hires List` with 20 rows of data prior to the exercise.

Source Code: Chapter_7_ExcelRowActivitiesExercise

Setup: Here are step-by-step implementation instructions:

1. In `StudioX`, add the `Use Excel File` activity and configure with the `C:\BookSamples\Chapter_07\EmployeeOnboarding.xlsx` file as demonstrated in the first exercise.
2. Then, drag the `Insert Rows` activity in the body of `Use Excel File` activity in the `Designer` panel.
3. Next, click the `Plus` icon in the `In what table or range` field, and navigate to the `EmployeeOnboarding` and select the `New Hires List` sheet.

4. Next, click the `Plus` icon in the `Number of rows` field, select `Number` option, and type in `2`.
5. Next, click the `Where` dropdown and select `Specific index`.
6. Next, click the `Plus` icon in the `Row Number` field, select `Number` option, and type in `5`.
7. Leave the default option of `Has headers` checked.

Once you have completed the exercise, the final configuration of the **Insert Rows** activity should resemble Figure 7-34.

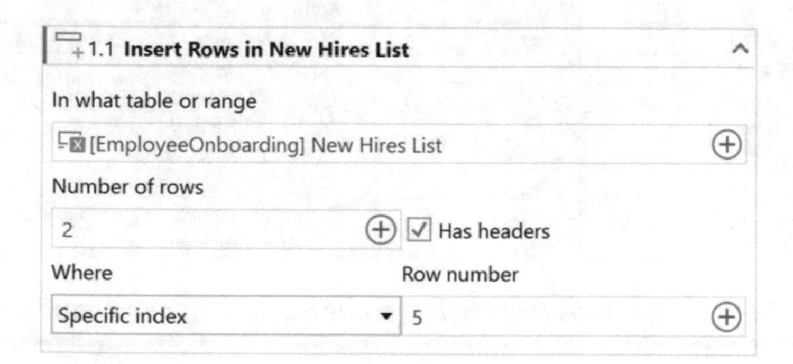

***Figure 7-34.** Displays the final configuration for the Insert Rows activity*

Figure 7-35 shows the output of the Insert Rows activity that has added two rows to the New Hires List sheet now extending from A1:F21 to A1:F23.

	A	B	C	D	E	F
1	ID	Last Name	First Name	Start Date	Department	Status
2	1	Brown	Sylvia	1/6/2020	Operations	
3	2	Carter	Samantha	4/10/2020	Accounting	
4	3	Clark	Jacob	7/20/2020	Human Resources	
5	4	Davis	Samuel	1/6/2020	Operations	
6						
7						
8	5	Davis	Sasha	7/20/2020	Information Technology	
9	6	Hill	Karen	1/6/2020	Operations	
10	7	Johnson	Elijah	7/20/2020	Information Technology	
11	8	Johnson	Adam	4/10/2020	Human Resources	
12	9	Jones	Daniel	7/20/2020	Information Technology	
13	10	Khan	Zain	1/6/2020	Operations	
14	11	Lane	Tamara	7/20/2020	Human Resources	
15	12	Lopez	Maya	4/10/2020	Accounting	
16	13	Miller	Raymond	7/20/2020	Information Technology	
17	14	Patel	Priya	1/6/2020	Accounting	
18	15	Sanchez	Gabriella	7/20/2020	Information Technology	
19	16	Singh	Aditya	4/10/2020	Human Resources	
20	17	Smith	Carolyn	1/6/2020	Accounting	
21	18	Smith	John	7/20/2020	Human Resources	
22	19	Williams	Jane	4/10/2020	Information Technology	
23	20	Young	Alyssa	1/6/2020	Accounting	

Figure 7-35. *New rows inserted in rows 6 and 7 of the New Hires List*

Delete Rows

The **Delete Rows** activity deletes one or multiple rows identified in a sheet, range, or table from a workbook.

Tip You can use a Filter activity to hide rows and then choose to delete all hidden rows in the Delete Rows activity.

Configuration

This section provides instructions on how to configure a **Delete Rows** activity, as shown in Figure 7-36.

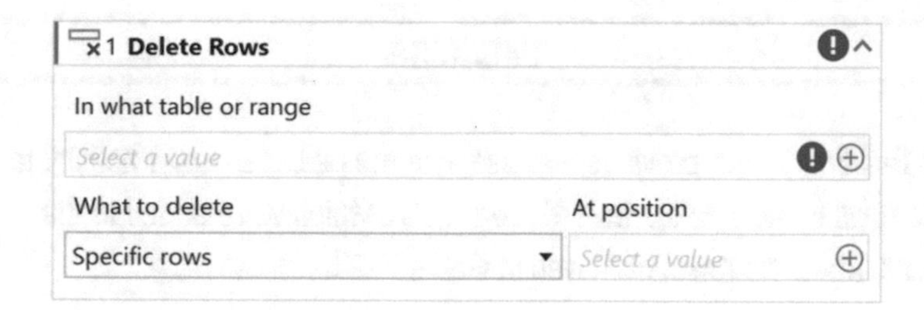

Figure 7-36. Delete Rows activity card

In what table or range: This is a required configuration available on the activity card. This configuration provides StudioX with the range, sheet, or table that contains the row(s) that need to be deleted.

What to delete: This is a required configuration available on the activity card. This configuration provides the type of row(s) that need to be deleted. The dropdown selection options are `Specify row`, `All visible rows`, `All hidden rows`, and `All duplicate rows`. For example, if there is a table where you have hidden three rows and you select the `All hidden rows` option, then those three hidden rows in the table are deleted.

At position: This is a required configuration available on the activity card. This configuration is required only when the `Specify rows` option is selected in `What to delete`. The most common way to configure the `At position` field is by using the `Text Builder` to enter the row numbers being deleted.

Has headers: This is an optional configuration available on the Properties panel. By default, this is selected, that is, the first row in the range will be considered as a header and will not be deleted.

EXERCISE

Goal: Building on our previous exercise, use the `Delete Rows` activity to delete rows 6 and 7 in the `New Hires List` which were added in the `Insert Rows` exercise, displayed in Figure 7-35.

Source Code: Chapter_7_ExcelRowActivitiesExercise

Setup: Here are step-by-step implementation instructions:

1. In `StudioX`, add the `Delete Rows` activity in the body of `Use Excel File` activity after `Insert Rows` activity.
2. Next, click the `Plus` icon in the `In what table or range` field, and navigate to the `EmployeeOnboarding` and select the `New Hires List` sheet.
3. Leave the `What to delete` dropdown with the default selection of `Specific rows`.
4. Next, click the `Plus` menu in the `At position` field, select the `Text` option, and type `6, 7`.

Once you have completed the exercise, the final configuration of the **Delete Rows** activity should resemble Figure 7-37.

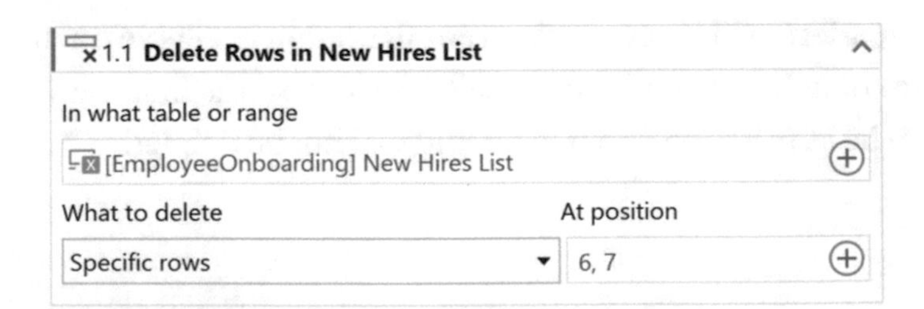

***Figure 7-37.** Displays the final configuration for the Delete Rows activity*

The output of this example, displayed in Figure 7-38, deletes the two blank rows at row numbers 6 and 7 in the New Hires List sheet.

	A	B	C	D	E	F
1	ID	Last Name	First Name	Start Date	Department	Status
2	1	Brown	Sylvia	1/6/2020	Operations	
3	2	Carter	Samantha	4/10/2020	Accounting	
4	3	Clark	Jacob	7/20/2020	Human Resources	
5	4	Davis	Samuel	1/6/2020	Operations	
6	5	Davis	Sasha	7/20/2020	Information Technology	
7	6	Hill	Karen	1/6/2020	Operations	
8	7	Johnson	Elijah	7/20/2020	Information Technology	
9	8	Johnson	Adam	4/10/2020	Human Resources	
10	9	Jones	Daniel	7/20/2020	Information Technology	
11	10	Khan	Zain	1/6/2020	Operations	
12	11	Lane	Tamara	7/20/2020	Human Resources	
13	12	Lopez	Maya	4/10/2020	Accounting	
14	13	Miller	Raymond	7/20/2020	Information Technology	
15	14	Patel	Priya	1/6/2020	Accounting	
16	15	Sanchez	Gabriella	7/20/2020	Information Technology	
17	16	Singh	Aditya	4/10/2020	Human Resources	
18	17	Smith	Carolyn	1/6/2020	Accounting	
19	18	Smith	John	7/20/2020	Human Resources	
20	19	Williams	Jane	4/10/2020	Information Technology	
21	20	Young	Alyssa	1/6/2020	Accounting	

Figure 7-38. *Rows 6 and 7 are deleted after the Delete Rows activity is executed*

Find First/Last Data Row

The **Find First/Last Data Row** activity allows you to find the first and last row number containing data in a sheet, table, or range in a workbook.

Tip This activity can be helpful when needing to know what the last row of data is to know where to append or write data to.

Configuration

This section provides instructions on how to configure a **Find First/Last Data Row** activity, shown in Figure 7-39.

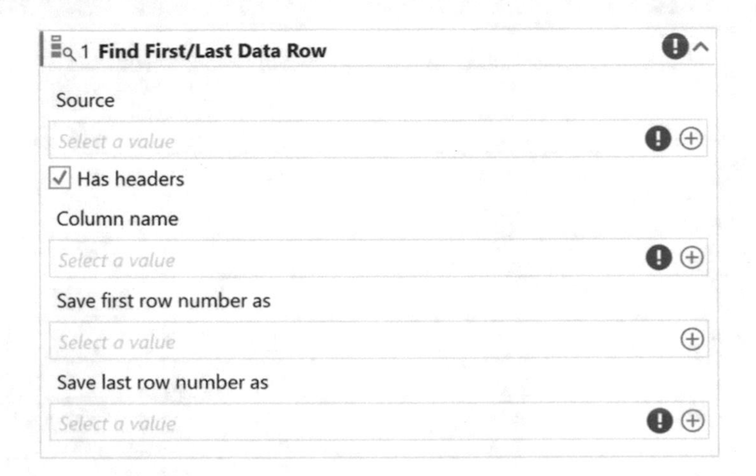

Figure 7-39. *Find First/Last Data Row activity card*

Source: This is a required configuration available on the activity card. This configuration provides the sheet, range, or table containing the data that needs to be searched for the first and/or last row number.

Has headers: This is an optional configuration available on the activity card. This configuration indicates that the first row for the `Source` is a header row that will not be counted in the output. For example, if there is a table in range A1:B20 and the `Has headers` option is checked, then the last data row count will be 19. By default, this option is checked, meaning that the first row will not be counted when calculating the output. If unselected, the first row will be counted the same as any other row in the identified range.

Column name: This is a required configuration available on the activity card. This configuration specifies the name of the column where the first and last row data needs to be searched. Commonly, this field is configured by selecting the column header name populated in the `Range` menu.

Save first row number as: This is an optional configuration available on the activity card. This configuration specifies where to store the number index of the first row containing data in the source range. Commonly this value can be saved in an Excel cell, saved for later use, or copied to the clipboard.

Save last row number as: This is a required configuration available on the activity card. This configuration specifies where to store the number index of the last row containing data in the source range. Commonly this value can be saved in an Excel cell, saved for later use, or copied to the clipboard.

Blank rows to skip: This is an optional configuration through the Properties panel under Input. This configuration specifies how many consecutive blank rows the automation should encounter before determining that the end of the range has been reached. This configuration is defaulted to 1.

First row offset: This is an optional configuration through the Properties panel under Input. This configuration allows the user to add a specified number of rows to offset the first row containing data for the activity output. For example, you could enter the number 5; then the activity will search for the number of the fifth row containing data. This configuration is defaulted to 0.

Last row offset: This is an optional configuration through the Properties panel under Input. This configuration allows the user to subtract a specified number of rows to offset the last row containing data for the activity output. For example, you could enter the number 5; then the activity will search for the fifth row from the bottom of the range that contains data. This configuration is defaulted to 0.

EXERCISE

Goal: Building on our previous exercise, use the `Find First/Last Data Row` activity to find the last row containing data in the ID column for the `New Hire Checklist`. Figure 7-40 displays the data that will be searched.

	A	B	C	D	E	F	G	H	I
1	ID	Last Name	First Name	Orientation	Employee Hanbook	Policy Training	Benefits Package	Direct Deposit Setup	Technology Setup
2	1	Brown	Sylvia	Complete	Complete	Complete	Complete	Complete	Complete
3	2	Carter	Samantha	Not Started	Not Started	In Progress	Not Started	Not Started	Not Started
4	3	Clark	Jacob	Not Started	Not Started	In Progress	In Progress	Not Started	In Progress
5	4	Davis	Samuel	Complete	Complete	Complete	Complete	Complete	Complete
6	5	Davis	Sasha	Complete	Complete	Complete	Complete	Complete	Complete
7	6	Hill	Karen	Complete	Complete	Complete	Complete	Complete	Complete
8	7	Johnson	Elijah	Not Started	In Progress	In Progress	Not Started	In Progress	Not Started
9	8	Johnson	Adam	Complete	Complete	Complete	Complete	Complete	Complete
10	9	Jones	Daniel	Not Started	Not Started	In Progress	Not Started	Not Started	Not Started
11	10	Khan	Zain	Complete	Complete	Complete	Complete	Complete	Complete
12	11	Lane	Tamara	Not Started	Not Started	In Progress	Not Started	Not Started	Not Started
13	12	Lopez	Maya	Complete	Complete	Complete	Complete	Complete	Complete
14	13	Miller	Raymond	Not Started	In Progress	Not Started	In Progress	Not Started	Not Started
15	14	Patel	Priya	Complete	Complete	Complete	Complete	Complete	Complete
16	15	Sanchez	Gabriella	Not Started	In Progress	In Progress	Not Started	In Progress	Not Started
17	16	Singh	Aditya	Complete	Complete	Complete	Complete	Complete	Complete
18	17	Smith	Carolyn	Complete	Complete	Complete	Complete	Complete	Complete
19	18	Smith	John	Not Started	Not Started	Not Started	In Progress	Not Started	Not Started
20	19	Williams	Jane	Complete	Complete	Complete	Complete	Complete	Complete
21	20	Young	Alyssa	Not Started	Not Started	Not Started	In Progress	Not Started	Not Started

***Figure 7-40.** New Hire Checklist data that will be searched*

Source Code: Chapter_7_ExcelRowActivitiesExercise

Setup: Here are step-by-step implementation instructions:

1. In `StudioX`, add the `Find First/Last Data Row` activity in the body of `Use Excel File` activity after `Delete Rows` activity.

2. Next, click the `Plus` icon in the `Source` field, and navigate to the `EmployeeOnboarding` and select the `New Hire Checklist` sheet.

3. Next, click the `Plus` icon in the `Column Name` field, and navigate to the `Range` and select the `ID` column header.

4. Next, click the `Plus` icon in the `Save last row number as` field, and select the `Copy to clipboard` option.

5. Leave the `Has headers` option checked.
6. Leave the `Save first row number as` field blank.
7. Next, add the `Write Line` activity to the body of `Use Excel File` activity after the `Find First/Last Data Row` activity.
8. In the `Text` field, click the `Plus` icon and select `Paste from clipboard` option.

Once you have completed the exercise, the final configuration of the **Find First/Last Data Row** activity should resemble Figure 7-41.

Figure 7-41. *Displays the configuration for the Find First/Last Data Row activity*

This example will output 21 as the last row for the New Hire Checklist as there are 20 new employees total. This can be seen in Figure 7-42.

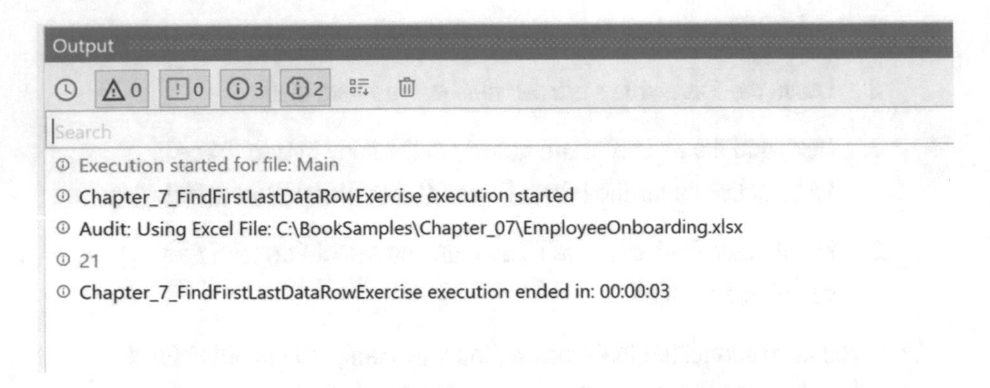

Figure 7-42. *Output of Find First/Last Data Row exercise*

For Each Excel Row

The **For Each Excel Row** activity allows you to execute one or more activities against each row in a sheet, table, or range within the same workbook.

Note Configuring at least one child activity within the For Each Excel Row activity is required to perform the iterative actions on multiple Excel rows.

Configuration

This section provides instructions on how to configure a **For Each Excel Row** activity, as shown in Figure 7-43.

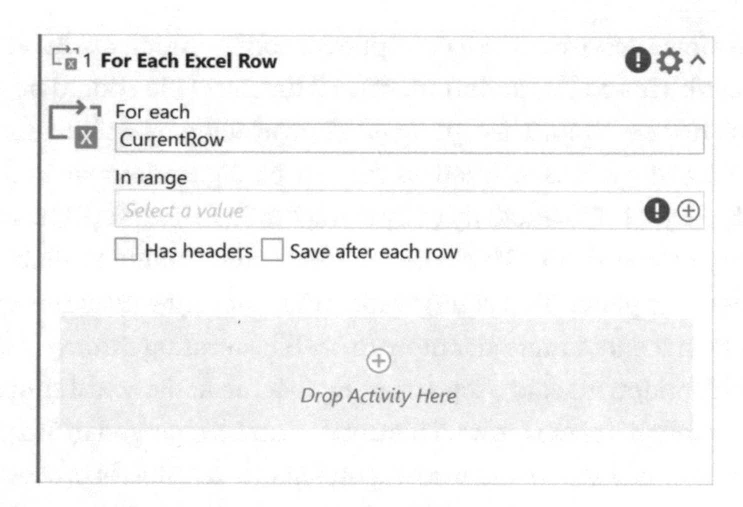

Figure 7-43. *For Each Excel Row activity card*

For each: This is a required configuration available on the activity card. This configuration allows you to provide a name to reference the current row which is being iterated through; this is especially important for referencing in the child activities. The default text is entered as `CurrentRow` and can be changed based on what data you are iterating through. For example, you can choose to name the current row as Employees when you are executing the same actions through a list of multiple employees. You can simply type in the desired name to configure this field or leave the default.

In range: This is a required configuration available on the activity card. This configuration provides the sheet, range, or table containing the rows of data that need to be iterated through.

Has headers: This is an optional configuration available on the activity card. This configuration indicates that the first row for the `In range` is a header row. Selecting this option will ease in referencing specific column names for the `CurrentRow` data in child activities. This option is unselected at default.

Save after each row: This is an optional configuration available on the activity card. This configuration specifies if the Excel file should be saved after each row iteration. This option is left unselected by default, meaning that the Excel file will save based on the save configurations set in the parent `Use Excel File` activity or by using the `Save Excel File` activity.

EmptyRowBehavior: This is an optional configuration available on the `Properties` panel. This configuration indicates how the automation will behave if it encounters an empty row while iterating through data. The default option is set to `Stop`, meaning that the activity will simply stop iterating through the next rows. This option can be changed to `Skip` so that the empty row is skipped and the next one in the iteration is processed, or `Process` so the empty row is included in the iteration data processing.

EXERCISE

Goal: Building on our previous exercise, use the `For Each Excel Row` activity to iterate through each employee row in the New Hires List data range. This activity will be executed with the `Write Cell` activity exercise.

Source Code: Chapter_7_ExcelRowActivitiesExercise

Setup: Here are step-by-step implementation instructions:

1. In `StudioX`, add the `For Each Excel Row` activity in the body of `Use Excel File` activity after `Find First/Last Data Row` activity from the previous exercise.
2. Next, in the `For each` field, type in the text `Employee`.
3. Next, click the `Plus` icon in the `In range` field, and navigate to the `EmployeeOnboarding` and select the `New Hires List` sheet.
4. Next, check the `Has headers` option.

5. Check the `Save after each row` option.
6. Then, click the `Properties` panel and select the `EmptyRowBehavior` option to change it to `Skip`. This step is optional.

Once you have completed the exercise, the final configuration of the **For Each Excel Row** activity should resemble Figure 7-44.

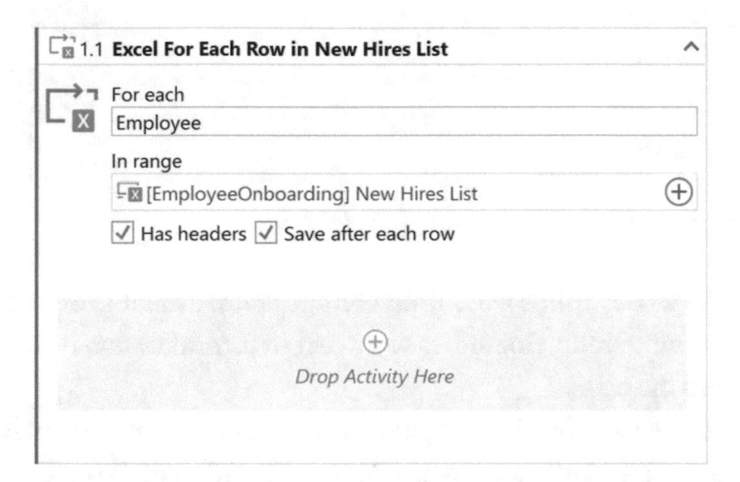

***Figure 7-44.** Displays the configuration for the For Each Excel Row activity*

In the upcoming Activity section, the automation will iterate through each Employee row in the New Hire List sheet and perform the Write Cell.

Write Cell

The **Write Cell** activity allows you to write a value or a formula into a given cell in a worksheet.

Common examples for this activity include writing to a single cell or using a **For Each** activity to write text or formulas into multiple cells incrementing with each row.

Configuration

This section provides instructions on how to configure a **Write Cell** activity, shown in Figure 7-45.

Figure 7-45. *Write Cell activity card*

What to write: This is a required configuration available on the activity card. This configuration identifies what text or formula value needs to be entered into the cell.

Where to write: This is a required configuration available on the activity card. This configuration provides the location of the cell where the `What to write` value will be entered.

Auto increment row: This is an optional configuration available on the activity card. This configuration, if checked, will automatically increase the row number for each iteration and write the cell value into multiple cells in the range. By default, this field is unchecked, meaning the value will only be entered into the single cell indicated in the `Where to write` field.

Note The Auto increment row option should be utilized with the For Each Excel Row, For Each Email, For Each File In Folder, or Repeat Number Of Times activities if writing to more than one cell.

EXERCISE

Goal: Building on our previous exercise, use the Write Cell activity with the For Each Excel Row activity to write the text Active as the status for each employee in the New Hires List. Figure 7-38 shows the New Hires List before the exercise.

Source Code: Chapter_7_ExcelRowActivitiesExercise

Setup: Here are step-by-step implementation instructions:

1. In StudioX, add the Write Cell activity to the body of For Each Excel Row activity from the previous exercise.
2. Next, select the Plus icon in What to write field, select Text option, and type Active.
3. Next, select the Plus icon in the Where to write field and hover over Employee option to select the Status field.
4. Next, check the box for the Auto increment row field.

Once you have completed the exercise, the final configuration of the **Write Cell** activity should resemble Figure 7-46.

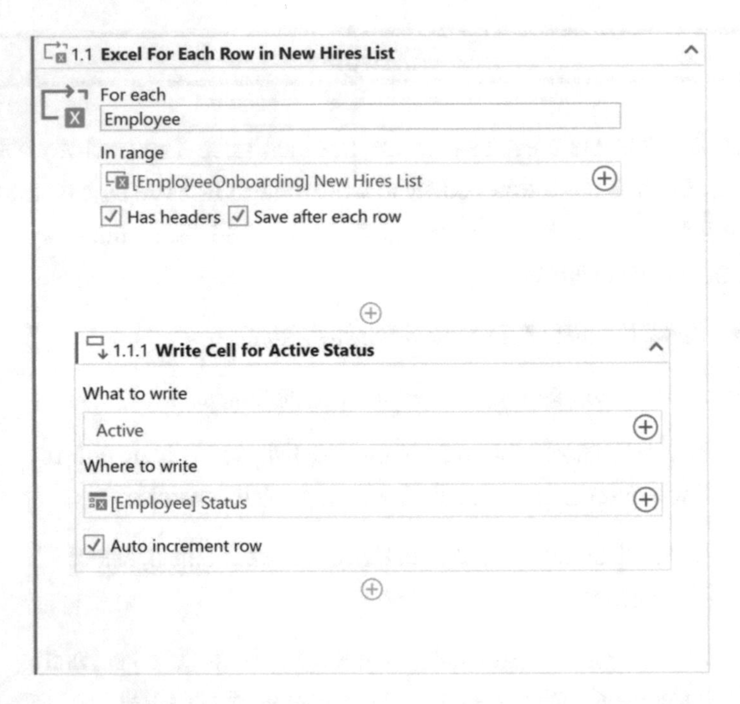

***Figure 7-46.** Displays the final configuration for the Write Cell activity*

Figure 7-47 shows the output of the Write Cell and For Each Excel Row activity that has added Active to the Status column for all employees.

	A	B	C	D	E	F
1	ID	Last Name	First Name	Start Date	Department	Status
2	1	Brown	Sylvia	1/6/2020	Operations	Active
3	2	Carter	Samantha	4/10/2020	Accounting	Active
4	3	Clark	Jacob	7/20/2020	Human Resources	Active
5	4	Davis	Samuel	1/6/2020	Operations	Active
6	5	Davis	Sasha	7/20/2020	Information Technology	Active
7	6	Hill	Karen	1/6/2020	Operations	Active
8	7	Johnson	Elijah	7/20/2020	Information Technology	Active
9	8	Johnson	Adam	4/10/2020	Human Resources	Active
10	9	Jones	Daniel	7/20/2020	Information Technology	Active
11	10	Khan	Zain	1/6/2020	Operations	Active
12	11	Lane	Tamara	7/20/2020	Human Resources	Active
13	12	Lopez	Maya	4/10/2020	Accounting	Active
14	13	Miller	Raymond	7/20/2020	Information Technology	Active
15	14	Patel	Priya	1/6/2020	Accounting	Active
16	15	Sanchez	Gabriella	7/20/2020	Information Technology	Active
17	16	Singh	Aditya	4/10/2020	Human Resources	Active
18	17	Smith	Carolyn	1/6/2020	Accounting	Active
19	18	Smith	John	7/20/2020	Human Resources	Active
20	19	Williams	Jane	4/10/2020	Information Technology	Active
21	20	Young	Alyssa	1/6/2020	Accounting	Active

Figure 7-47. *Displays the text Active added for all the employees' Statuses*

Create Pivot Table

The **Create Pivot Table** activity creates a pivot table that helps you organize, analyze, and summarize data. Additionally, the **Create Pivot Table** activity is accompanied by the **Add Pivot Field** child activity allowing you to add rows, columns, filters, or values to the new pivot table.

Configuration

This section provides instructions on how to configure a **Create Pivot Table** activity, shown in Figure 7-48.

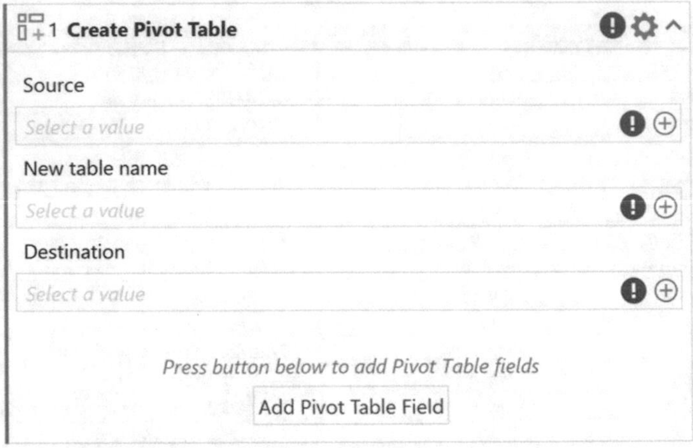

Figure 7-48. *Create Pivot Table activity card*

Source: This is a required configuration available on the activity card. This configuration provides the source range or table that provides the data for the new pivot table. The common method to configure this field is by choosing a named table as the source.

New table name: This is a required configuration available on the activity card. This configuration provides the name for the pivot table. The most common way to configure this field is by selecting the `Text Builder` option to type in the desired name manually.

Destination: This is a required configuration available on the activity card. This field identifies the sheet, table, or range where the new pivot table should be created. This field is commonly configured by selecting a named sheet or table or by using the `Indicate in Excel` option.

Pivot Field: The child activity allows you to add rows, columns, values, and filters for the new pivot table being created in the parent `Create Pivot Table` activity.

Field: The `Field` value is required for the `Add Pivot Field` child activity configuration. This configuration specifies what field from the `Source` should be added to the new pivot table. You can configure this by selecting a `column header` from the `Range` selector or using `Indicate in Excel` to select the desired column header.

Is a: This is a required configuration available on the activity card of `Pivot Field` child activity. This field allows you to select if the field being added to the new pivot table is a `row`, `column`, `filter`, or `value`.

Function: This is an optional configuration available on the activity card. This configuration only appears if the `Is a` field is set to the option `Value`. The `Function` field specifies what function is used to determine the value, with the following dropdown options seen in Figure 7-49.

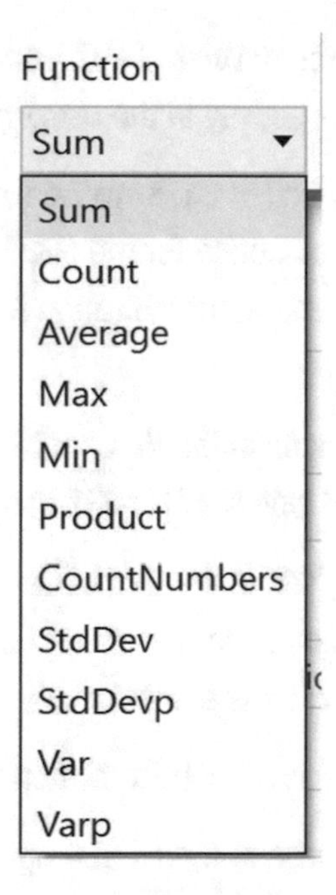

***Figure 7-49.** Dropdown options for Function field*

EXERCISE

Goal: Use the Use Excel File and the Create Pivot Table activities to create a pivot table in the EmployeeOnboarding Excel file to add the count of employee ID by job function for the first ten new hires in the New Hires List.

Source Code: Chapter_7_ExcelTableActivitiesExercise

Setup: Here are step-by-step implementation instructions:

1. In StudioX, add the Use Excel File activity and configure with the C:\BookSamples\Chapter_07\EmployeeOnboarding.xlsx file as demonstrated in the first exercise.
2. Then, drag the Create Pivot Table activity card in the body of Use Excel File activity in the Designer panel.
3. Next, click the Plus icon in the Source field and select the Indicate in Excel option for the EmployeeOnboarding file. Once Excel is open, select the range A1:F11 in the New Hires List sheet.
4. Next, click the Plus icon in the New table name field, select the Text option, and type NewHireStats_PivotTable.
5. Next, click the Plus icon in the Destination field and navigate to the EmployeeOnboarding Excel file and select the New Hire Statistics sheet.
6. Next, click the Add Pivot Field button to add the first field.
7. Next, click the Plus icon in the Field option, navigate to the Range option, and select Department.
8. Next, leave the Is a option set to the default value of Row.

9. Next, click the `Add Pivot Field` button to add the second field.
10. Next, click the `Plus` icon in the `Field` option, navigate to the `Range` option, and select `ID`.
11. Next, click the dropdown for the `Is a` option and select the value `Value`.
12. Next, click the dropdown for the `Function` option and select the value `Count`.

Once you have completed the exercise, the final configuration of the Create Pivot Table activity should resemble Figure 7-50.

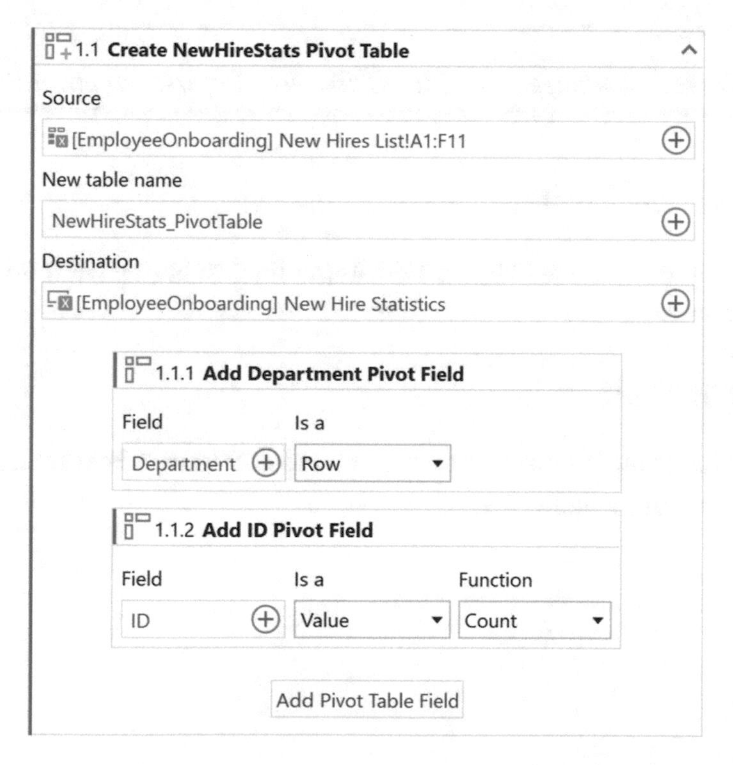

***Figure 7-50.** Create Pivot Table activity final configuration*

The output of this example will create a NewHireStats pivot table with the Department and ID columns displaying how many employees were hired by department. Figure 7-51 displays the new pivot table in the EmployeeOnboarding Excel file.

	A	B	C
1	**Row Labels**	**Count of ID**	
2	Accounting	1	
3	Human Resources	2	
4	Information Technology	3	
5	Operations	4	
6	**Grand Total**	**10**	
7			

Figure 7-51. *NewHireStats_PivotTable for the first ten employees*

Format as Table

The **Format as Table** activity formats a specified range as a new table with a given name.

Configuration

This section provides instructions on how to configure a **Format as Table** activity, shown in Figure 7-52.

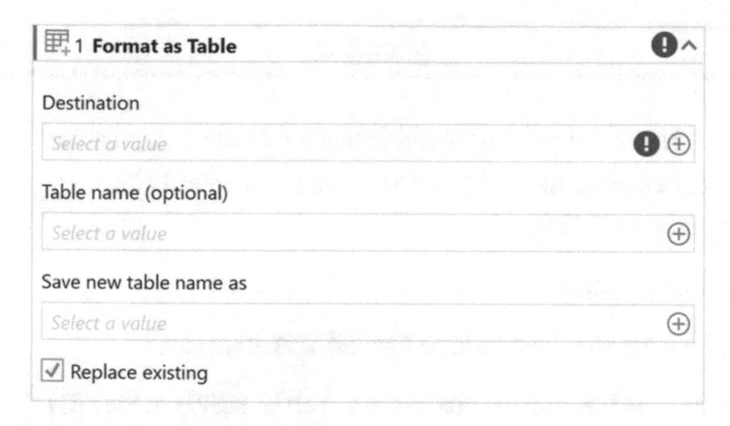

Figure 7-52. *Format as Table activity card*

Destination: This is a required configuration available on the activity card. This configuration identifies the range of cells that need to be converted from a range to a table. This field is commonly configured by utilizing the `Indicate in Excel` option.

Table name (optional): This is an optional configuration available on the activity card. This configuration provides the name for the table. The most common way to configure this field is by selecting the `Text Builder` option to type in the desired name manually. If left blank, the first available name such as "Table1" will automatically be assigned to the table.

Save new table name as: This is an optional configuration available on the activity card. This configuration provides the name to reference the table for later use and is helpful if the `Table name` configuration is left blank, so you can retrieve the name that was assigned by the automation. The common way to configure this field is by selecting the `Save for later` option and assign a variable name to output for future activities.

Replace existing: This is an optional configuration available on the activity card. This configuration deletes and replaces an existing table if one exists with the same name that is configured in `Save new table name as`. By default, this configuration is checked.

EXERCISE

Goal: Building on our previous exercise, use the `Format as Table` activity to format the `New Hires List` range as a table in the `EmployeeOnboarding.xlsx` file.

Source Code: Chapter_7_ExcelTableActivitiesExercise

Setup: Here are step-by-step implementation instructions:

1. In `StudioX`, add the `Format as Table` activity in the body of `Use Excel File` activity after `Change Pivot Table` activity from the previous exercise.
2. Click the `Plus` icon in the `Destination` field. Navigate to the `EmployeeOnboarding` workbook option, and select `Indicate in Excel`. Once Excel is open, select the range `A1:F21` in the `New Hires List` sheet.
3. Next, click the `Plus` icon in the `New table name` field, select the `Text` option, and type `NewHiresList_Table`.
4. Leave the `Save reference name` as blank.

Once you have completed the exercise, the final configuration of the **Format as Table** activity should resemble Figure 7-53.

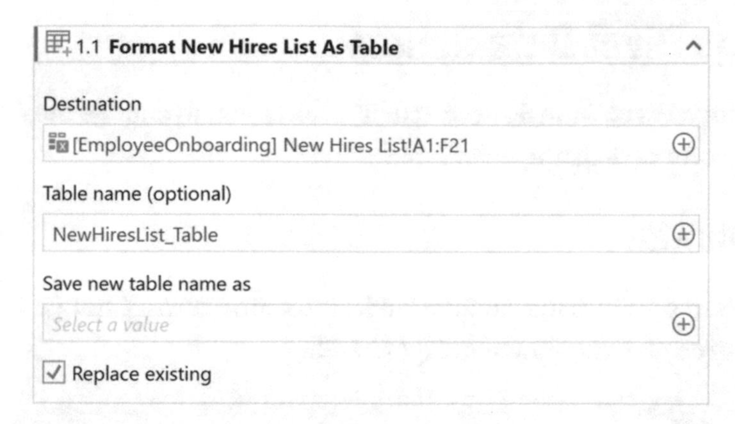

***Figure 7-53.** Configuration for Format as Table activity*

Figure 7-54 shows the range A1:F21 which is now converted to the new table.

	A	B	C	D	E	F
1	ID	Last Name	First Name	Start Date	Department	Status
2	1	Brown	Sylvia	1/6/2020	Operations	Active
3	2	Carter	Samantha	4/10/2020	Accounting	Active
4	3	Clark	Jacob	7/20/2020	Human Resources	Active
5	4	Davis	Samuel	1/6/2020	Operations	Active
6	5	Davis	Sasha	7/20/2020	Information Technology	Active
7	6	Hill	Karen	1/6/2020	Operations	Active
8	7	Johnson	Elijah	7/20/2020	Information Technology	Active
9	8	Johnson	Adam	4/10/2020	Human Resources	Active
10	9	Jones	Daniel	7/20/2020	Information Technology	Active
11	10	Khan	Zain	1/6/2020	Operations	Active
12	11	Lane	Tamara	7/20/2020	Human Resources	Active
13	12	Lopez	Maya	4/10/2020	Accounting	Active
14	13	Miller	Raymond	7/20/2020	Information Technology	Active
15	14	Patel	Priya	1/6/2020	Accounting	Active
16	15	Sanchez	Gabriella	7/20/2020	Information Technology	Active
17	16	Singh	Aditya	4/10/2020	Human Resources	Active
18	17	Smith	Carolyn	1/6/2020	Accounting	Active
19	18	Smith	John	7/20/2020	Human Resources	Active
20	19	Williams	Jane	4/10/2020	Information Technology	Active
21	20	Young	Alyssa	1/6/2020	Accounting	Active

***Figure 7-54.** Range A1:F21 converted from Range to Table*

Change Pivot Data Source

The **Change Pivot Data Source** activity allows you to change the source range data for an existing pivot table.

Configuration

This section provides instructions on how to configure a **Change Pivot Data Source** activity, shown in Figure 7-55.

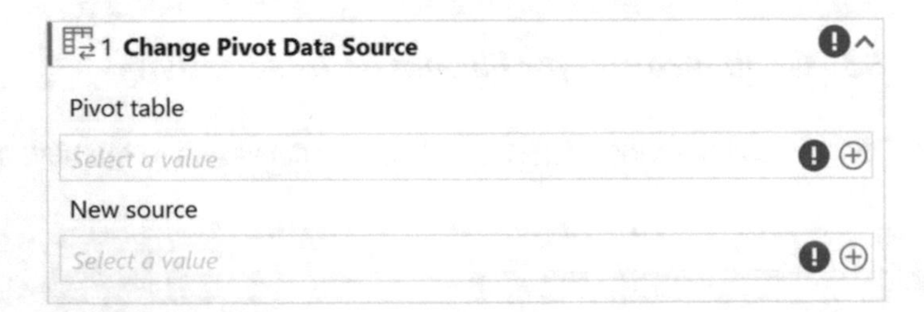

Figure 7-55. *Change Pivot Data Source activity card*

Pivot table: This is a required configuration available on the activity card. This configuration is used to identify the pivot table from the Excel file that will be updated with the new source data.

New source: This is a required configuration available on the activity card. This configuration is used to identify the new source table or range that will update the data in the pivot table.

EXERCISE

Goal: Building on our previous exercise, use the `Change Pivot Data Source` activity to change the `NewHireStats_PivotTable` from the range `A1:F11` to the `NewHiresList_Table` added in the `Format as Table` activity exercise. Figure 7-51 displays the `pivot table` prior to the exercise.

Source Code: Chapter_7_ExcelTableActivitiesExercise

Setup: Here are step-by-step implementation instructions:

1. In StudioX, add the Change Pivot Data Source activity in the body of Use Excel File activity after Format as Table activity from the previous exercise.
2. Click the Plus icon in the Pivot table field. Navigate through the EmployeeOnboarding workbook, New Hire Statistics sheet, and select the NewHireStats_PivotTable [Pivot Table].
3. Click the Plus icon in the New source field. Navigate through the EmployeeOnboarding workbook, New Hires List sheet, and select the NewHiresList_Table [Table].

Once you have completed the exercise, the final configuration of the **Change Pivot Data Source** activity should resemble Figure 7-56.

Figure 7-56. *Change Pivot Data Source configuration*

Once the Change Pivot Data Source activity is executed, Figure 7-57 shows the output of the updated data for the NewHireStats_PivotTable now accounting for all 20 employees.

	A	B	C
1	**Row Labels**	**Count of ID**	
2	Accounting	5	
3	Human Resources	5	
4	Information Technology	6	
5	Operations	4	
6	**Grand Total**	**20**	
7			

Figure 7-57. *Change Pivot Data Source activity output*

Refresh Pivot Table

The **Refresh Pivot Table** activity utilizes the Refresh feature in Excel to allow you to update the pivot table with any data changes that have occurred in the source data range.

Configuration

This section provides instructions on how to configure a **Refresh Pivot Table** activity, shown in 7-58.

Figure 7-58. *Refresh Pivot Table activity card*

Pivot table to refresh: This is a required configuration available on the activity card. This configuration provides StudioX with the pivot table that needs to be updated due to changes in its source data table or range. You can configure this field by selecting a named pivot table by clicking the `Plus` icon.

EXERCISE

Goal: Building on our previous exercise, use the `Refresh Pivot Table` activity to update the count of New Hires in the `NewHireStats_PivotTable`. This will be based on a `Write Cell` activity that updates the Department of the first employee from `Operations` to `Accounting`. Figure 7-57 shows the `NewHireStats_PivotTable` prior to the exercise.

Source Code: Chapter_7_ExcelTableActivitiesExercise

Setup: Here are step-by-step implementation instructions:

1. In `StudioX`, add the `Write Cell` activity in the body of `Use Excel File` activity after `Change Pivot Data Source` activity.
2. Next, in the `What to write` field, click `Plus` icon, select `Text` option, and type `Accounting`.
3. Next, in the `Where to write` field, select cell `E2` in the `New Hires List` sheet.
4. Next, drag the `Refresh Pivot Table` activity in the body of `Use Excel File` activity after the `Write Cell` activity in the `Designer` panel.
5. Next, click the `Plus` icon in the `Pivot table to refresh` field, and navigate to the `EmployeeOnboarding` workbook, the `New Hire Statistics` sheet, and select `NewHireStats_PivotTable`.

Once you have completed the exercise, the final configuration of the **Write Cell** and **Refresh Pivot Table** activities should resemble Figure 7-59.

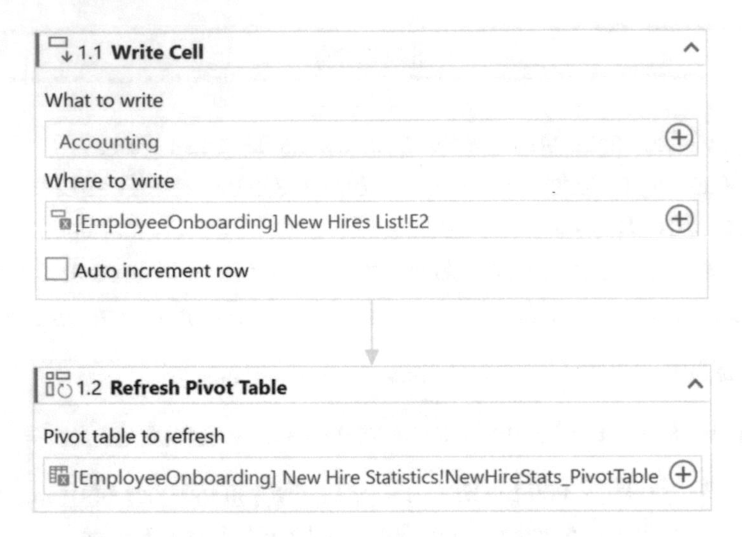

***Figure 7-59.** Displays the final configuration for the Refresh Pivot Table activity*

Figure 7-60 shows the output of the Refresh Pivot Table activity that has updated the counts for new employees in Accounting and Operations departments.

	A	B	C
1	**Row Labels**	**Count of ID**	
2	Accounting	6	
3	Human Resources	5	
4	Information Technology	6	
5	Operations	3	
6	**Grand Total**	**20**	
7			

***Figure 7-60.** Displays the refreshed pivot table*

Append Range

The **Append Range** activity copies data from a specific sheet, range, or table and pastes/appends it after data in another sheet range or table.

Configuration

This section provides instructions on how to configure an **Append Range** activity, shown in Figure 7-61.

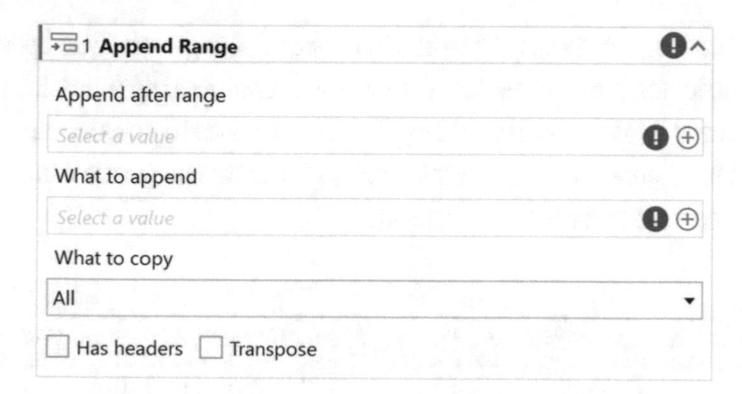

Figure 7-61. *Append Range activity card*

Append after range: This is a required configuration available on the activity card. This configuration provides the location of the sheet, range, or table that the data will be appended after.

What to append: This is a required configuration available on the activity card. This configuration provides the location of the sheet, range, or table that the data will be appended after.

What to copy: This is an optional configuration available on the activity card. This field determines how to paste the data during this activity, for example, selecting `All` copies values, formatting, and any formulas from the data selected in `What to append`. This field is defaulted to the option `All` and can be changed to `Values`, `Formulas`, or `Formats`.

Has headers: This is an optional configuration available on the activity card. If selected, it will consider the first row in the `What to append` range a header and not copy it. This is unselected at default.

Transpose: This is an optional configuration available on the activity card. `Transpose` will rotate the data being copied from columns to rows or vice versa. By default, this option is unselected.

EXERCISE

Goal: Use the `Use Excel File` and the `Append Range` activity to append the data for the `5` new hires from the `EmployeeOnboardingInput` Excel file and add it to the `NewHiresList_Table` in the `Employee Onboarding` parent file. Figure 7-62 displays the EmployeeOnboardingInput file with the `5` employee rows that need to be appended.

	A	B	C	D	E
1	ID	Last Name	First Name	Start Date	Department
2	21	George	Lana	7/7/2020	Accounting
3	22	Evans	Chad	7/8/2020	Human Resources
4	23	Nguyen	Linda	7/9/2020	Information Technology
5	24	Martin	Frank	7/10/2020	Operations
6	25	Lee	Aera	7/11/2020	Operations

***Figure 7-62.** Displays the Employee Onboarding Input data that needs to get appended*

Source Code: Chapter_7_ExcelRangeActivitiesExercise

Setup: Here are step-by-step implementation instructions:

1. In `StudioX`, add the `Use Excel File` activity and configure with the `C:\BookSamples\Chapter_07\EmployeeOnboarding.xlsx` file as demonstrated in the first exercise.

2. Then, drag a second `Use Excel File` activity in the body of the first one for the `C:\BookSamples\Chapter_07\EmployeeOnboardingInput.xlsx` file path and `EmployeeOnboardingInput` for `Reference as`.
3. Next, add the `Append Range` activity card to the body of the nested `Use Excel File` activity.
4. Click the `Plus` icon in the `Append after range` field. Navigate to the `EmployeeOnboarding` workbook option, and select the `NewHiresList_Table`.
5. Click the `Plus` icon in the `What to` append field. Navigate to the `EmployeeOnboardingInput` workbook option, and select the range `New Employees Input sheet`.
6. Check the `Has headers` option.

Once you have completed the exercise, the final configuration of the **Append Range** activity should resemble Figure 7-63.

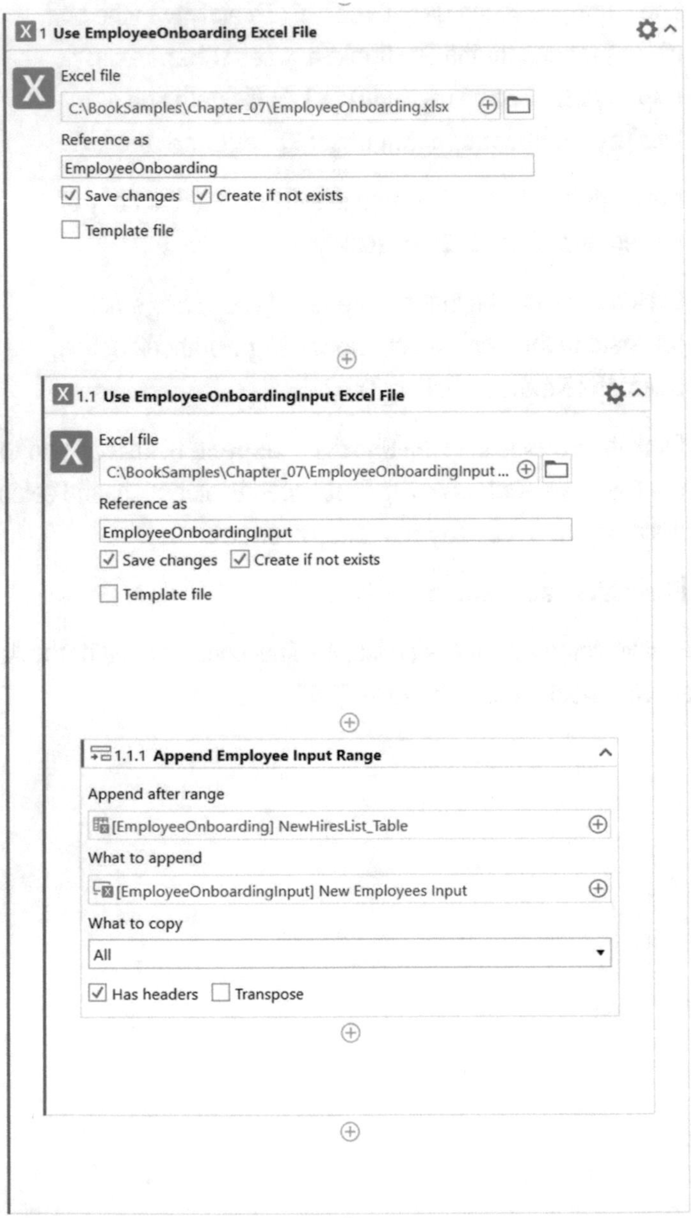

1 Use EmployeeOnboarding Excel File

Excel file

C:\BookSamples\Chapter_07\EmployeeOnboarding.xlsx

Reference as

EmployeeOnboarding

Save changes Create if not exists

Template file

1.1 Use EmployeeOnboardingInput Excel File

Excel file

C:\BookSamples\Chapter_07\EmployeeOnboardingInput ...

Reference as

EmployeeOnboardingInput

Save changes Create if not exists

Template file

1.1.1 Append Employee Input Range

Append after range

[EmployeeOnboarding] NewHiresList_Table

What to append

[EmployeeOnboardingInput] New Employees Input

What to copy

All

Has headers Transpose

Figure 7-63. *Final configuration for Append Range activity card*

The output of this activity, shown in Figure 7-64, has appended the five new rows from the input file to the NewHiresList_Table in the EmployeeOnboarding.xlsx file.

	A	B	C	D	E	F
1	ID	Last Name	First Name	Start Date	Department	Status
2	1	Brown	Sylvia	1/6/2020	Accounting	Active
3	2	Carter	Samantha	4/10/2020	Accounting	Active
4	3	Clark	Jacob	7/20/2020	Human Resources	Active
5	4	Davis	Samuel	1/6/2020	Operations	Active
6	5	Davis	Sasha	7/20/2020	Information Technology	Active
7	6	Hill	Karen	1/6/2020	Operations	Active
8	7	Johnson	Elijah	7/20/2020	Information Technology	Active
9	8	Johnson	Adam	4/10/2020	Human Resources	Active
10	9	Jones	Daniel	7/20/2020	Information Technology	Active
11	10	Khan	Zain	1/6/2020	Operations	Active
12	11	Lane	Tamara	7/20/2020	Human Resources	Active
13	12	Lopez	Maya	4/10/2020	Accounting	Active
14	13	Miller	Raymond	7/20/2020	Information Technology	Active
15	14	Patel	Priya	1/6/2020	Accounting	Active
16	15	Sanchez	Gabriella	7/20/2020	Information Technology	Active
17	16	Singh	Aditya	4/10/2020	Human Resources	Active
18	17	Smith	Carolyn	1/6/2020	Accounting	Active
19	18	Smith	John	7/20/2020	Human Resources	Active
20	19	Williams	Jane	4/10/2020	Information Technology	Active
21	20	Young	Alyssa	1/6/2020	Accounting	Active
22	21	George	Lana	7/7/2020	Accounting	
23	22	Evans	Chad	7/8/2020	Human Resources	
24	23	Nguyen	Linda	7/9/2020	Information Technology	
25	24	Martin	Frank	7/10/2020	Operations	
26	25	Lee	Aera	7/11/2020	Operations	

Figure 7-64. *NewHiresList_Table with the appended range in A22:F26*

Copy Range

The **Copy Range** activity allows you to copy and paste a range, sheet, or table from one location to another in the Excel workbooks available to the automation.

Configuration

This section provides instructions on how to configure a **Copy Range** activity, shown in Figure 7-65.

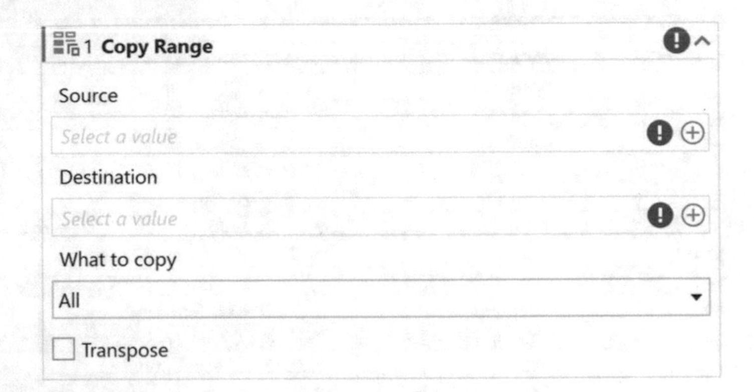

Figure 7-65. *Copy Range activity card*

Note Unlike the Append Range activity that pastes data starting at the first empty row, the Copy Range activity will overwrite data when it is pasted.

Source: This is a required configuration available on the activity card. This field identifies the sheet, table, or range source data that needs to be copied. This field is commonly configured by selecting a named sheet or table or by using the `Indicate in Excel` option.

Destination: This is a required configuration available on the activity card. This field identifies the sheet, table, or range where the `Source` data should be pasted. This field is commonly configured by selecting a named sheet or table or by using the `Indicate in Excel` option.

What to copy: This is an optional configuration available on the activity card. `What to copy` field determines how to paste the data during this activity, for example, selecting All copies values, formatting, and any formulas from the data selected in What to append. This field is defaulted to the option `All` and can be changed to `Values`, `Formulas`, or `Formats`.

Transpose: This is an optional configuration available on the activity card. `Transpose` will rotate the data being copied from columns to rows or vice versa. By default, this option is unselected.

EXERCISE

Goal: Building on our previous exercise, use the `Copy Range` activity to copy the employee `ID`, `Last Name`, and `First Name` from `NewHiresList_Table` rows `22-26` to the `NewHireChecklist Table`. New Hire Checklist table prior to the Copy Range exercise is displayed in Figure 7-66.

	A	B	C	D	E	F	G	H	I
1	ID	Last Name	First Name	Orientation	Employee Hanbook	Policy Training	Benefits Package	Direct Deposit Setup	Technology Setup
2	1	Brown	Sylvia	Complete	Complete	Complete	Complete	Complete	Complete
3	2	Carter	Samantha	Not Started	Not Started	In Progress	Not Started	Not Started	Not Started
4	3	Clark	Jacob	Not Started	Not Started	In Progress	In Progress	Not Started	In Progress
5	4	Davis	Samuel	Complete	Complete	Complete	Complete	Complete	Complete
6	5	Davis	Sasha	Complete	Complete	Complete	Complete	Complete	Complete
7	6	Hill	Karen	Complete	Complete	Complete	Complete	Complete	Complete
8	7	Johnson	Elijah	Not Started	In Progress	In Progress	Not Started	In Progress	Not Started
9	8	Johnson	Adam	Complete	Complete	Complete	Complete	Complete	Complete
10	9	Jones	Daniel	Not Started	Not Started	In Progress	Not Started	Not Started	Not Started
11	10	Khan	Zain	Complete	Complete	Complete	Complete	Complete	Complete
12	11	Lane	Tamara	Not Started	Not Started	In Progress	Not Started	Not Started	Not Started
13	12	Lopez	Maya	Complete	Complete	Complete	Complete	Complete	Complete
14	13	Miller	Raymond	Not Started	In Progress	Not Started	In Progress	Not Started	Not Started
15	14	Patel	Priya	Complete	Complete	Complete	Complete	Complete	Complete
16	15	Sanchez	Gabriella	Not Started	In Progress	In Progress	Not Started	In Progress	Not Started
17	16	Singh	Aditya	Complete	Complete	Complete	Complete	Complete	Complete
18	17	Smith	Carolyn	Complete	Complete	Complete	Complete	Complete	Complete
19	18	Smith	John	Not Started	Not Started	Not Started	In Progress	Not Started	Not Started
20	19	Williams	Jane	Complete	Complete	Complete	Complete	Complete	Complete
21	20	Young	Alyssa	Not Started	Not Started	Not Started	In Progress	Not Started	Not Started

***Figure 7-66.** New Hire Checklist, prior to the Copy Range activity*

Source Code: Chapter_7_ExcelRangeActivitiesExercise

Setup: Here are step-by-step implementation instructions:

1. In StudioX, add the Copy Range activity in the body of the nested Use Excel File activity after the Append Range activity from the previous exercise.
2. Next, click the Plus icon in the Source field, navigate to the EmployeeOnboarding file, and select Indicate in Excel. Once Excel is open, select the range A22:C26 in the New Hires List sheet.
3. Next, click the Plus icon in the Destination field, navigate to the EmployeeOnboarding file, and select Indicate in Excel. Once Excel is open, select cell A22:C26 in the New Hire Checklist sheet.
4. Leave the What to copy and Transpose options with their default selections.

Once you have completed the exercise, the final configuration of the **Copy Range** activity should resemble Figure 7-67.

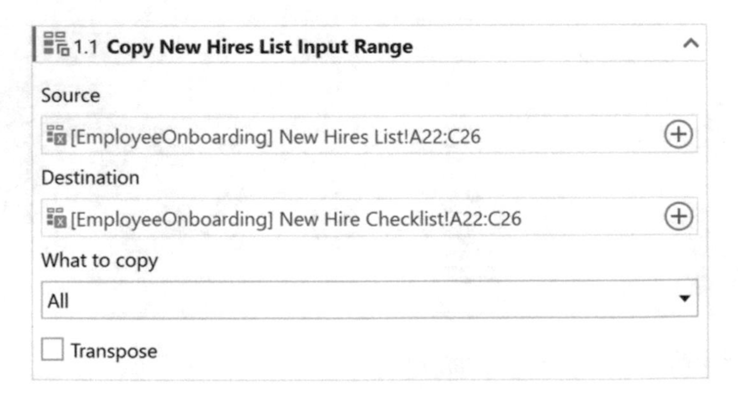

***Figure 7-67.** Copy Range activity configuration*

The output of this example will copy range A22:C26 from the New Hires List and paste it to range A22:C26 in the New Hire Checklist as shown in Figure 7-68.

	A	B	C	D	E	F	G	H	I
1	ID	Last Name	First Name	Orientation	Employee Hanbook	Policy Training	Benefits Package	Direct Deposit Setup	Technology Setup
2	1	Brown	Sylvia	Complete	Complete	Complete	Complete	Complete	Complete
3	2	Carter	Samantha	Not Started	Not Started	In Progress	Not Started	Not Started	Not Started
4	3	Clark	Jacob	Not Started	Not Started	In Progress	In Progress	Not Started	In Progress
5	4	Davis	Samuel	Complete	Complete	Complete	Complete	Complete	Complete
6	5	Davis	Sasha	Complete	Complete	Complete	Complete	Complete	Complete
7	6	Hill	Karen	Complete	Complete	Complete	Complete	Complete	Complete
8	7	Johnson	Elijah	Not Started	In Progress	In Progress	Not Started	In Progress	Not Started
9	8	Johnson	Adam	Complete	Complete	Complete	Complete	Complete	Complete
10	9	Jones	Daniel	Not Started	Not Started	In Progress	Not Started	Not Started	Not Started
11	10	Khan	Zain	Complete	Complete	Complete	Complete	Complete	Complete
12	11	Lane	Tamara	Not Started	Not Started	In Progress	Not Started	Not Started	Not Started
13	12	Lopez	Maya	Complete	Complete	Complete	Complete	Complete	Complete
14	13	Miller	Raymond	Not Started	In Progress	Not Started	In Progress	Not Started	Not Started
15	14	Patel	Priya	Complete	Complete	Complete	Complete	Complete	Complete
16	15	Sanchez	Gabriella	Not Started	In Progress	In Progress	Not Started	In Progress	Not Started
17	16	Singh	Aditya	Complete	Complete	Complete	Complete	Complete	Complete
18	17	Smith	Carolyn	Complete	Complete	Complete	Complete	Complete	Complete
19	18	Smith	John	Not Started	Not Started	Not Started	In Progress	Not Started	Not Started
20	19	Williams	Jane	Complete	Complete	Complete	Complete	Complete	Complete
21	20	Young	Alyssa	Not Started	Not Started	Not Started	In Progress	Not Started	Not Started
22	21	George	Lana						
23	22	Evans	Chad						
24	23	Nguyen	Linda						
25	24	Martin	Frank						
26	25	Lee	Aera						

Figure 7-68. *Copy Range activity output*

Clear Sheet/Range/Table

The **Clear Sheet/Range/Table** activity allows you to clear data for a specified sheet, range, or table in the Excel workbooks available to the automation.

Configuration

This section provides instructions on how to configure a **Clear Sheet/Range/Table** activity, shown in Figure 7-69.

1 Clear Sheet/Range/Table

Range to clear

Select a value

Has headers

Figure 7-69. *Clear Sheet/Range/Table activity card*

Range to clear: This is a required configuration available on the activity card. This field identifies the sheet, table, or range source data that needs to be cleared. This field is commonly configured by selecting a named sheet or table or by using the `Indicate in Excel` option.

Has headers: This is an optional configuration available on the activity card. This configuration is unselected at default, meaning that all the data will be cleared. If selected, then the header row will not be cleared in the identified range.

EXERCISE

Goal: Building on our previous exercise, use the `Clear Sheet/Range/Table` activity to clear the InputTable data from the `EmployeeOnboardingInput` Excel file. Figure 7-62 displays this data prior to being cleared.

Source Code: Chapter_7_ExcelRangeActivitiesExercise

Setup: Here are step-by-step implementation instructions:

1. In `StudioX`, add the `Clear Sheet/Range/Table` activity in the body of the inner `Use Excel File` activity after the `Copy Range` activity.
2. Next, click the `Plus` icon in the `Range to clear` field, navigate to the `EmployeeOnboardingInput` file, the `New Employees Input [Sheet]`, and select the `InputTable [Table]`.
3. Then, check the `Has headers` option so that the header row is not cleared.

Once you have completed the exercise, the final configuration of the **Clear Sheet/Range/Table** activity should resemble Figure 7-70.

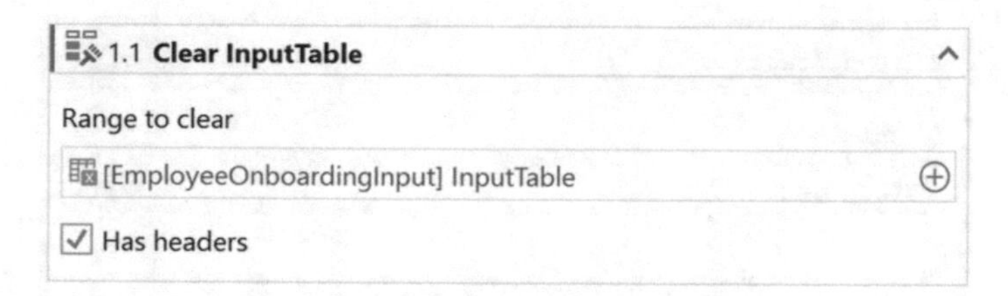

***Figure 7-70.** Clear Sheet/Range/Table activity configuration*

The output of this example will clear the InputTable from the EmployeeOnboardingInput file and leave the header row as is; this is shown in Figure 7-71.

	A	B	C	D	E	F
1	ID	Last Name	First Name	Start Date	Department	
2						
3						
4						
5						
6						
7						

***Figure 7-71.** InputTable data cleared*

Sort Range

The **Sort Range** activity allows you to sort the data in a sheet, range, or table by one or multiple columns in Excel.

Configuration

This section provides instructions on how to configure a **Sort Range** activity, shown in Figure 7-72.

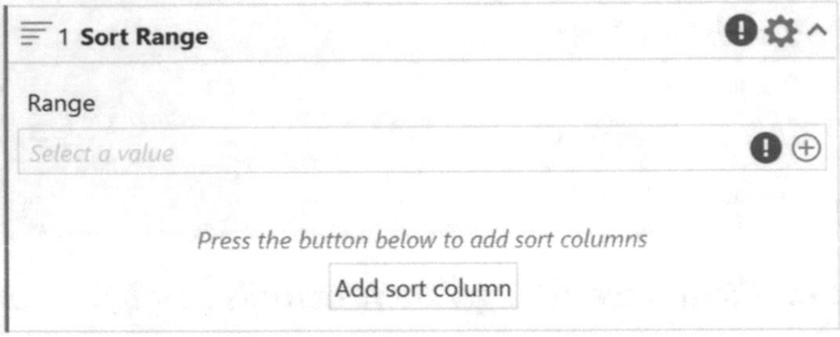

Figure 7-72. Sort Range activity card

Range: This is a required configuration available on the activity card. The `Range` field identifies which sheet, table, or range contains the column(s) that need to be sorted or the target location.

Sort By Column: At least one `Sort By Column` child activity is required to be configured for the `Sort Range` activity. A `Sort By Column` child activity can be added by clicking `Add Sort Column` button on the activity card of `Sort Range` activity. The `Add Sort Column` option can be done multiple times against the same range to sort by as many columns as required based on requirements. The `Sort By Column` child activity requires that `Column` and `Direction` fields be configured.

Note If multiple columns are indicated through the Add Sort Column option, then the automation will sort the data in the order the columns are placed in the parent Sort Range activity.

Column: This is a required field on the `Sort By Column` child activity card. This field identifies which column in the range to sort the data by. Commonly, this field is identified through the `Range` option by selecting the column header name automatically populated from the `Range` field or `Indicate in Excel` to select any cell in the target column. Figure 7-73 displays additional options to configure this field.

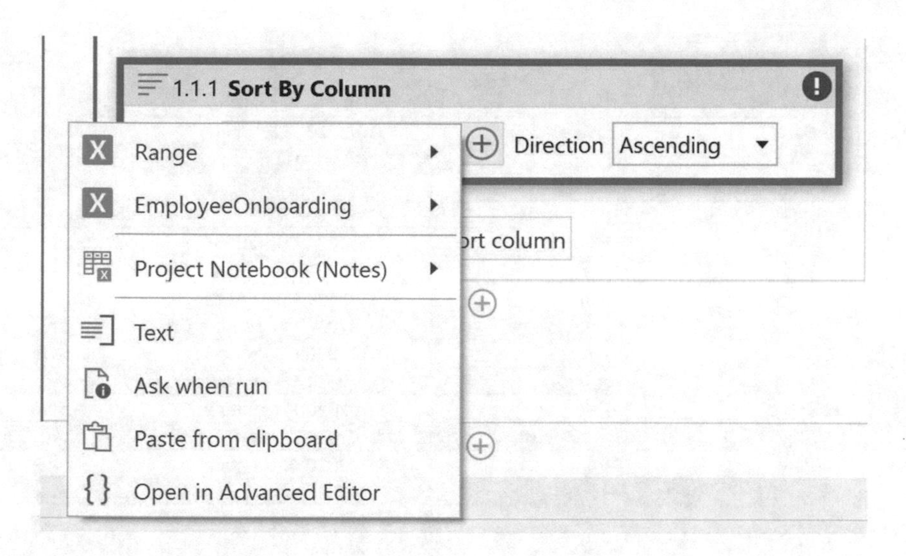

Figure 7-73. Displays the Column field configuration options

Direction: This is a required field on the `Sort By Column` child activity card. The `Direction` field provides the direction to sort the identified column. The default selection for this field is `Ascending` and can be changed to `Descending` through the dropdown menu.

EXERCISE

Goal: Building on our previous exercise, use the `Sort Range` activity to sort the Employees in alphabetical order by `Last Name`. Figure 7-74 displays the `NewHiresList_Table` prior to the `Sort Range` activity.

	A	B	C	D	E	F
1	ID	Last Name	First Name	Start Date	Department	Status
2	1	Brown	Sylvia	1/6/2020	Accounting	Active
3	2	Carter	Samantha	4/10/2020	Accounting	Active
4	3	Clark	Jacob	7/20/2020	Human Resources	Active
5	4	Davis	Samuel	1/6/2020	Operations	Active
6	5	Davis	Sasha	7/20/2020	Information Technology	Active
7	6	Hill	Karen	1/6/2020	Operations	Active
8	7	Johnson	Elijah	7/20/2020	Information Technology	Active
9	8	Johnson	Adam	4/10/2020	Human Resources	Active
10	9	Jones	Daniel	7/20/2020	Information Technology	Active
11	10	Khan	Zain	1/6/2020	Operations	Active
12	11	Lane	Tamara	7/20/2020	Human Resources	Active
13	12	Lopez	Maya	4/10/2020	Accounting	Active
14	13	Miller	Raymond	7/20/2020	Information Technology	Active
15	14	Patel	Priya	1/6/2020	Accounting	Active
16	15	Sanchez	Gabriella	7/20/2020	Information Technology	Active
17	16	Singh	Aditya	4/10/2020	Human Resources	Active
18	17	Smith	Carolyn	1/6/2020	Accounting	Active
19	18	Smith	John	7/20/2020	Human Resources	Active
20	19	Williams	Jane	4/10/2020	Information Technology	Active
21	20	Young	Alyssa	1/6/2020	Accounting	Active
22	21	George	Lana	7/7/2020	Accounting	
23	22	Evans	Chad	7/8/2020	Human Resources	
24	23	Nguyen	Linda	7/9/2020	Information Technology	
25	24	Martin	Frank	7/10/2020	Operations	
26	25	Lee	Aera	7/11/2020	Operations	

Figure 7-74. *New Hires List Table before the Sort Range activity exercise*

Source Code: Chapter_7_ExcelRangeActivitiesExercise

Setup: Here are step-by-step implementation instructions:

1. In `StudioX`, add the `Sort Range` activity in the body of the inner `Use Excel File` activity after the `Clear Sheet/Range/Table` activity from the previous exercise.
2. Next, select the `Plus` icon in the `Range` field and navigate to the `EmployeeOnboarding` Excel file to select the `New Hires List` ➤ `NewHiresList_Table`.

3. Then, click the `Add sort column` button to add the `Sort By Column` child activity.
4. Next, select the `Plus` icon in the `Column` field, and navigate to the `Range` option to select `Last Name` as the column header.
5. Leave the `Direction` field as the default option of `Ascending`.

Once you have completed the exercise, the final configuration of the **Sort Range** activity should resemble Figure 7-75.

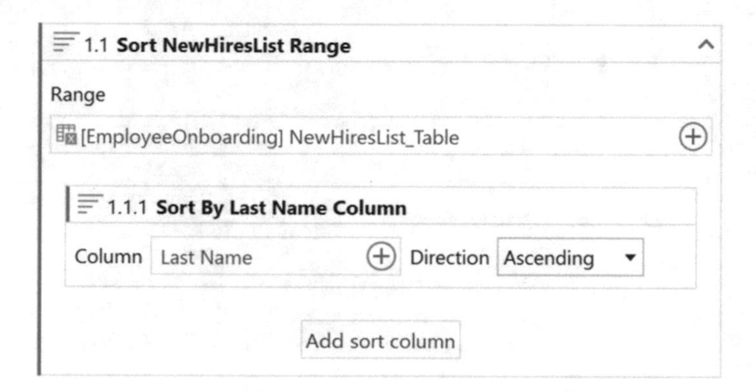

Figure 7-75. *Displays the final configuration for the Sort Range activity*

Figure 7-76 shows the output of the Sort Range activity that has sorted the New Hires List in alphabetical order of Last Names.

	A	B	C	D	E	F
1	ID	Last Name	First Name	Start Date	Department	Status
2	1	Brown	Sylvia	1/6/2020	Accounting	Active
3	2	Carter	Samantha	4/10/2020	Accounting	Active
4	3	Clark	Jacob	7/20/2020	Human Resources	Active
5	4	Davis	Samuel	1/6/2020	Operations	Active
6	5	Davis	Sasha	7/20/2020	Information Technology	Active
7	22	Evans	Chad	7/8/2020	Human Resources	
8	21	George	Lana	7/7/2020	Accounting	
9	6	Hill	Karen	1/6/2020	Operations	Active
10	7	Johnson	Elijah	7/20/2020	Information Technology	Active
11	8	Johnson	Adam	4/10/2020	Human Resources	Active
12	9	Jones	Daniel	7/20/2020	Information Technology	Active
13	10	Khan	Zain	1/6/2020	Operations	Active
14	11	Lane	Tamara	7/20/2020	Human Resources	Active
15	25	Lee	Aera	7/11/2020	Operations	
16	12	Lopez	Maya	4/10/2020	Accounting	Active
17	24	Martin	Frank	7/10/2020	Operations	
18	13	Miller	Raymond	7/20/2020	Information Technology	Active
19	23	Nguyen	Linda	7/9/2020	Information Technology	
20	14	Patel	Priya	1/6/2020	Accounting	Active
21	15	Sanchez	Gabriella	7/20/2020	Information Technology	Active
22	16	Singh	Aditya	4/10/2020	Human Resources	Active
23	17	Smith	Carolyn	1/6/2020	Accounting	Active
24	18	Smith	John	7/20/2020	Human Resources	Active
25	19	Williams	Jane	4/10/2020	Information Technology	Active
26	20	Young	Alyssa	1/6/2020	Accounting	Active

Figure 7-76. *New Hires List sorted by Last Name*

Auto Fill

The **Auto Fill** activity utilizes the auto fill feature in Excel to fill data in adjacent cells (cells in the same column) based on existing data, similar to using the Fill option in Excel or dragging the `Plus` icon on the corner of a cell to other surrounding cells.

Popular use cases for automatically filling data include date, text, number, or formula fields that need to be automatically filled into their adjacent fields in Excel.

Configuration

This section provides instructions on how to configure an **Auto Fill** activity, shown in Figure 7-77.

Figure 7-77. *Auto Fill activity card*

Select source: This is a required configuration available on the activity card. This field identifies the reference cell or range field(s) that will be used to fill in data in the adjacent fields. For example, if the date 1/1/2020 is in cell A1 and needs to be auto filled in the cells below with the date 1/2/2020 and so on, then cell A1 will be the source.

EXERCISE

Goal: Building on our previous exercise, use the `Auto Fill` activity to automatically fill in the `Active` status in the blank `Status` fields in the `New Hires List` sheet.

Source Code: Chapter_7_ExcelRangeActivitiesExercise

Setup: Here are step-by-step implementation instructions:

1. In `StudioX`, add the `Auto Fill` activity in the body of the inner `Use Excel File` activity after the `Sort Range` activity from the previous exercise.

2. In the `Select source` field, click the `Plus` icon, navigate to the `EmployeeOnboarding` workbook, and select the `Indicate in Excel` option. Once Excel is open, select the cell `F2` in the `New Hires List` sheet and press `Confirm` to populate.

Once you have completed the exercise, the final configuration of the **Auto Fill** activity should resemble Figure 7-78.

Figure 7-78. *Auto Fill activity final configuration*

After this automation has been executed, the missing status data is automatically populated with Active based on cell F2. Figure 7-79 displays the New Hires List sheet with the statuses auto filled.

	A	B	C	D	E	F
1	ID	Last Name	First Name	Start Date	Department	Status
2	1	Brown	Sylvia	1/6/2020	Accounting	Active
3	2	Carter	Samantha	4/10/2020	Accounting	Active
4	3	Clark	Jacob	7/20/2020	Human Resources	Active
5	4	Davis	Samuel	1/6/2020	Operations	Active
6	5	Davis	Sasha	7/20/2020	Information Technology	Active
7	22	Evans	Chad	7/8/2020	Human Resources	Active
8	21	George	Lana	7/7/2020	Accounting	Active
9	6	Hill	Karen	1/6/2020	Operations	Active
10	7	Johnson	Elijah	7/20/2020	Information Technology	Active
11	8	Johnson	Adam	4/10/2020	Human Resources	Active
12	9	Jones	Daniel	7/20/2020	Information Technology	Active
13	10	Khan	Zain	1/6/2020	Operations	Active
14	11	Lane	Tamara	7/20/2020	Human Resources	Active
15	25	Lee	Aera	7/11/2020	Operations	Active
16	12	Lopez	Maya	4/10/2020	Accounting	Active
17	24	Martin	Frank	7/10/2020	Operations	Active
18	13	Miller	Raymond	7/20/2020	Information Technology	Active
19	23	Nguyen	Linda	7/9/2020	Information Technology	Active
20	14	Patel	Priya	1/6/2020	Accounting	Active
21	15	Sanchez	Gabriella	7/20/2020	Information Technology	Active
22	16	Singh	Aditya	4/10/2020	Human Resources	Active
23	17	Smith	Carolyn	1/6/2020	Accounting	Active
24	18	Smith	John	7/20/2020	Human Resources	Active
25	19	Williams	Jane	4/10/2020	Information Technology	Active
26	20	Young	Alyssa	1/6/2020	Accounting	Active

Figure 7-79. *Status column auto filled with Active*

Fill Range

The **Fill Range** activity located allows you to insert text or formula into a specified range/table in the parent Use Excel File activity or Project Notebook.

Configuration

This section provides instructions on how to configure a **Fill Range** activity, shown in Figure 7-80.

Figure 7-80. Fill Range activity card

Where to write: This is a required configuration available on the activity card. This configuration provides StudioX with the range or table where data will be written. The common way to configure this field is to select a named table by clicking the `Plus` icon.

What to write: This is a required configuration available on the activity card. This configuration determines the value or formula that needs to be added to the cells in the range or table. The common way to configure this field is to type in the value using the `Text` or `Number` builder.

Note If you choose to write a formula in a range, then the cells being filled will use Excel's auto incrementation functionality.

EXERCISE

Goal: Building on our previous exercise, use the `Fill Range` activity to write in a formula in the Bonus column for `10%` of the `Salary` amount from cell `B2` in the `New Hire Salaries` sheet.

Source Code: Chapter_7_ExcelRangeActivitiesExercise

Setup: Here are step-by-step implementation instructions:

1. In `StudioX`, add the `Fill Range` activity in the body of the inner `Use Excel File` activity after the `Auto Fill` activity from the previous exercise.
2. Next, click the `Plus` icon in the `Where to write` field, and navigate to the `EmployeeOnboarding` workbook to select the `Indicate in Excel` option. Once Excel is open, click the `New Hire Salaries` sheet, and select the cell `C2`.
3. Next, click the `Plus` icon in the `What to write` field, select `Text` option, and type in the formula `=B2*0.10`.

Once you have completed the exercise, the final configuration of the **Fill Range** activity should resemble Figure 7-81.

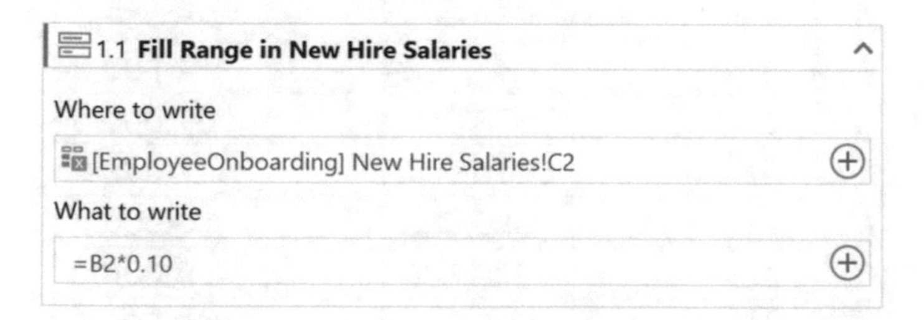

***Figure 7-81.** Displays the final configuration for the Fill Range activity*

	A	B	C
1	ID	Salary	Estimated Bonus
2	1	$45,000	$4,500.0
3	2	$100,000	$10,000.0
4	3	$75,000	$7,500.0
5	4	$50,000	$5,000.0
6	5	$65,000	$6,500.0
7	6	$125,000	$12,500.0
8	7	$80,000	$8,000.0
9	8	$30,000	$3,000.0
10	9	$85,000	$8,500.0
11	10	$40,000	$4,000.0
12	11	$105,000	$10,500.0
13	12	$55,000	$5,500.0
14	13	$60,000	$6,000.0
15	14	$45,000	$4,500.0
16	15	$70,000	$7,000.0
17	16	$110,000	$11,000.0
18	17	$80,000	$8,000.0
19	18	$30,000	$3,000.0
20	19	$85,000	$8,500.0
21	20	$60,000	$6,000.0

***Figure 7-82.** Displays the New Hire Salaries with the Bonus column filled*

Figure 7-82 shows the output of this activity that has added the formula =B2*0.10 to cell C2:C21 in the New Hire Salaries sheet which shows the calculation of 10% of Salary.

Write Range

The **Write Range** activity allows you to write data values from a range, table, sheet, or saved in the automation to another specified range in the Excel workbooks available to the automation.

Note Unlike the Copy and Append Range activities that will only copy the specific range from Excel, the Write Range activity can write a range from other sources, such as saved values from data scraping.

Configuration

This section provides instructions on how to configure a **Write Range** activity, shown in Figure 7-83.

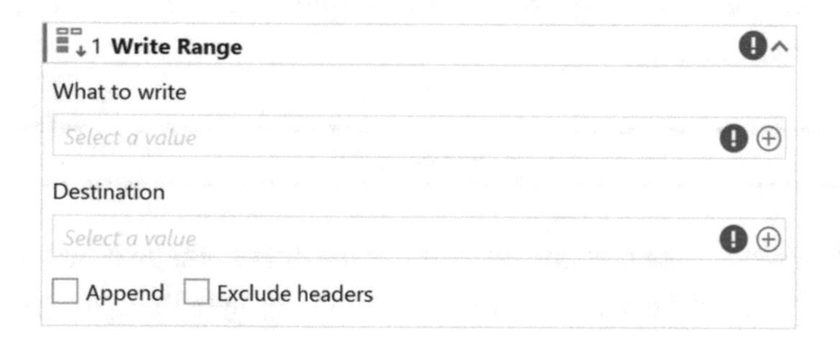

Figure 7-83. *Write Range activity card*

What to write: This is a required configuration available on the activity card. This field identifies the sheet, table, or range source data that needs to be written. This field is commonly configured by selecting a sheet, range, or table using the `Indicate in Excel` option. Alternatively, a saved value from Excel or any other compatible source can also be indicated.

Destination: This is a required configuration available on the activity card. This field identifies the sheet, table, or range where the `What to write` data should be written. This field is commonly configured by selecting a named sheet or table or by using the `Indicate in Excel` option.

Append: This is an optional configuration available on the activity card. This field determines if the data being written should be appended to the destination range. If checked, the data will be written starting at the first empty cell in the range. If left unchecked, the data will be written to the destination range overwriting any data that is existing already. This option is unchecked at default.

Exclude headers: This is an optional configuration available on the activity card. This field determines if the source data being written should include or exclude headers. If checked, the headers from the `What to write` source data will not be written to the `Destination` range. If unchecked, the data being written will include the headers. This option is unchecked at default.

EXERCISE

Goal: Building on our previous exercise, use the `Write Range` activity to write the Salaries from the Input Excel file to the Employee Onboarding Excel file. Figure 7-84 displays the new employee salaries.

	A	B
1	ID	Salary
2	21	$55,000
3	22	$200,000
4	23	$80,000
5	24	$97,000
6	25	$60,000

Figure 7-84. *Input salaries Excel file that will be written to the Employee Onboarding Excel file*

Source Code: Chapter_7_ExcelRangeActivitiesExercise

Setup: Here are step-by-step implementation instructions:

1. In `StudioX`, add the `Write Range` activity in the body of the inner `Use Excel File` activity after the `Fill Range` activity from the previous exercise.
2. Next, click the `Plus` icon in the `What to write` field, navigate to the `EmployeeOnboardingInput` file, and select `Indicate in Excel`. Once Excel is open, select cell `A2` in the `New Employees Salary` sheet.
3. Next, click the `Plus` icon in the `Destination` field, navigate to the `EmployeeOnboarding` file, and select the `New Hire Salaries` sheet.
4. Then, check the `Append` and `Exclude headers` options.

Once you have completed the exercise, the final configuration of the **Write Range** activity should resemble Figure 7-85.

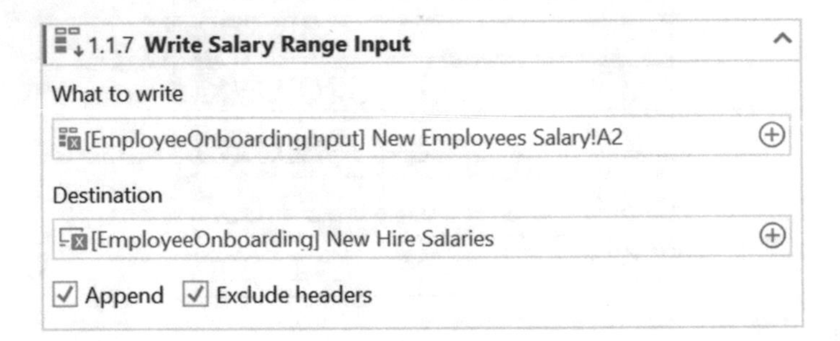

***Figure 7-85.** Write Range activity configuration*

The output of this example will write range A2:B6 from the Salary Input Data to the New Hire Salaries data in the Employee Onboarding Excel file. This will result in extending the New Hire Salaries data to 25 employees as shown in Figure 7-86. The data for column C has also been auto-populated as this was a range filled with the bonus values in the previous exercise.

	A	B	C
1	ID	Salary	Estimated Bonus
2	1	$45,000	$4,500.0
3	2	$100,000	$10,000.0
4	3	$75,000	$7,500.0
5	4	$50,000	$5,000.0
6	5	$65,000	$6,500.0
7	6	$125,000	$12,500.0
8	7	$80,000	$8,000.0
9	8	$30,000	$3,000.0
10	9	$85,000	$8,500.0
11	10	$40,000	$4,000.0
12	11	$105,000	$10,500.0
13	12	$55,000	$5,500.0
14	13	$60,000	$6,000.0
15	14	$45,000	$4,500.0
16	15	$70,000	$7,000.0
17	16	$110,000	$11,000.0
18	17	$80,000	$8,000.0
19	18	$30,000	$3,000.0
20	19	$85,000	$8,500.0
21	20	$60,000	$6,000.0
22	21	$55,000	$5,500.0
23	22	$200,000	$20,000.0
24	23	$80,000	$8,000.0
25	24	$97,000	$9,700.0
26	25	$60,000	$6,000.0

Figure 7-86. *Write Range activity output*

Read Cell Formula

The **Read Cell Formula** activity allows you to read a formula from a specified cell in an Excel workbook used in the automation.

Configuration

This section provides instructions on how to configure a **Read Cell Formula** activity, shown in Figure 7-87.

Figure 7-87. *Read Cell Formula activity card*

Cell: This is a required configuration available on the activity card. This configuration provides the location of the cell to read the formula from. The common way to configure this field is through the `Indicate in Excel` option.

Save to: This is a required configuration available on the activity card. This configuration determines where the formula from the `Cell` will be saved.

EXERCISE

Goal: Use the `Use Excel File` and the `Read Cell Formula` activity to read and print the formula in cell C2 from the `New Hire Salaries` sheet.

Source Code: Chapter_7_ExcelCellActivitiesExercise

Setup: Here are step-by-step implementation instructions:

1. In `StudioX`, add the `Use Excel File` activity and configure with the `C:\BookSamples\Chapter_07\EmployeeOnboarding.xlsx` file as demonstrated in the first exercise.
2. Then add a `Read Cell Formula` activity to the body of the `Use Excel File` activity.
3. Next, click the `Plus` icon in the `Cell` field, and navigate to the `EmployeeOnboarding` workbook to select the `Indicate in Excel` option. Once Excel is open, click the `New Hire Salaries` sheet, and select the cell `C2`.
4. Next, click the `Plus` icon in the `Save to` field, select the `Save for later use` option, and name your value as `CellFormulaValue`.
5. Then add a `Write Line` activity after the `Read Cell Formula` activity.
6. Next, click the `Plus` icon in the `Text` field, hover over `Use Saved Value`, and select `CellFormulaValue`.

Once you have completed the exercise, the final configuration of the **Read Cell Formula** activity should resemble Figure 7-88.

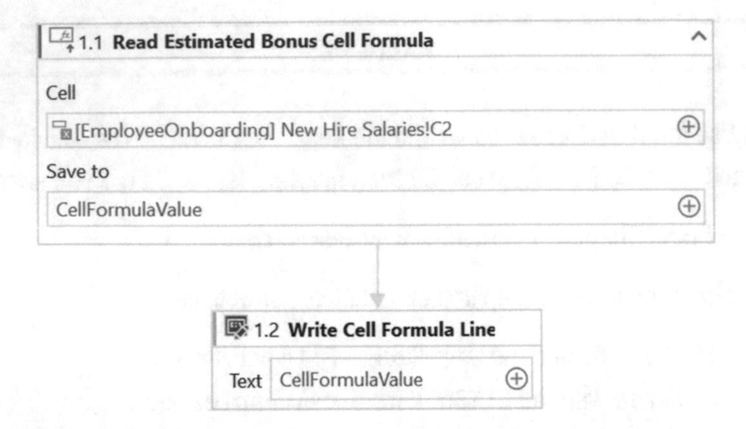

Figure 7-88. *Displays the final configuration for the Read Cell Formula activity*

> Figure 7-91 shows the output of this activity that has returned the formula =B2*0.10 to the Output panel.

Read Cell Value

The **Read Cell Value** activity allows you to read the value from a specified cell in an Excel workbook used in the automation.

Configuration

This section provides instructions on how to configure a **Read Cell Value** activity, shown in Figure 7-89.

1 Read Cell Value
Cell
Select a value
Save to
Select a value
Get formatted text

Figure 7-89. *Read Cell Value activity card*

Cell: This is a required configuration available on the activity card. This configuration provides the location of the cell to read the value from. The common way to configure this field is through the `Indicate in Excel` option.

Save to: This is a required configuration available on the activity card. This configuration determines where the value from the `Cell` will be saved.

Get formatted text: This is an optional configuration available on the activity card. This option is selected by default, meaning that the cell's value and number format will both be read, for example, date, currency, or percentage. If unselected, only the cell's value will be read.

EXERCISE

Goal: Building on our previous exercise, use the `Read Cell Value` activity to read and print the value in cell C2 from the `New Hire Salaries` sheet. This is similar to the exercise for Read Cell Formula, except instead of printing the formula B2*0.10, this exercise will print the cell value of $4,500.

Source Code: Chapter_7_ExcelCellActivitiesExercise

Setup: Here are step-by-step implementation instructions:

1. In `StudioX`, add the `Read Cell Value` activity in the body of `Use Excel File` activity after the `Read Cell Formula` activity from the previous exercise.
2. Next, click the `Plus` icon in the `Cell` field, and navigate to the `EmployeeOnboarding` workbook to select the `Indicate in Excel` option. Once Excel is open, click the `New Hire Salaries` sheet, and select the cell `C2`.
3. Next, click the `Plus` icon in the `Save to` field, select the `Save for later use` option, and name your value as `CellValue`.

4. Then add a `Write Line` activity after the `Read Cell Value` activity.

5. Next, click the `Plus` icon in the `Text` field, hover over `Use Saved Value`, and select `CellValue`.

Once you have completed the exercise, the final configuration of the **Read Cell Value** activity should resemble Figure 7-90.

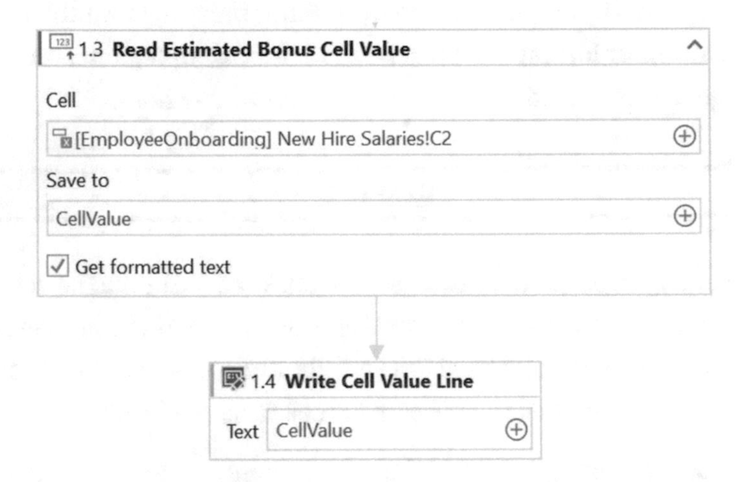

***Figure 7-90.** Displays the final configuration for the Read Cell Value activity*

Figure 7-91 shows the output of this activity that has returned the value $4,500 from cell C2 to the Output panel.

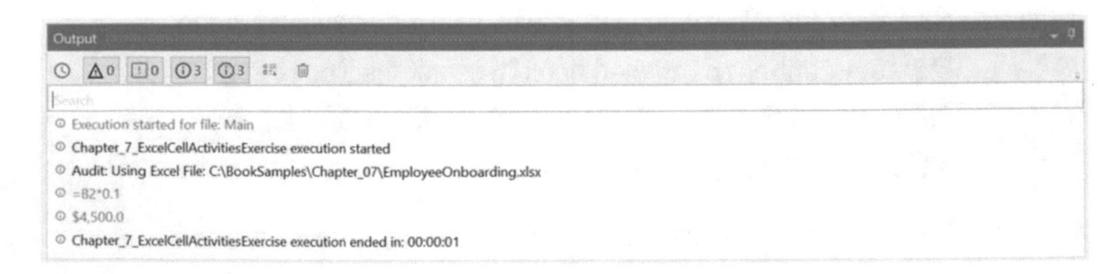

***Figure 7-91.** Output of Read Cell Formula and Read Cell Value activities*

Format Cells

The **Format Cells** activity updates the format of cells in a specified range.

Note Format Cells activity will format all the cells in the specified range and override any existing formatting.

Configuration

This section provides instructions on how to configure a **Format Cells** activity, shown in Figure 7-92.

Figure 7-92. *Format Cells activity card*

Source: This is a required configuration available on the activity card. This configuration provides StudioX with the range that contains the cells that will be formatted.

Format data as type: This is a required configuration available on the activity card. This configuration determines the format of the fields in the `Source` range. You can provide this configuration by clicking the `Set Format` button and selecting the appropriate `Category`. The `Category` dropdown includes `General`, `Number`, `Date`, `Time`, `Percentage`, `Currency`, `Text`, and `Custom` options.

EXERCISE

Goal: Building on our previous exercise, use the Format Cells activity to format the Salary and Estimated Bonus column in the New Hire Salaries sheet to a Currency format.

Source Code: Chapter_7_ExcelCellActivitiesExercise

Setup: Here are step-by-step implementation instructions:

1. In StudioX, add the Format Cells activity in the body of Use Excel File activity after the Read Cell Value activity from the previous exercise.
2. Next, click the Plus icon in the Source field, navigate to the EmployeeOnboarding workbook, and select Indicate in Excel option. Once Excel opens, click the New Hire Salaries sheet, and select the range B2:C21.
3. Next, click the Set Format button in the Format data as type field.
4. Once the Format data as type dialog opens, select Currency from Category dropdown, type 2 in Decimals field, check Use 1000 Separator, type $ in Symbol field, and uncheck Set at the end. Figure 7-93 shows this configuration.

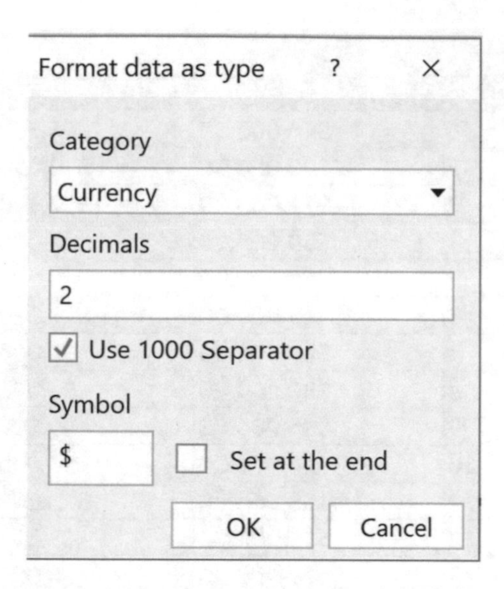

***Figure 7-93.** Displays the Format data as type field configurations*

Once you have completed the exercise, the final configuration of the **Format Cells** activity should resemble Figure 7-94.

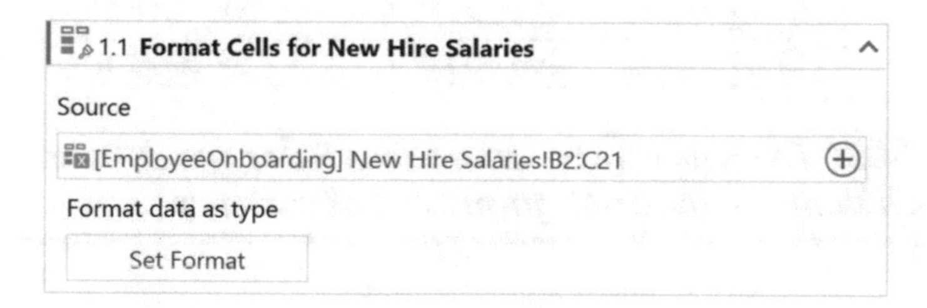

***Figure 7-94.** Displays the final configuration for the Format Cells activity*

Figure 7-95 shows the output of the Format Cells activity that has updated the cells in the Salary and Estimated Bonus column range B2:C21 to a Currency format.

	A	B	C
1	ID	Salary	Estimated Bonus
2	1	$45,000.00	$4,500.00
3	2	$100,000.00	$10,000.00
4	3	$75,000.00	$7,500.00
5	4	$50,000.00	$5,000.00
6	5	$65,000.00	$6,500.00
7	6	$125,000.00	$12,500.00
8	7	$80,000.00	$8,000.00
9	8	$30,000.00	$3,000.00
10	9	$85,000.00	$8,500.00
11	10	$40,000.00	$4,000.00
12	11	$105,000.00	$10,500.00
13	12	$55,000.00	$5,500.00
14	13	$60,000.00	$6,000.00
15	14	$45,000.00	$4,500.00
16	15	$70,000.00	$7,000.00
17	16	$110,000.00	$11,000.00
18	17	$80,000.00	$8,000.00
19	18	$30,000.00	$3,000.00
20	19	$85,000.00	$8,500.00
21	20	$60,000.00	$6,000.00

Figure 7-95. *The New Hire Salaries sheet, Salary column and Estimated Bonus column now formatted as currency*

Export to CSV

The **Export to CSV** activity allows you to export a range, sheet, or table from an Excel file to a CSV file.

Configuration

This section provides instructions on how to configure an **Export to CSV** activity, shown in Figure 7-96.

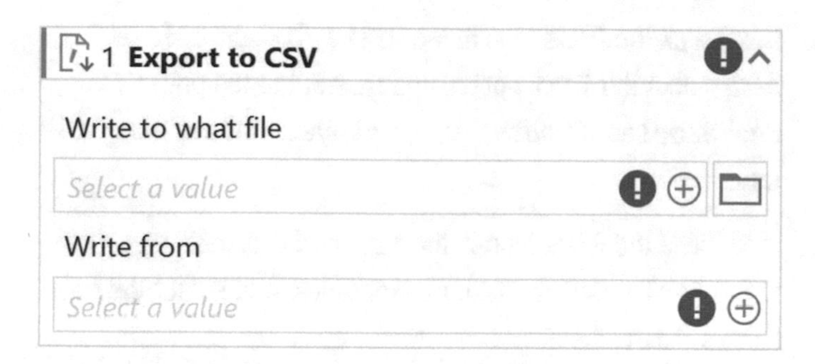

Figure 7-96. Export to CSV activity card

Write to what file: This is a required configuration available on the activity card. This configuration specifies the file path of the CSV that the range, sheet, or table will be exported to.

Write from: This is a required configuration available on the activity card. This configuration determines which range, table, or sheet will be exported from the Excel file to the CSV file.

EXERCISE

Goal: Use the `Use Excel File` and the `Export to CSV` activities to export the `New Hires List [Sheet]` from the `EmployeeOnboarding` Excel file to the `Employee Onboarding Summary.csv` file.

Source Code: Chapter_7_ExcelMiscActivitiesExercise

Setup: Here are step-by-step implementation instructions:

1. In `StudioX`, add the `Use Excel File` activity and configure with the `C:\BookSamples\Chapter_07\EmployeeOnboarding.xlsx` file as demonstrated in the first exercise.
2. Then, drag the `Export to CSV` activity card in the body of `Use Excel File` activity in the `Designer` panel.

3. Next, click the `Plus` icon in the `Write to what file` field, select the `Text` option, and type in the file path `C:\BookSamples\Chapter_07\Employee Onboarding CSV.csv`.

4. Next, click the `Plus` icon in the `Write from` field, navigate to the `EmployeeOnboarding` workbook, and select the `New Hires List` sheet.

Once you have completed the exercise, the final configuration of the **Export to CSV** activity should resemble Figure 7-97.

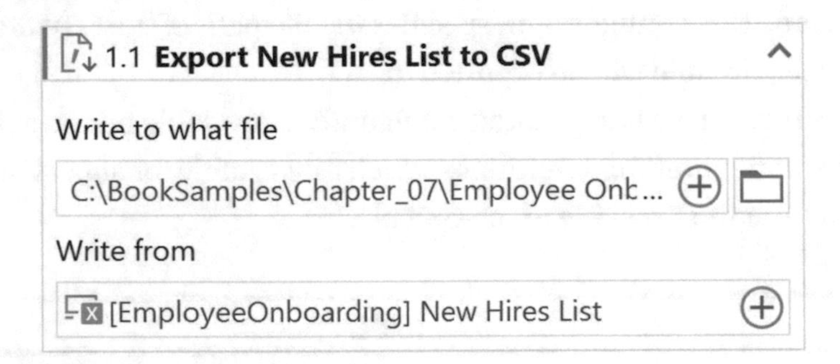

***Figure 7-97.** Displays the final configuration for the Export to CSV activity*

The output of this example exports the New Hires List sheet from the EmployeeOnboarding Excel to the Employee Onboarding Summary CSV file as demonstrated in Figure 7-98.

	A	B	C	D	E	F
1	ID	Last Name	First Name	Start Date	Department	Status
2	1	Brown	Sylvia	1/6/2020	Accounting	Active
3	2	Carter	Samantha	4/10/2020	Accounting	Active
4	3	Clark	Jacob	7/20/2020	Human Resources	Active
5	4	Davis	Samuel	1/6/2020	Operations	Active
6	5	Davis	Sasha	7/20/2020	Information Technology	Active
7	22	Evans	Chad	7/8/2020	Human Resources	Active
8	21	George	Lana	7/7/2020	Accounting	Active
9	6	Hill	Karen	1/6/2020	Operations	Active
10	7	Johnson	Elijah	7/20/2020	Information Technology	Active
11	8	Johnson	Adam	4/10/2020	Human Resources	Active
12	9	Jones	Daniel	7/20/2020	Information Technology	Active
13	10	Khan	Zain	1/6/2020	Operations	Active
14	11	Lane	Tamara	7/20/2020	Human Resources	Active
15	25	Lee	Aera	7/11/2020	Operations	Active
16	12	Lopez	Maya	4/10/2020	Accounting	Active
17	24	Martin	Frank	7/10/2020	Operations	Active
18	13	Miller	Raymond	7/20/2020	Information Technology	Active
19	23	Nguyen	Linda	7/9/2020	Information Technology	Active
20	14	Patel	Priya	1/6/2020	Accounting	Active
21	15	Sanchez	Gabriella	7/20/2020	Information Technology	Active
22	16	Singh	Aditya	4/10/2020	Human Resources	Active
23	17	Smith	Carolyn	1/6/2020	Accounting	Active
24	18	Smith	John	7/20/2020	Human Resources	Active
25	19	Williams	Jane	4/10/2020	Information Technology	Active
26	20	Young	Alyssa	1/6/2020	Accounting	Active

Employee Onboarding CSV

Figure 7-98. *Employee Onboarding CSV populated with New Hires List*

Save Excel File

The **Save Excel File** activity allows you to save the referenced Excel file after Excel automation activities have been executed in the UiPath project.

Configuration

This section provides instructions on how to configure a **Save Excel File** activity, shown in Figure 7-99.

Figure 7-99. *Save Excel File activity card*

File: This is a required configuration available on the activity card. This configuration identifies the workbook that needs to be saved. By default, this field is populated with the workbook from the parent `Use Excel File` activity.

Tip The Save Excel File activity is useful for incrementally saving changes in case of any errors that may occur.

EXERCISE

Goal: Building on our previous exercise, use the `Save Excel File` activity to save the `EmployeeOnboarding` Excel file.

Source Code: Chapter_7_ExcelMiscActivitiesExercise

Setup: Here are step-by-step implementation instructions:

1. In `StudioX`, add the `Save Excel File` activity in the body of `Use Excel File` activity after the `Export to CSV` activity from the previous exercise.
2. The `File` field will be pre-populated with the `EmployeeOnboarding` Excel file from the parent activity.

Once you have completed the exercise, the final configuration of the **Save Excel File** activity should resemble Figure 7-100.

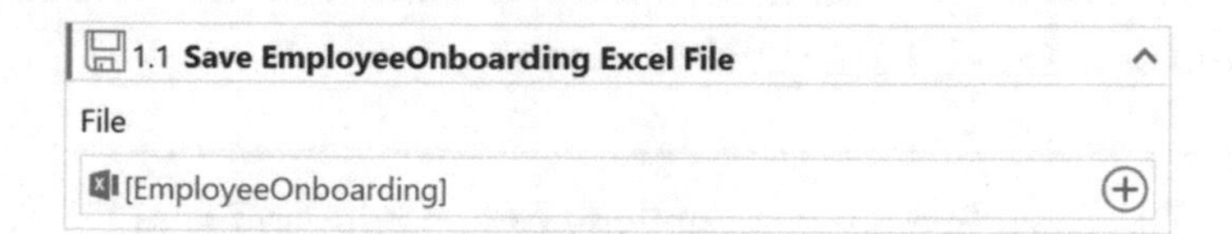

Figure 7-100. *Displays the final configuration for the Save Excel File activity*

Save Excel File As

The **Save Excel File As** activity allows you to save the referenced Excel file as a different Excel file type including .xlsx, .xlsb, .xlsm, and .xls.

Configuration

This section provides instructions on how to configure a **Save Excel File As** activity, shown in Figure 7-101.

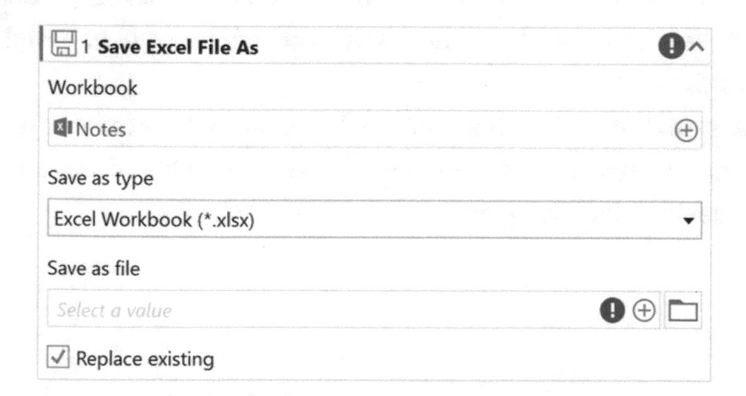

Figure 7-101. *Save Excel File As activity card*

Workbook: This is a required configuration available on the activity card. This field identifies the workbook that needs to be saved with a different Excel file type. This field is pre-populated with the workbook from the parent `Use Excel File` activity.

Note The Save Excel File As activity only saves an Excel file to another Excel file type extension limited to .xslx, .xls, .xlsm, and .xlsb.

Save as type: This is a required configuration available on the activity card. This field identifies the new Excel format that the workbook needs to be saved as through the activity. The following are the dropdown options available:

- `Excel Workbook (*.xlsx)`
- `Excel Binary Workbook (*.xlsb)`
- `Excel Macro-Enabled Workbook (*.xlsm)`
- `Excel 97-2003 Workbook (*.xls)`

Save as file: This is a required configuration available on the activity card. This field provides the name of the Excel file that will be saved as a new Excel format.

Replace existing: This is an optional configuration available on the activity card. If checked, this will replace the existing file of the same name in the target location. By default, this field is checked.

EXERCISE

Goal: Building on our previous exercise, use the `Save Excel File As` activity to save the `EmployeeOnboarding.xlsx` file as `Employee Onboarding Macro Enabled.xlsm` file.

Source Code: Chapter_7_ExcelMiscActivitiesExercise

Setup: Here are step-by-step implementation instructions:

1. In `StudioX`, add the `Save Excel File As` activity in the body of `Use Excel File` activity after the `Save Excel File` activity from the previous exercise.
2. The `Workbook` field will be pre-populated with the `EmployeeOnboarding` Excel file from the parent activity.
3. Next, click the `Save as type` dropdown to select `Excel Macro-Enabled Workbook (*.xlsm)` option.
4. Next, click the `Plus` icon in the `Save as file` field, select the `Text` option, and type in the file path `C:\BookSamples\Chapter_07\Employee Onboarding Macro Enabled.xlsm`.
5. Leave the `Replace existing` option checked.

Once you have completed the exercise, the final configuration of the **Save Excel File As** activity should resemble Figure 7-102.

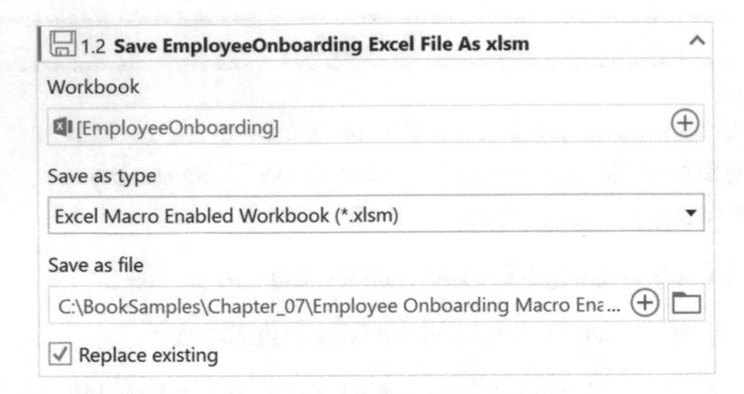

***Figure 7-102.** Displays the final configuration for the Save Excel File As activity*

Figure 7-103 shows the output of the Save Excel File As activity that has added the .xlsm file to the Chapter_07 folder.

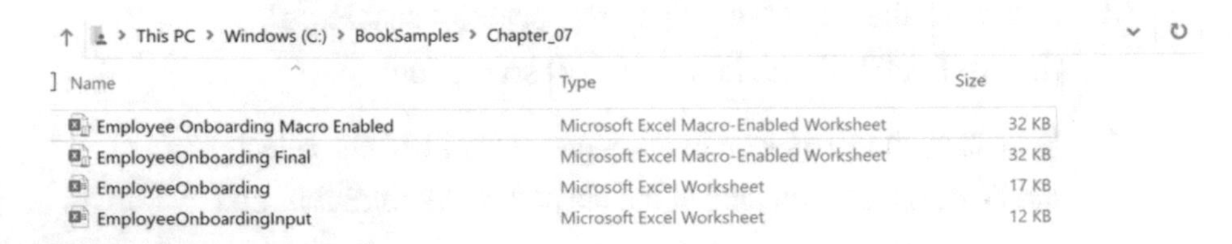

***Figure 7-103.** Employee Onboarding Macro Enabled xlsm file added*

Save Excel File As PDF

The **Save Excel File As PDF** activity allows you to save the referenced Excel file or the Project Notebook as a PDF file.

Configuration

This section provides instructions on how to configure a **Save Excel File As PDF** activity, shown in Figure 7-104.

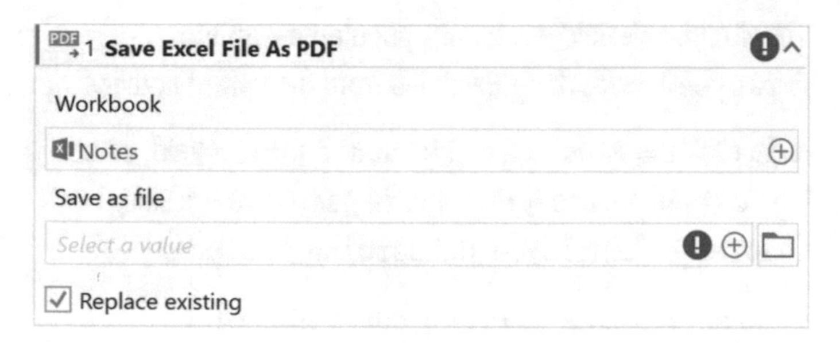

Figure 7-104. Save Excel File As PDF activity card

Workbook: This is a required configuration available on the activity card. This field identifies the workbook that needs to be saved as a PDF. This field is pre-populated with the workbook from the parent `Use Excel File` activity.

Save as file: This is a required configuration available on the activity card. This field provides the file path to save the new PDF file created from the activity.

Replace existing: This is an optional configuration available on the activity card. If checked, this will replace the existing file of the same name in the target location. By default, this field is checked.

EXERCISE

Goal: Building on our previous exercise, use the `Save Excel File As PDF` activity to save the `EmployeeOnboarding.xlsx` file as `EmployeeOnboarding.pdf` file.

Source Code: Chapter_7_ExcelMiscActivitiesExercise

Setup: Here are step-by-step implementation instructions:

1. In `StudioX`, add the `Save Excel File As PDF` activity in the body of `Use Excel File` activity after the `Save Excel File As` activity from the previous exercise.

2. The `Workbook` field will be pre-populated with the `EmployeeOnboarding` Excel file from the parent activity.

3. Next, click the `Plus` icon in the `Save as file` field, select the `Text` option, and type in the file path `C:\BookSamples\Chapter_07\Employee Onboarding PDF.pdf`.

4. Leave the `Replace existing` option checked.

Once you have completed the exercise, the final configuration of the **Save Excel File As PDF** activity should resemble Figure 7-105.

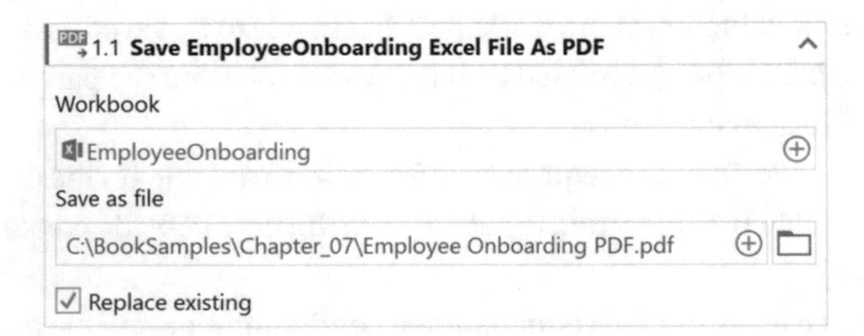

***Figure 7-105.** Displays the final configuration for the Save Excel File As PDF activity*

Figure 7-106 shows the output of the Save Excel File As PDF activity that has saved the Employee Onboarding PDF file in the Chapter_07 folder.

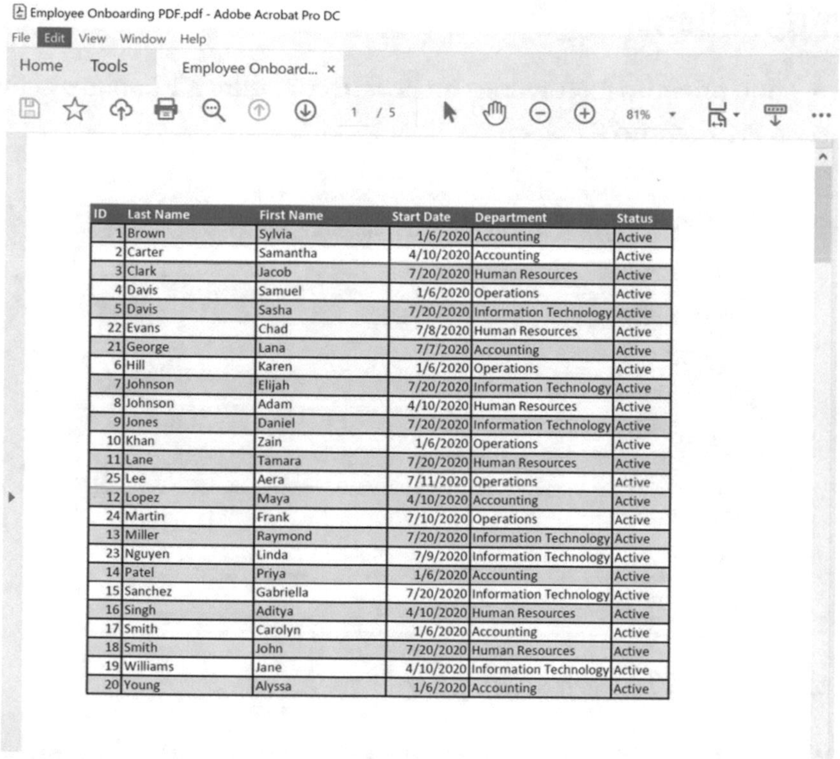

Employee Onboarding PDF.pdf - Adobe Acrobat Pro DC

File Edit View Window Help

Home Tools Employee Onboard... ×

1 / 5 81%

ID	Last Name	First Name	Start Date	Department	Status
1	Brown	Sylvia	1/6/2020	Accounting	Active
2	Carter	Samantha	4/10/2020	Accounting	Active
3	Clark	Jacob	7/20/2020	Human Resources	Active
4	Davis	Samuel	1/6/2020	Operations	Active
5	Davis	Sasha	7/20/2020	Information Technology	Active
22	Evans	Chad	7/8/2020	Human Resources	Active
21	George	Lana	7/7/2020	Accounting	Active
6	Hill	Karen	1/6/2020	Operations	Active
7	Johnson	Elijah	7/20/2020	Information Technology	Active
8	Johnson	Adam	4/10/2020	Human Resources	Active
9	Jones	Daniel	7/20/2020	Information Technology	Active
10	Khan	Zain	1/6/2020	Operations	Active
11	Lane	Tamara	7/20/2020	Human Resources	Active
25	Lee	Aera	7/11/2020	Operations	Active
12	Lopez	Maya	4/10/2020	Accounting	Active
24	Martin	Frank	7/10/2020	Operations	Active
13	Miller	Raymond	7/20/2020	Information Technology	Active
23	Nguyen	Linda	7/9/2020	Information Technology	Active
14	Patel	Priya	1/6/2020	Accounting	Active
15	Sanchez	Gabriella	7/20/2020	Information Technology	Active
16	Singh	Aditya	4/10/2020	Human Resources	Active
17	Smith	Carolyn	1/6/2020	Accounting	Active
18	Smith	John	7/20/2020	Human Resources	Active
19	Williams	Jane	4/10/2020	Information Technology	Active
20	Young	Alyssa	1/6/2020	Accounting	Active

***Figure 7-106.** Employee Onboarding PDF saved*

VLookup

The **VLookup** activity utilizes Excel's VLookup function to find data in a range, sheet, or table from the referenced Excel file.

Common examples for the **VLookup** activity include finding an inventory item and then locating and returning the exact price of the item or finding an employee and then locating the start date for the specific employee.

Configuration

This section provides instructions on how to configure a **VLookup** activity, shown in Figure 7-107.

Figure 7-107. *VLookup activity card*

Value to lookup: This is a required configuration available on the activity card. The `Value to lookup` field identifies what value you are searching for in the target range. For example, to find the start date of John Smith, the Value to lookup field will be the cell address containing the last name Smith.

Note The Value to lookup field must always be from the first column in the identified range, the same as it is for the VLookup function in Excel.

In range: This is a required configuration available on the activity card. The `In range` selection provides the sheet, table, or range containing the value that needs to be looked up.

Column Index: This is a required configuration available on the activity card. This field identifies which column in the range contains the value that needs to be looked up and returned. For example, if the start date of John Smith is in column D of a table, the Column Index will be the number 4 to indicate that the value is in the fourth column of the range.

Exact match: This is an optional configuration available on the activity card. Checking this field means the VLookup activity will return only exact matches; if unchecked, the activity will return approximate matches. The default selection for this field is checked and is usually left as is.

Save to: This is an optional configuration available on the activity card; however, this is an output field that should be defined for the `VLookup` activity. The `Save to` field provides the cell where the returned value from the `VLookup` activity should be saved.

EXERCISE

Goal: Building on our previous exercise, use the `VLookup` activity to save and print the `Department` for `Aera Lee` from the `NewHiresList_Table`.

Source Code: Chapter_7_ExcelMiscActivitiesExercise

Setup: Here are step-by-step implementation instructions:

1. In `StudioX`, add the `VLookup` activity in the body of `Use Excel File` activity after the `Save Excel File As PDF` activity from the previous exercise.
2. Next, click the `Plus` icon in `Value to lookup` field and navigate to the `EmployeeOnboarding` Excel file to select the `Indicate in Excel` option. Once Excel is open, select cell `A15` in the `New Hires List` sheet containing the `ID` number for `Aera Lee`.
3. Next, click the `Plus` icon in the `In range` field, navigate to the `EmployeeOnboarding` Excel file, `New Hires List [Sheet]`, and select `NewHiresList_Table [Table]`.

4. Next, click the `Plus` icon in the `Column Index` field and choose the `Number` option. Once the Number editor is open, type in the number `5`.

5. Leave the `Exact match` field checked, as default.

6. Next, click the `Plus` icon in the `Save to` field and select the `Save for later use` option. Once the dialog box opens, type in `Aera Lee- Department` as the saved value name.

7. Then, add a `Write Line` activity under the VLookup activity.

8. Next, click the `Plus` icon in the `Text` field, navigate to `Use Saved Value` option, and select `Aera Lee- Department`.

Once you have completed the exercise, the final configuration of the **VLookup** and Write Line activity should resemble Figure 7-108.

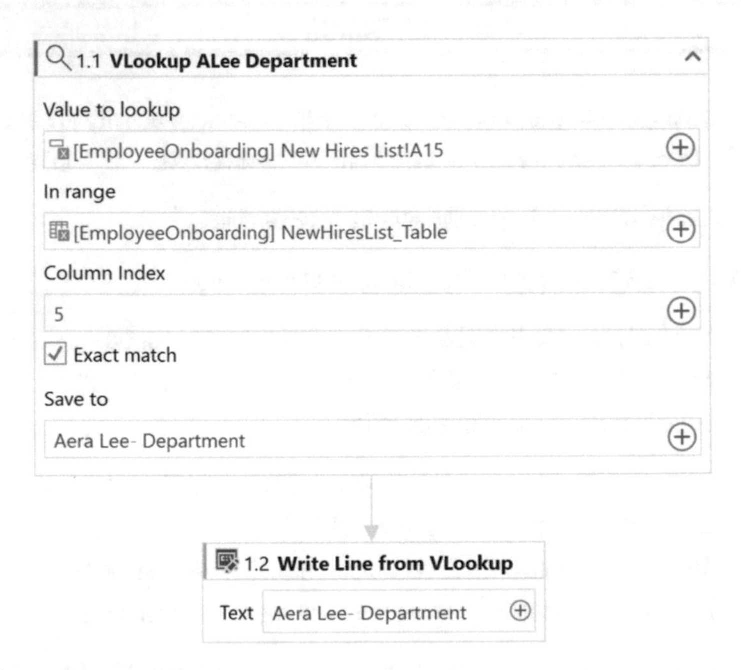

Figure 7-108. *Displays the final configuration for the VLookup and Write Line activity*

Figure 7-109 shows the output of the VLookup activity that has saved the department value of Operations into the Aera Lee- Department value.

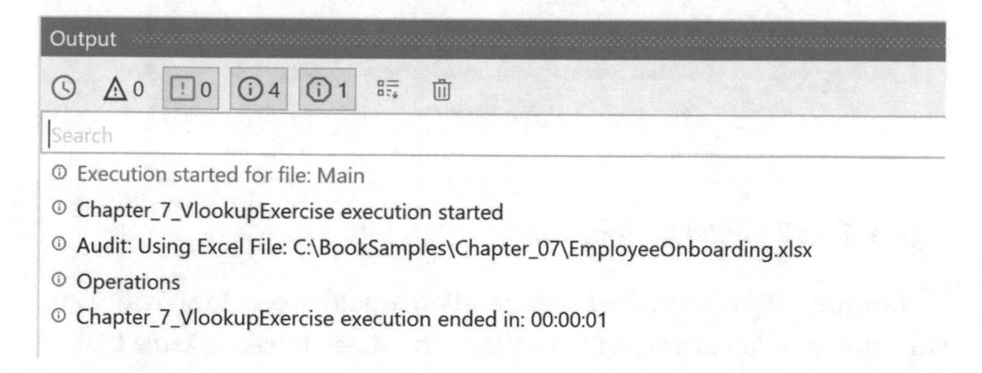

Figure 7-109. *Shows the new Saved Value from the VLookup activity*

Filter

The **Filter** activity utilizes the data filter option in Excel to allow you to filter a column in a range, sheet, or table based on a given value in the referenced Excel file.

Configuration

This section provides instructions on how to configure a **Filter** activity, shown in Figure 7-110.

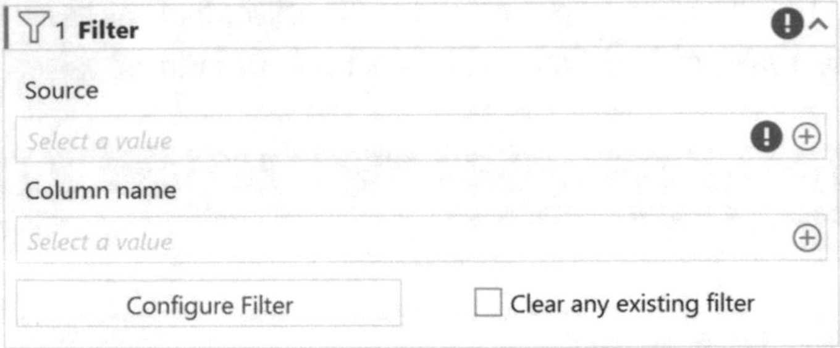

***Figure 7-110.** Filter activity card*

Source: This is a required configuration available on the activity card. This configuration provides StudioX with the sheet, range, or table that needs to be filtered.

Column name: This is a required configuration available on the activity card. This configuration determines column header in the range, sheet, or table that needs to be filtered. For example, if you want to filter on Completed transactions, then you would select Transaction Status as the column header.

Configure Filter: This is a required configuration available on the activity card. This configuration provides value that the column should be filtered on. For example, for finding the Completed transactions in the preceding example for the Column name field, your `Configure Filter` value would be Completed.

To configure this field, use the `Basic filter` option which allows you to enter one or more values to filter on using the same configuration options mentioned in the `Column name` field. You can filter on one or multiple values using this option by pressing the `Add` to filter on a second value in the same column.

Additional options are available through the `Advanced filter` which lets you specify conditional filters like the Custom AutoFilter option in Excel. For example, you can utilize the `Advanced filter` to filter out values that are greater than 100 in a Quantity column.

Clear any existing filter: This is an optional configuration available on the activity card. This configuration allows you to clear any existing filters from the range, sheet, or table defined in the Source field. By default, this option is not checked.

EXERCISE

Goal: Building on our previous exercise, use the Filter activity to filter the NewHiresList table for Employees with the Department of Accounting or Human Resources.

Source Code: Chapter_7_ExcelMiscActivitiesExercise

Setup: Here are step-by-step implementation instructions:

1. In StudioX, add the Filter activity in the body of Use Excel File activity after the VLookup activity from the previous exercise.
2. Next, click the Plus icon in the Source field, and navigate to the EmployeeOnboarding workbook, New Hires List [Sheet], and select the NewHiresList_Table [Table].
3. Next, click the Plus icon in the Column name field, and hover over Range to select the Department column.
4. Next, click Configure Filter button, and with the Basic filter option selected, click the Plus icon in the Is equal to field in the Value section. Use the Text option to type Accounting.
5. Click the Add button to add another filter. Click the Plus icon, select Text option, and type in Human Resources for the second filter. Once completed, click OK. Your filters should resemble Figure 7-111.

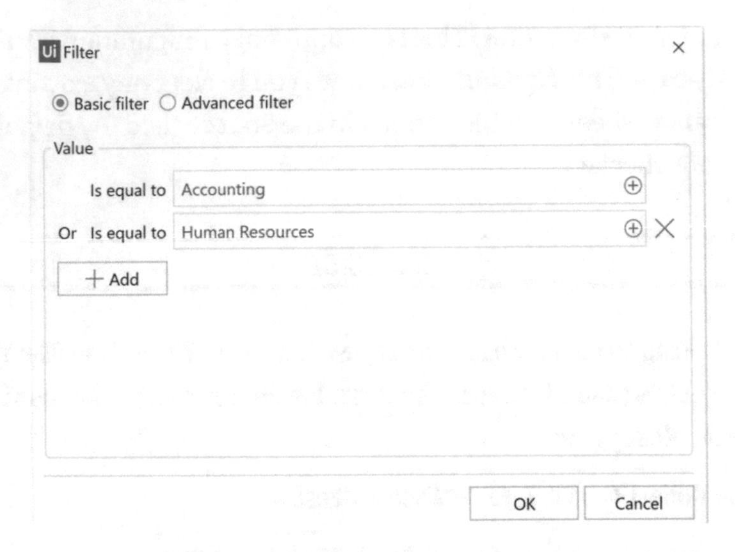

***Figure 7-111.** Configure Filter selections*

Once you have completed the exercise, the final configuration of the **Filter** activity should resemble Figure 7-112.

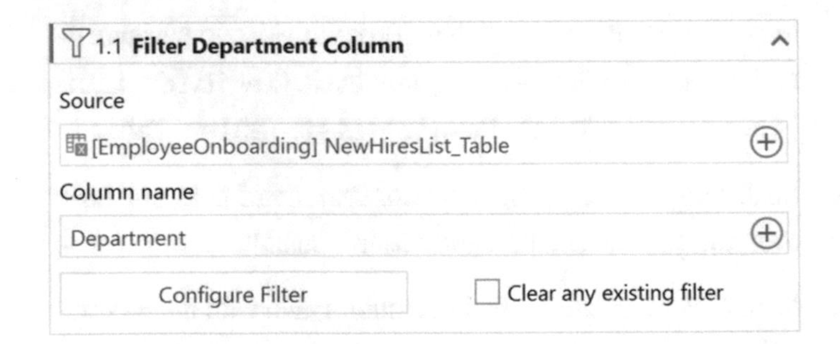

***Figure 7-112.** Displays the final configuration for the Filter activity*

Figure 7-113 shows the output of the Output activity with the filtered down NewHiresList_Table, from 25 rows to 13 rows with only Accounting and Human Resources functions.

	A	B	C	D	E	F
1	ID	Last Name	First Name	Start Date	Department	Status
2	1	Brown	Sylvia	1/6/2020	Accounting	Active
3	2	Carter	Samantha	4/10/2020	Accounting	Active
4	3	Clark	Jacob	7/20/2020	Human Resources	Active
7	22	Evans	Chad	7/8/2020	Human Resources	Active
8	21	George	Lana	7/7/2020	Accounting	Active
11	8	Johnson	Adam	4/10/2020	Human Resources	Active
14	11	Lane	Tamara	7/20/2020	Human Resources	Active
16	12	Lopez	Maya	4/10/2020	Accounting	Active
20	14	Patel	Priya	1/6/2020	Accounting	Active
22	16	Singh	Aditya	4/10/2020	Human Resources	Active
23	17	Smith	Carolyn	1/6/2020	Accounting	Active
24	18	Smith	John	7/20/2020	Human Resources	Active
26	20	Young	Alyssa	1/6/2020	Accounting	Active

Figure 7-113. *The NewHiresList_Table filtered to only show the 13 rows with Accounting or Human Resources employees*

Run Spreadsheet Macro

The **Run Spreadsheet Macro** activity allows you to run a selected macro in the referenced Excel file.

Configuration

This section provides instructions on how to configure a **Run Spreadsheet Macro** activity, shown in Figure 7-114.

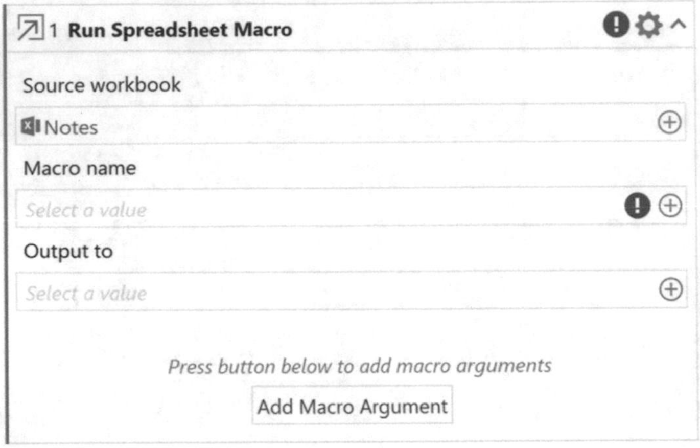

1 Run Spreadsheet Macro

Source workbook

Notes

Macro name

Select a value

Output to

Select a value

Press button below to add macro arguments

Add Macro Argument

Figure 7-114. *Run Spreadsheet Macro activity card*

Source workbook: This is a required configuration available on the activity card. This configuration identifies the workbook where the macro needs to be executed. By default, this field is pre-populated with the workbook from the parent `Use Excel File` activity.

Note The Run Macro activity must be executed in a Macro Enabled Excel file.

Macro name: This is a required configuration available on the activity card. This configuration provides the name of the macro that will be executed through this activity. You can use `Text Builder` to type in the Macro name manually.

Output to: This is an optional configuration available on the activity card. This configuration provides the location to save any values returned by the Macro if required.

Add Macro Argument: This is an optional configuration available on the activity card. This configuration allows you to pass an input `Argument value` when the `Run Macro` activity is executed. Multiple argument values can be added through this child activity if required for the Macro.

EXERCISE

Goal: Use the `Use Excel File` and the `Run Spreadsheet Macro` activity to run the `NewHiresListMacro` macro in the EmployeeOnboarding for Run Macro Exercise.xlsm file. Figure 7-115 shows the New Hires List sheet prior to execution.

ID	Last Name	First Name	Start Date	Department	Status
1	Brown	Sylvia	1/6/2020	Accounting	Active
2	Carter	Samantha	4/10/2020	Accounting	Active
3	Clark	Jacob	7/20/2020	Human Resources	Active
4	Davis	Samuel	1/6/2020	Operations	Active
5	Davis	Sasha	7/20/2020	Information Technology	Active
22	Evans	Chad	7/8/2020	Human Resources	Active
21	George	Lana	7/7/2020	Accounting	Active
6	Hill	Karen	1/6/2020	Operations	Active
7	Johnson	Elijah	7/20/2020	Information Technology	Active
8	Johnson	Adam	4/10/2020	Human Resources	Active
9	Jones	Daniel	7/20/2020	Information Technology	Active
10	Khan	Zain	1/6/2020	Operations	Active
11	Lane	Tamara	7/20/2020	Human Resources	Active
25	Lee	Aera	7/11/2020	Operations	Active
12	Lopez	Maya	4/10/2020	Accounting	Active
24	Martin	Frank	7/10/2020	Operations	Active
13	Miller	Raymond	7/20/2020	Information Technology	Active
23	Nguyen	Linda	7/9/2020	Information Technology	Active
14	Patel	Priya	1/6/2020	Accounting	Active
15	Sanchez	Gabriella	7/20/2020	Information Technology	Active
16	Singh	Aditya	4/10/2020	Human Resources	Active
17	Smith	Carolyn	1/6/2020	Accounting	Active
18	Smith	John	7/20/2020	Human Resources	Active
19	Williams	Jane	4/10/2020	Information Technology	Active
20	Young	Alyssa	1/6/2020	Accounting	Active

***Figure 7-115.** Displays the EmployeeOnboarding sheets before the automation is executed*

Source Code: Chapter_7_ExcelMacroActivitiesExercise

Setup: Here are step-by-step implementation instructions:

1. In StudioX, add the Use Excel File activity and configure with the C:\BookSamples\Chapter_07\EmployeeOnboarding Final.xlsm file and Reference as as EmployeeOnboardingMacro.
2. Then, drag the Run Spreadsheet Macro activity card in the body of Use Excel File activity.
3. The Source workbook field will be auto-populated with the EmployeeOnboardingMacro Excel file from the parent activity.
4. Next, click the Text option in the Macro name field, and type in NewHiresListMacro in the Text builder.

Once you are done, the final configuration for the **Run Macro** activity card is shown in Figure 7-116.

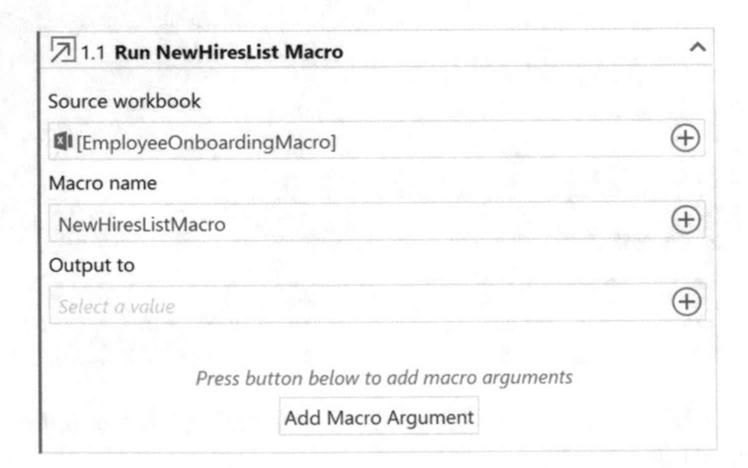

***Figure 7-116.** Displays the final configuration for the Run Macro activity*

Figure 7-117 shows the output of the Run Macro activity that has executed the NewHiresListMacro in the EmployeeOnboarding Excel file. As a result, the font has been changed to Times New Roman, the sort is by ID, and the New Hire Statistics sheet has moved right after the New Hires List sheet.

	A	B	C	D	E	F
1	ID	Last Name	First Name	Start Date	Department	Status
2	1	Brown	Sylvia	1/6/2020	Accounting	Active
3	2	Carter	Samantha	4/10/2020	Accounting	Active
4	3	Clark	Jacob	7/20/2020	Human Resources	Active
5	4	Davis	Samuel	1/6/2020	Operations	Active
6	5	Davis	Sasha	7/20/2020	Information Technology	Active
7	6	Hill	Karen	1/6/2020	Operations	Active
8	7	Johnson	Elijah	7/20/2020	Information Technology	Active
9	8	Johnson	Adam	4/10/2020	Human Resources	Active
10	9	Jones	Daniel	7/20/2020	Information Technology	Active
11	10	Khan	Zain	1/6/2020	Operations	Active
12	11	Lane	Tamara	7/20/2020	Human Resources	Active
13	12	Lopez	Maya	4/10/2020	Accounting	Active
14	13	Miller	Raymond	7/20/2020	Information Technology	Active
15	14	Patel	Priya	1/6/2020	Accounting	Active
16	15	Sanchez	Gabriella	7/20/2020	Information Technology	Active
17	16	Singh	Aditya	4/10/2020	Human Resources	Active
18	17	Smith	Carolyn	1/6/2020	Accounting	Active
19	18	Smith	John	7/20/2020	Human Resources	Active
20	19	Williams	Jane	4/10/2020	Information Technology	Active
21	20	Young	Alyssa	1/6/2020	Accounting	Active
22	21	George	Lana	7/7/2020	Accounting	Active
23	22	Evans	Chad	7/8/2020	Human Resources	Active
24	23	Nguyen	Linda	7/9/2020	Information Technology	Active
25	24	Martin	Frank	7/10/2020	Operations	Active
26	25	Lee	Aera	7/11/2020	Operations	Active
27						

New Hires List | New Hire Statistics | New Hire Checklist | New Hire Salaries

***Figure 7-117.** Displays the NewHiresListMacro executed*

CHAPTER 8

CSV Automation

CSV is a comma-separated value file that saves data in a single tabular format and allows organizations to manage and transfer large databases in a compressed format. Common use cases include storing large amounts of customer or product data for use across an organization, making CSV automation a critical capability for data-focused organizations. UiPath StudioX CSV automation allows citizen developers to take critical actions such as reading, appending, and writing to CSV files using the simple to use StudioX designer.

Learning Objectives

At the end of this chapter, you will learn how to

- Write data to a CSV file
- Append data to a CSV file
- Read a CSV file and use the data

Sample Overview

Throughout this chapter, we will be using Employee Data to showcase all the CSV automation activities. For this use case, the Human Resources (HR) team at a small firm retrieves data daily for new employees that have

A. Javed et al., *Robotic Process Automation using UiPath StudioX*,
https://doi.org/10.1007/978-1-4842-6794-3_8

recently joined the firm. This data is then appended and compiled into an Employee Summary CSV data file that allows the organization to maintain information for their employees and keep the data accurate and up to date for processing.

This section will familiarize you with the prerequisites for all exercises in this chapter.

Download the source code from the book's site, and make sure you move the entire BookSamples folder to your C:\ drive. All exercises in this chapter assume the folder paths will be **C:\BookSamples\Chapter_08**. Figure 8-1 shows the physical folder structure required for this sample. This folder structure comes with the source code.

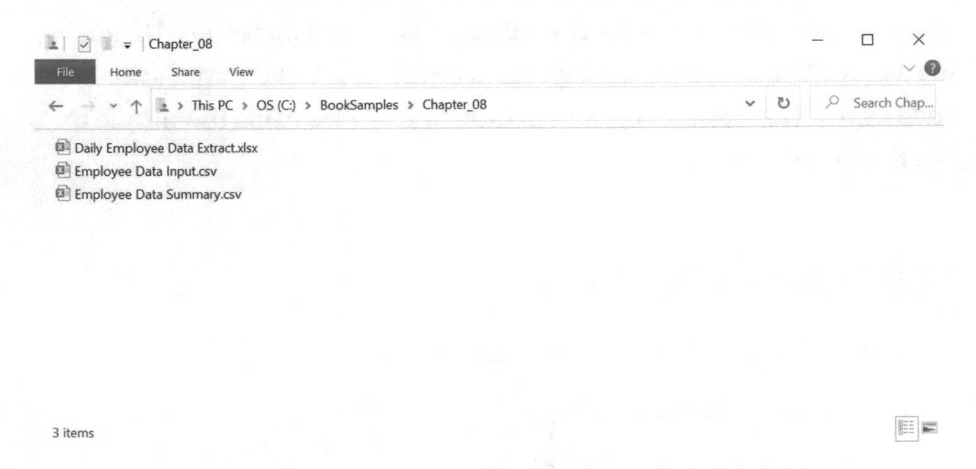

Figure 8-1. *Folder structure used for the CSV exercises*

Employee Data Summary.csv: The Summary CSV captures an aggregate of the Employee IDs, first names, and last names.

Employee Data Input.csv: The Input CSV captures new employees that need to get added to the summary.

Daily Employee Data Extract.xlsx: This Excel file is used as the source of Employee Data that is copied or appended to the CSV files.

Activities Reference

As shown in Figure 8-2, all CSV automation activities can be found under the CSV category. The following sections will provide instructions on how to configure and use each activity.

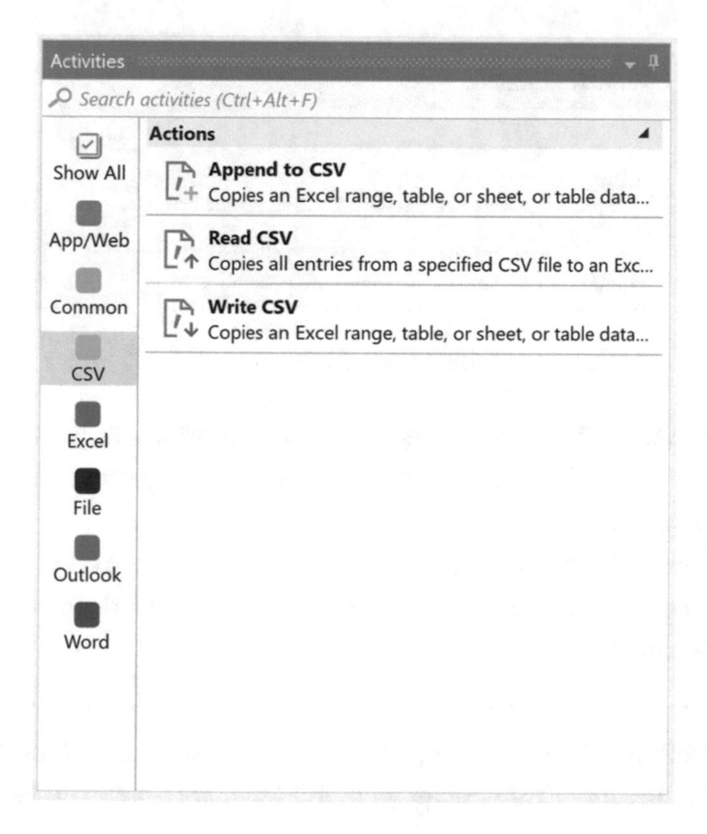

Figure 8-2. *Activities for CSV automation*

Write CSV

The **Write CSV** activity allows you to copy a range, sheet, or table from an Excel file and write or overwrite the data to a specified CSV.

Configuration

This section provides instructions on how to configure a **Write CSV** activity, shown in Figure 8-3.

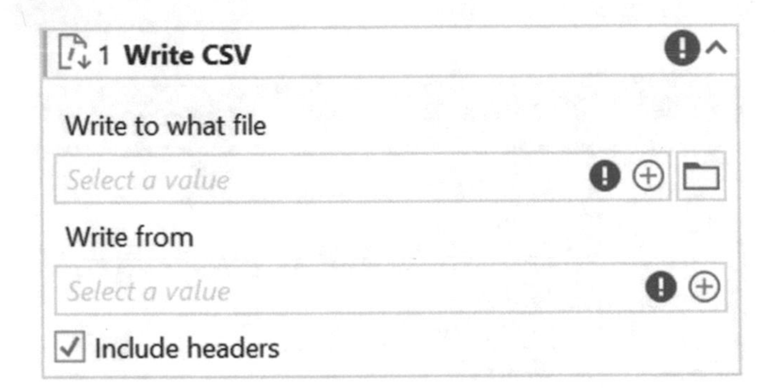

Figure 8-3. *Activity card for Write CSV*

Write to what file: This is a required configuration available on the activity card. This configuration identifies the full file path of the CSV file to read and copy over.

Write from: This is a required configuration available on the activity card. This configuration provides the DataTable that needs to be copied from the Excel file.

Tip If the DataTable is being copied from a parent Excel file to the CSV, add the Write CSV activity as a child activity under the Use Excel File activity card.

Include headers: This is an optional configuration available on the activity card. This configuration specifies that the first row in the source Excel file is a header row and whether it should be copied to the CSV or not. By default, this option is checked, that is, the header will be copied.

Delimiter: This is an optional configuration available on the Properties panel. This configuration specifies the delimiter used to separate the values in the CSV file. By default, the Delimiter value is set to Comma. It can be changed in the dropdown to Tab, Semicolon, Caret, or Pipe as shown in Figure 8-4.

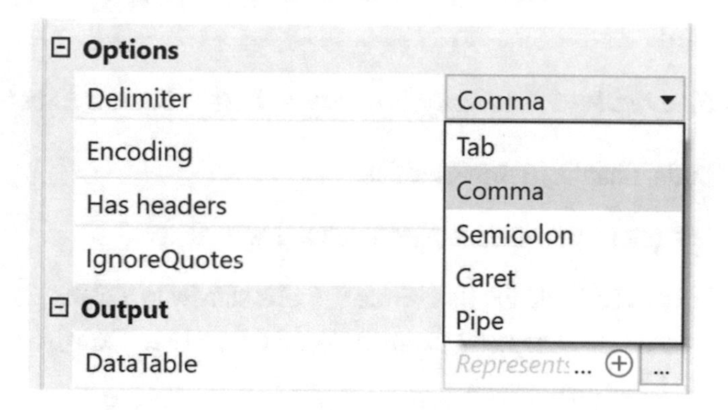

***Figure 8-4.** Delimiter field dropdown values*

Encoding: This is an optional configuration available on the Properties panel. This configuration allows you to specify the Encoding[1,2] in case the content contains any characters other than standard English.

EXERCISE

Goal: Use the Write CSV activity to write the Employee ID, First Name, and Last Name from the Daily Employee Data Extract.xlsx file to the Employee Data Input.csv file. Figure 8-5 displays the source data in the Excel file that needs to be written to CSV.

[1] Learn more about character encoding from W3C website – www.w3.org/International/questions/qa-what-is-encoding

[2] Complete list of encoding types supported by UiPath – https://docs.uipath.com/activities/docs/supported-character-encoding

	A	B	C	D	E	F
1	ID	Last Name	First Name	Start Date	Function	
2	21	George	Lana	7/7/2020	Accounting	
3	22	Evans	Chad	7/8/2020	Human Resources	
4	23	Nguyen	Linda	7/9/2020	Information Technology	
5	24	Martin	Frank	7/10/2020	Operations	
6	25	Lee	Aera	7/11/2020	Operations	
7						

***Figure 8-5.** Displays the Daily Employee Data Extract Excel*

Source Code: Chapter_8_CSVActivities

Setup: Here are step-by-step implementation instructions:

1. In `StudioX`, add the `Use Excel File` activity to a blank process (see Chapter 7 for instructions on the `Use Excel File` activity).

2. Next, click the `Folder` icon in the `Excel file` field of the `Use Excel File` activity and navigate through the file explorer to select `C:\BookSamples\Chapter_08\Daily Employee Data Extract.xlsx`.

3. Next, in the `Reference as` field, type in `EmployeeExcelData`.

4. Next, add a `Write CSV` activity to the body of `Use Excel File` activity.

5. Next, click the `Folder` icon in the `Write to what file` field on the `Write CSV` activity. Navigate through the file explorer to select `C:\BookSamples\Chapter_08\Employee Data Input.csv`.

6. Next, click the `Plus` icon in the `Write from` field and hover over `EmployeeExcelData` to select `Indicate in Excel`. Once Excel is open, select the range `A1:C6`.

7. Uncheck the `Include headers` option.

Once you have completed the exercise, the final configuration of the **Write CSV** activity should resemble Figure 8-6.

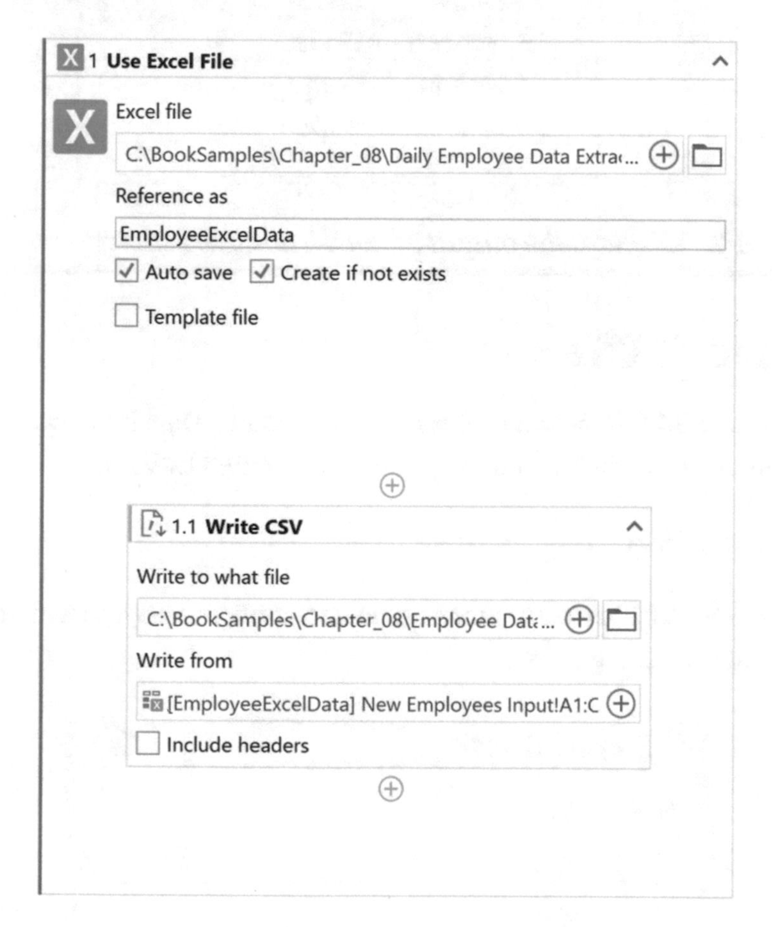

***Figure 8-6.** Displays the final configuration for the Write CSV activity*

Figure 8-7 shows the output of the **Write CSV** activity that has written the Excel data range to the Employee Data Input CSV.

	A	B	C	D
1	ID	Last Name	First Name	
2	21	George	Lana	
3	22	Evans	Chad	
4	23	Nguyen	Linda	
5	24	Martin	Frank	
6	25	Lee	Aera	
7				

***Figure 8-7.** Displays the output of the Write CSV activity*

Append To CSV

The **Append To CSV** activity allows you to append a DataTable from a range, sheet, or table from an Excel file to a specified CSV file.

Configuration

This section provides instructions on how to configure an **Append To CSV** activity, shown in Figure 8-8.

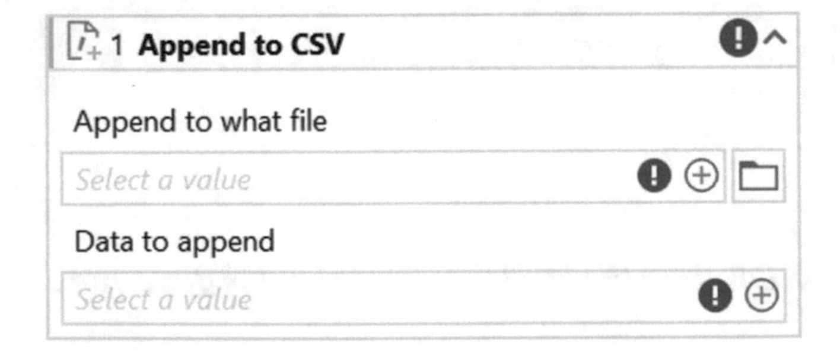

***Figure 8-8.** Activity card for Append To CSV*

Append to what file: This is a required configuration available on the activity card. This configuration identifies the full file path of the CSV file to read and copy over.

Data to append: This is a required configuration available on the activity card. This configuration provides the DataTable that needs to be appended to the CSV file.

Tip If the DataTable that needs to be appended is from a parent Excel file to the CSV, add the Append To CSV activity as a child activity under the Use Excel File activity card.

Delimiter: This is an optional configuration available on the Properties panel. This configuration specifies the delimiter used to separate the values in the CSV file. By default, the Delimiter value is set to Comma. It can be changed in the dropdown to Tab, Semicolon, Caret, or Pipe.

Encoding: This is an optional configuration available on the Properties panel. This configuration allows you to specify the Encoding[3,4] in case the content contains any characters other than standard English.

EXERCISE

Goal: Use the Append To CSV activity to append the Employee ID, First Name, and Last Name from the Daily Employee Data Extract.xlsx file to the Employee Data Summary.csv file. Figure 8-9 displays the Employee Data Summary CSV file before this exercise.

[3]Learn more about character encoding from W3C website – www.w3.org/International/questions/qa-what-is-encoding

[4]Complete list of encoding types supported by UiPath – https://docs.uipath.com/activities/docs/supported-character-encoding

	A	B	C	D
1	ID	Last Name	First Name	
2	1	Brown	Sylvia	
3	2	Carter	Samantha	
4	3	Clark	Jacob	
5	4	Davis	Samuel	
6	5	Davis	Sasha	
7	6	Hill	Karen	
8	7	Johnson	Elijah	
9	8	Johnson	Adam	
10	9	Jones	Daniel	
11	10	Khan	Zain	
12	11	Lane	Tamara	
13	12	Lopez	Maya	
14	13	Miller	Raymond	
15	14	Patel	Priya	
16	15	Sanchez	Gabriella	
17	16	Singh	Aditya	
18	17	Smith	Carolyn	
19	18	Smith	John	
20	19	Williams	Jane	
21	20	Young	Alyssa	
22				

Figure 8-9. *Displays the Employee Data Summary CSV before the Append To CSV activity*

Source Code: Chapter_8_CSVActivities

Setup: Here are step-by-step implementation instructions:

1. In `StudioX`, add the `Append To CSV` activity to the `Designer` panel, nested under the `Use Excel File`, after the `Write CSV` activity referenced in the previous exercise.

2. Next, click the `Folder` icon in the `Append to what file` field. Navigate through the file explorer to select `C:\BookSamples\Chapter_08\Employee Data Summary.csv`.

3. Next, click the `Plus` icon in the `Data to append` field and hover over `EmployeeExcelData` to select `Indicate in Excel`. Once Excel is open, select the range `A2:C6`.

Once you have completed the exercise, the final configuration of the **Append To CSV** activity should resemble Figure 8-10.

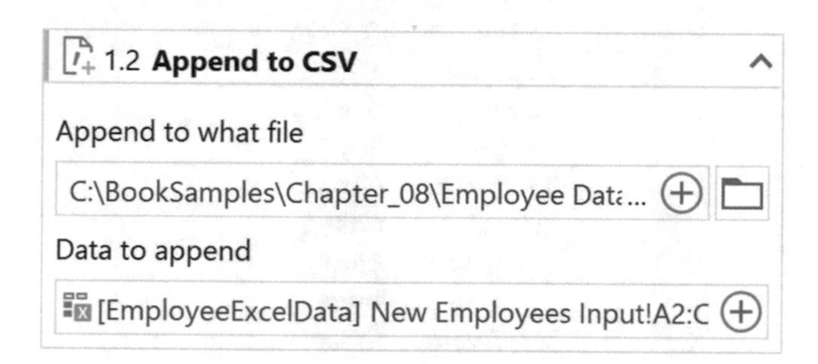

Figure 8-10. *Displays the final configuration for the Append To CSV activity*

Figure 8-11 shows the output of the **Append To CSV** activity that has appended the data from the Input file to the Summary file.

	A	B	C	D
1	ID	Last Name	First Name	
2	1	Brown	Sylvia	
3	2	Carter	Samantha	
4	3	Clark	Jacob	
5	4	Davis	Samuel	
6	5	Davis	Sasha	
7	6	Hill	Karen	
8	7	Johnson	Elijah	
9	8	Johnson	Adam	
10	9	Jones	Daniel	
11	10	Khan	Zain	
12	11	Lane	Tamara	
13	12	Lopez	Maya	
14	13	Miller	Raymond	
15	14	Patel	Priya	
16	15	Sanchez	Gabriella	
17	16	Singh	Aditya	
18	17	Smith	Carolyn	
19	18	Smith	John	
20	19	Williams	Jane	
21	20	Young	Alyssa	
22	21	George	Lana	
23	22	Evans	Chad	
24	23	Nguyen	Linda	
25	24	Martin	Frank	
26	25	Lee	Aera	
27				

Figure 8-11. *Output of Append To CSV activity*

Read CSV

The **Read CSV** activity allows you to read and copy data from a CSV file and write it to an Excel range, sheet, or table or save the value to the clipboard for later use.

Configuration

This section provides instructions on how to configure a **Read CSV** activity, shown in Figure 8-12.

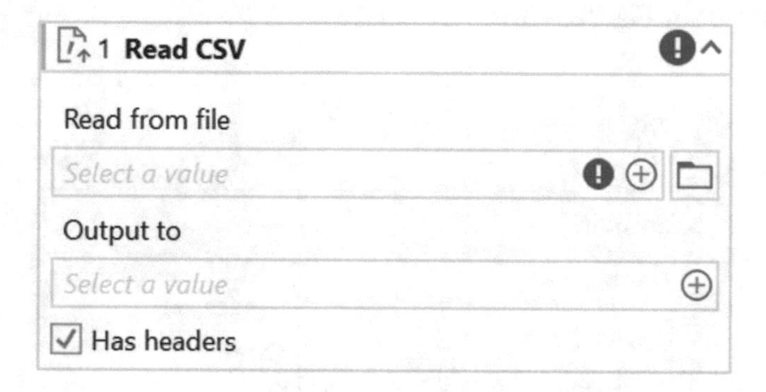

Figure 8-12. *Activity card for Read CSV*

Read from file: This is a required configuration available on the activity card. This configuration identifies the full file path of the CSV file to read.

Output to: This is a required configuration available on the activity card. This configuration provides the output data table location where the CSV data should be copied.

Tip If the CSV data should be copied to a parent Excel file, add the Read CSV activity as a child activity under the Use Excel File activity card.

Has headers: This is an optional configuration available on the activity card. This configuration is selected by default, meaning the first row in the CSV file is a header row. If unselected, the header row will not be read.

Delimiter: This is an optional configuration available on the `Properties` panel. This configuration specifies the delimiter used to separate the values in the CSV file. By default, the Delimiter value is set to `Comma`. It can be changed in the dropdown to `Tab`, `Semicolon`, `Caret`, or `Pipe` as shown in Figure 8-13.

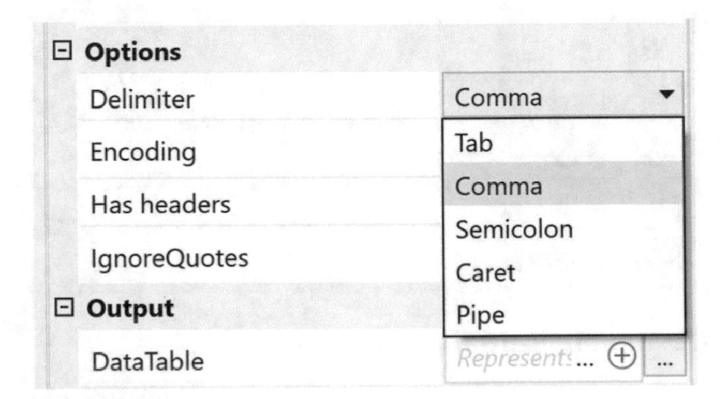

***Figure 8-13.** Delimiter field dropdown values*

Encoding: This is an optional configuration available on the `Properties` panel. This configuration allows you to specify the `Encoding`[5,6] in case the content contains any characters other than standard English.

IgnoreQuotes: This is an optional configuration available on the `Properties` panel. This configuration specifies if quotes should be ignored while reading the CSV file. By default, this field is blank, and if selected, it means that quotes are also copied over by the activity.

---

[5]Learn more about character encoding from W3C website – `www.w3.org/International/questions/qa-what-is-encoding`

[6]Complete list of encoding types supported by UiPath – `https://docs.uipath.com/activities/docs/supported-character-encoding`

EXERCISE

Goal: Use the Read CSV activity to read the Employee Data Summary.csv file and print the contents in the Output panel.

Source Code: Chapter_8_CSVActivities

Setup: Here are step-by-step implementation instructions:

1. In StudioX, add the Read CSV activity to the Designer panel after the Use Excel File activity.
2. Next, click the Folder icon in the Read from file field. Navigate through the file explorer C:\BookSamples\Chapter_08\Employee Data Summary.csv.
3. Next, click the Plus icon in the Output to field and select the Clipboard option.
4. Leave the Has headers option checked.
5. Then, add a Write Line activity under the Read CSV activity card.
6. Next, click the Plus icon in the Text field, and select the Clipboard option.

Once you have completed the exercise, the final configuration of the **Read CSV** activity should resemble Figure 8-14.

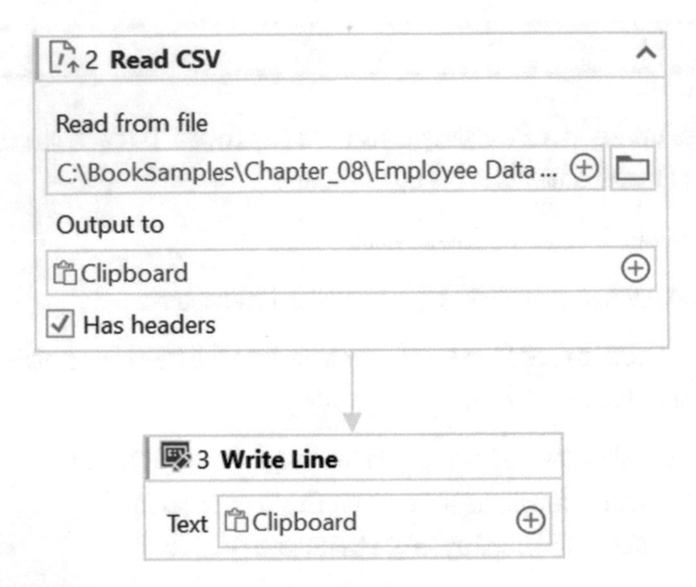

***Figure 8-14.** Displays the final configuration for the Read CSV activity*

Figure 8-15 shows the output of the Read CSV activity that has added the 25 rows of Employee Data to the Output panel.

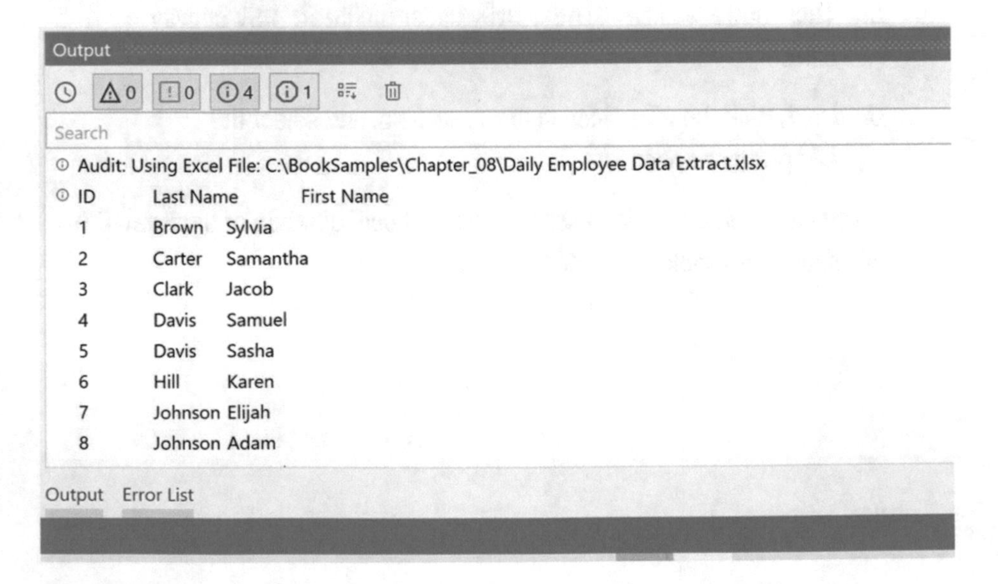

***Figure 8-15.** Displays the Employee Data from Read CSV saved*

CHAPTER 9

File Automation

Managing data, whether physical or digital, is a critical part of our daily jobs. An effective system to manage files and folders is becoming more important than ever with the growing amount of data that is available in today's working environment. We use folders to help us organize information so that it can be easily searched and retrieved later. Folders can also be used as a to-do list, that is, storing files in folders based on their status. The use of files is also quite common in our day-to-day jobs, for example, exporting data from systems in Excel/CSV, moving data between systems, and Word or Excel file templates to generate orders and invoices. As a result, UiPath StudioX File automation capabilities that organize, manage, and edit files and folders become a useful and essential tool. Folders help us organize information so that it can be easily searched and retrieved later. Folders can also be used as a to-do list, that is, storing files in folders based on their status.

Learning Objectives

At the end of this chapter, you will learn how to

- Get properties of a folder
- Check if a folder already exists
- Create, delete, copy, and move folders

A. Javed et al., *Robotic Process Automation using UiPath StudioX*,
https://doi.org/10.1007/978-1-4842-6794-3_9

- Fetch all files in a folder
- Get properties of a file
- Check if a file already exists
- Create, delete, copy, and move files
- Extract contents of a compressed file
- Create a compressed file
- Read contents of a text file
- Write and append content to a text file

Sample Overview

Throughout this chapter, we will be using a simplified version of a recruitment process to showcase all File automation activities.

This section will familiarize you with the prerequisites for all exercises in this chapter.

Download the source code from the book's site, and make sure you move the entire BookSamples folder to your C:\ drive. All exercises in this chapter assume the folder paths will be **C:\BookSamples\Chapter_09**. Figure 9-1 shows the folder structure required for this sample. This folder structure comes with the source code.

Note To run the exercises multiple times, make sure that the required files are available; otherwise, UiPath StudioX will throw an error.

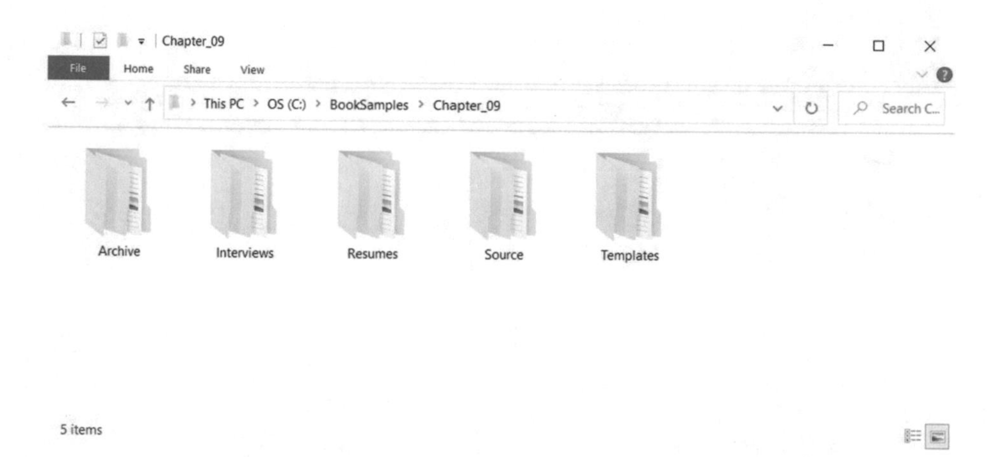

Figure 9-1. *Folder structure of sample recruitment management process*

Archive: This folder is used for archiving old folders and files. One sample subfolder already exists for a past date that is used in an exercise.

Interviews: This folder is used for keeping track of interviews. This folder contains two subfolders, Pending and Scheduled. At the end of each day, resumes from the MoveForward folder are moved to the Interviews\Pending folder. Once interviews are scheduled, resumes are moved to the Interviews\Scheduled folder.

Resumes: This folder is used for keeping track of resumes received every day. As shown in Figure 9-2, a new folder is created for each day. Three sample subfolders already exist for past dates that are used in a few exercises.

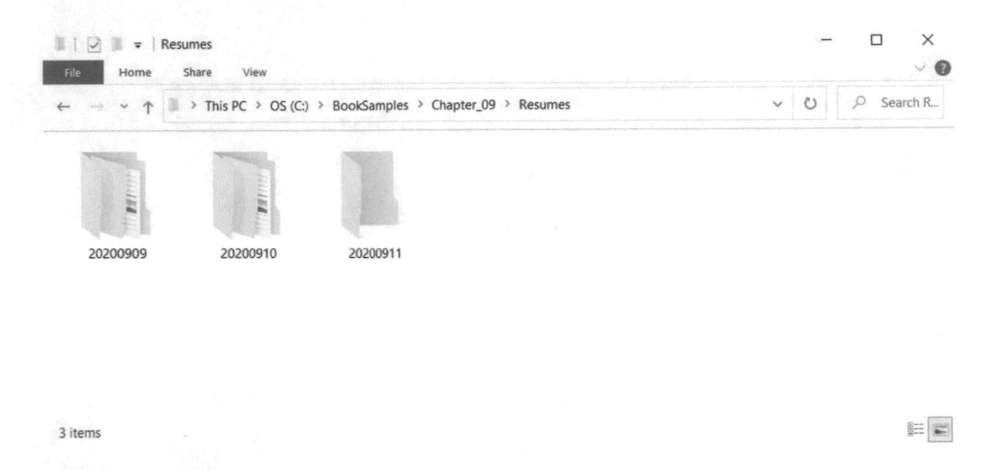

Figure 9-2. *Contents of Resumes folder*

Each daily folder contains three subfolders, New, MoveForward, and Reject. Whenever a new daily folder is created, these subfolders are copied from the Templates folder. All resumes are added to the New subfolder and, after review, a resume is either moved to MoveForward or Reject subfolder.

Source: This folder, as shown in Figure 9-3, is used for receiving compressed files from external staffing agencies. These compressed files contain candidate resumes. By default, this folder contains three sample extract files used in a few exercises.

Figure 9-3. *Contents of Source folder*

Templates: This is a static folder. As shown in Figure 9-4, it contains templates of folders and files that are copied to other folders during the process. This helps avoid creating the same folder structures repeatedly.

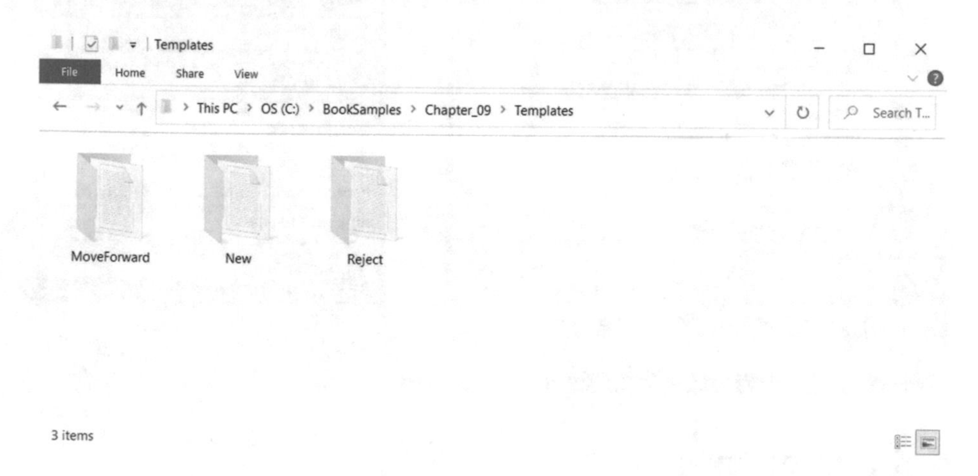

Figure 9-4. *Contents of Templates folder*

Activities Reference

As shown in Figure 9-5, all folder automation activities can be found under the File category. The following sections will provide instructions on how to configure and use each activity.

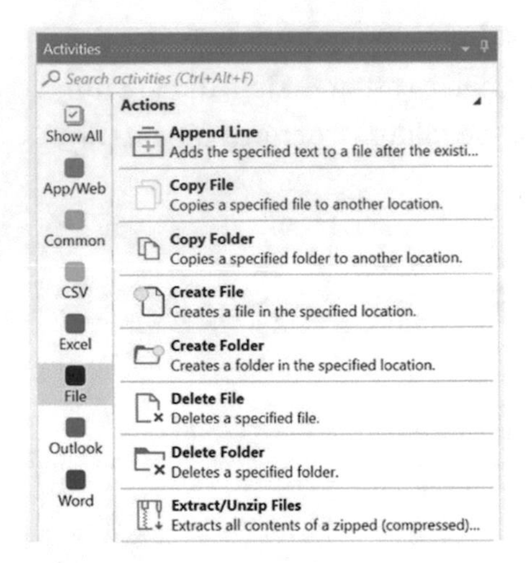

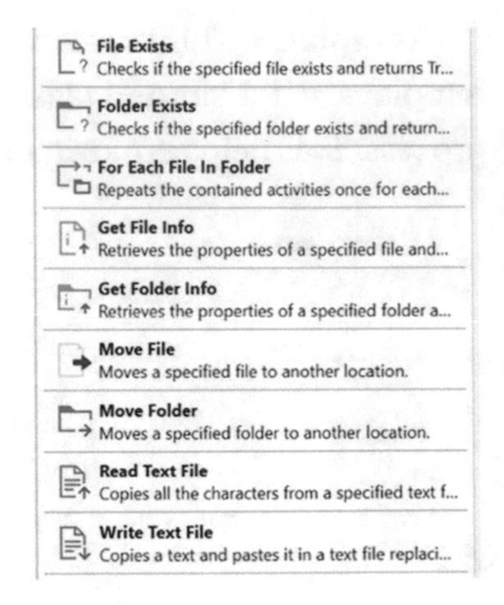

Figure 9-5. *Activities for File automation*

Get Folder Info

The **Get Folder Info** activity allows you to retrieve and save properties of a specified folder.

Configuration

This section provides instructions on how to configure a **Get Folder Info** activity, shown in Figure 9-6.

Figure 9-6. *Activity card for Get Folder Info*

Folder path: This is a required configuration available on the activity card. This configuration allows you to specify the full path of the folder for which you want to retrieve information.

Output to: This is an optional configuration available on the activity card. This configuration allows you to save folder information for later use, shown in Figure 9-7.

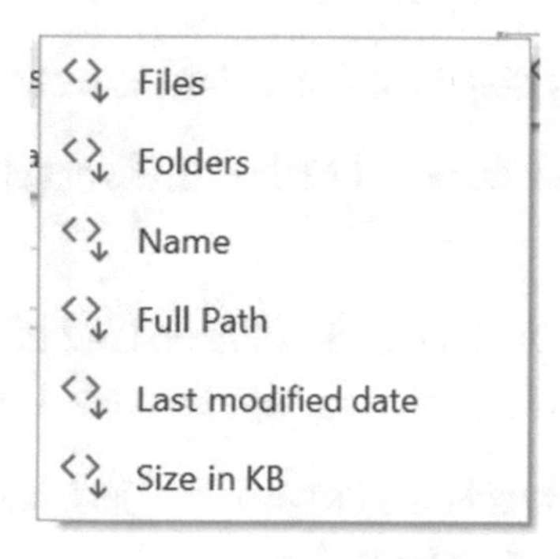

Figure 9-7. *Properties returned by Get Folder Info activity*

Table 9-1 provides a brief description of all properties returned by this activity.

Table 9-1. *Properties returned by the Get Folder Info activity*

Property	Description
Files	Number of files in all subfolders
Folders	Number of subfolders
Name	Name of the specified folder
Full path	The full path of the specified folder
Last modified date	Data and time when the specified folder was last modified
Size in KB	Size of the specified folder in kilobytes (KB)

EXERCISE

Goal: Use the Get Folder Info activity to get properties of C:\BookSamples\Chapter_09\Resumes\20200910 folder, and print the last modified date.

Source Code: Chapter_9_FolderActivitiesExercise

Setup: Here are step-by-step implementation instructions:

1. In StudioX, add the Get Folder Info activity to a blank process.
2. In the Folder path field, click the Folder icon and select C:\BookSamples\Chapter_09\Resumes\20200910 folder.
3. Click the Plus icon in the Output to field, name the value as FolderInfo, and click Ok.
4. Next, add a Write Line activity in the Designer panel after the Get Folder Info activity.
5. In the Text field, click the Plus icon, and select the Text option.
6. In the Text Builder, type Folder.
7. While the Text Builder is still open, click the Plus icon, select Use Saved Value option, and hover over FolderInfo to select Name.
8. Next, in the Text Builder, type was last modified on:.
9. Next, in the Text Builder, click the Plus icon, select Use Saved Value option, and hover over FolderInfo to select LastModifiedDate. The complete text should read as Folder FolderInfo ➤ Name was last modified on: FolderInfo ä LastModifiedDate.
10. Click Save.

Once you have completed the exercise, the final configuration of the **Get Folder Info** activity should resemble Figure 9-8. Figure 9-9 shows the output of this exercise.

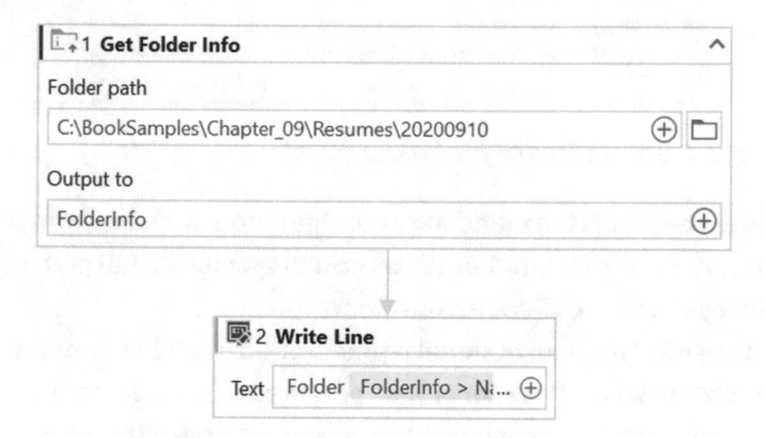

Figure 9-8. *Final configuration of Get Folder Info activity*

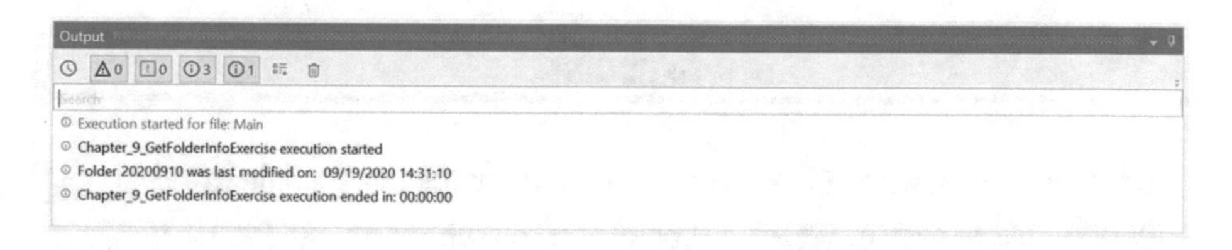

Figure 9-9. *Output of Get Folder Info activity*

Folder Exists

The **Folder Exists** activity allows you to check if a folder already exists in the target location.

Configuration

This section provides instructions on how to configure a **Folder Exists** activity, shown in Figure 9-10.

Figure 9-10. Activity card for Folder Exists

Folder path: This is a mandatory configuration available from the activity card. This configuration allows you to specify the full path of the folder that you want to check exists or not.

Save result: This is an optional configuration available from the activity card. This configuration allows you to specify where to save the result for later use. This activity returns a value of `True` if the folder exists and `False` if the folder does not exist.

EXERCISE

Goal: Use the `Folder Exists` activity to check if the current date folder (YYYYMMDD) already exists in `C:\BookSamples\Chapter_09\Resumes` location, and print the result.

Source Code: Chapter_9_FolderActivitiesExercise

Setup: Here are step-by-step implementation instructions:

1. In `StudioX`, add the `Folder Exists` activity to a blank process.
2. We are going to dynamically generate the path of the folder we want to check exists or not. In the `Folder path` field, click the `Plus` icon, select `Text` option, and type `C:\BookSamples\Chapter_09\Resumes\`.

3. Then click the Plus icon in Text Builder and hover over Project Notebook (Notes) to select [Notes] Date!YYYYMMDD. Figure 9-11 (on the next page) shows how the text should look. Click Save.

4. Click the Plus icon in the Save result field, name the value as FolderExists, and click Ok.

5. Add the Write Line activity to the Designer panel after the Folder Exists activity.

6. In the Text field, click the Plus icon and select the Text option. In the Text Builder, type Folder exists:.

7. While the Text Builder is still open, click the Plus icon and hover over Use Saved Value to select FolderExists. Click Save.

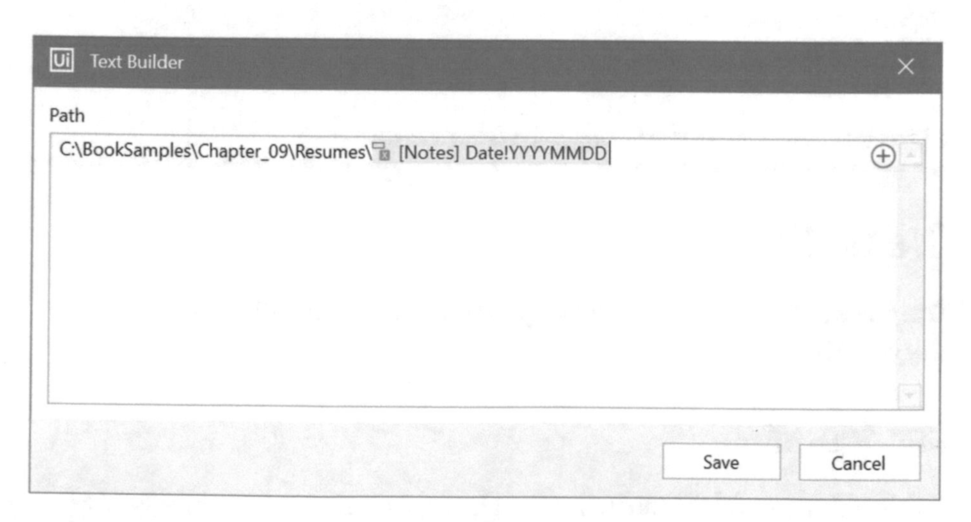

Figure 9-11. *Dynamic path of folder for Folder Exists activity*

Once you have completed the exercise, the final configuration of the **Folder Exists** activity should resemble Figure 9-12. Figure 9-13 shows the output of this exercise.

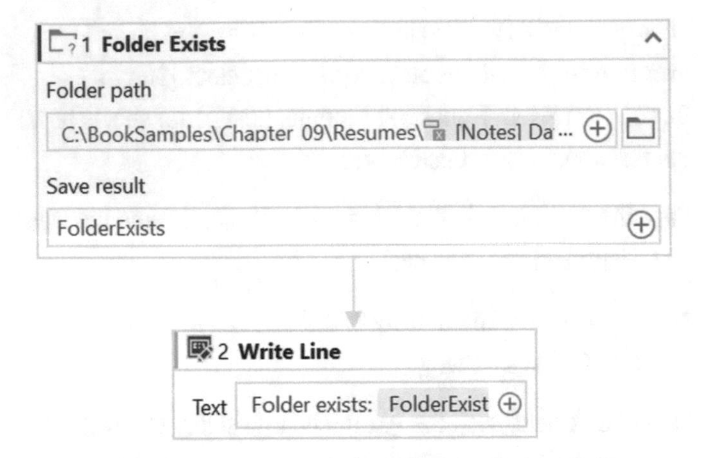

***Figure 9-12.** Final configuration of Folder Exists exercise*

***Figure 9-13.** Output of Folder Exists exercise*

Create Folder

The **Create Folder** activity allows you to create a new folder in the specified location.

Configuration

This section provides instructions on how to configure a **Create Folder** activity, shown in Figure 9-14.

***Figure 9-14.** Activity card for Create Folder*

Folder name: This is a required configuration available on the activity card. This configuration allows you to specify the full path of the folder that you want to create.

EXERCISE

Goal: Use the `Create Folder` activity to create a new folder for the current date (YYYYMMDD) in `C:\BookSamples\Chapter_09\Resumes` location.

Source Code: Chapter_9_FolderActivitiesExercise

Setup: Here are step-by-step implementation instructions:

1. In `StudioX`, add the `Create Folder` activity to a blank process.
2. We are going to generate the path of the folder dynamically. In the `Folder path` field, click the `Plus` icon, select `Text` option, and type `C:\BookSamples\Chapter_09\Resumes\`.
3. Next, click the `Plus` icon in the `Text Builder` and hover over `Project Notebook (Notes)` to select `[Notes] Date!`YYYYMMDD. Figure 9-15 shows the final text entered in the `Text Builder`. Click `Save`.

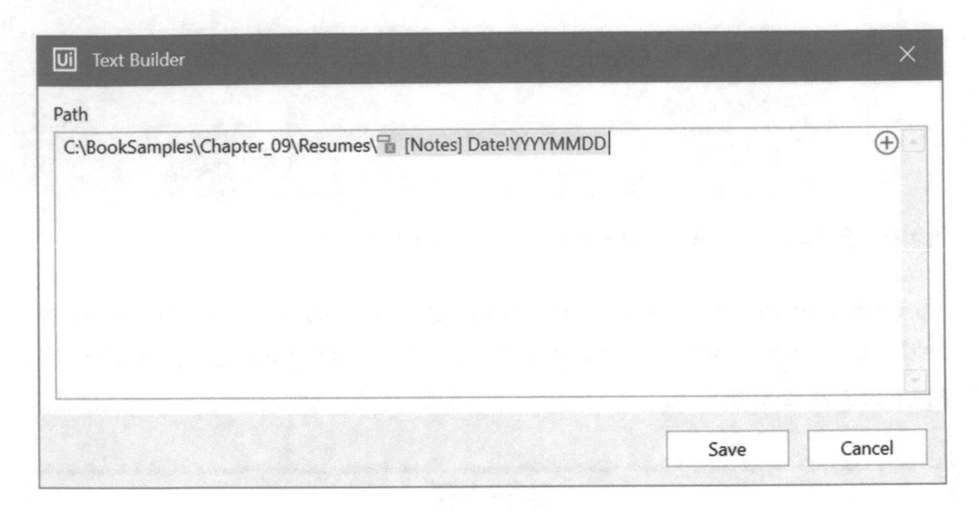

Figure 9-15. *Dynamic path of folder for Create Folder activity*

Once you have completed the exercise, the final configuration of the **Create Folder** activity should resemble Figure 9-16. Figure 9-17 shows the folder structure of the target location before and after this exercise.

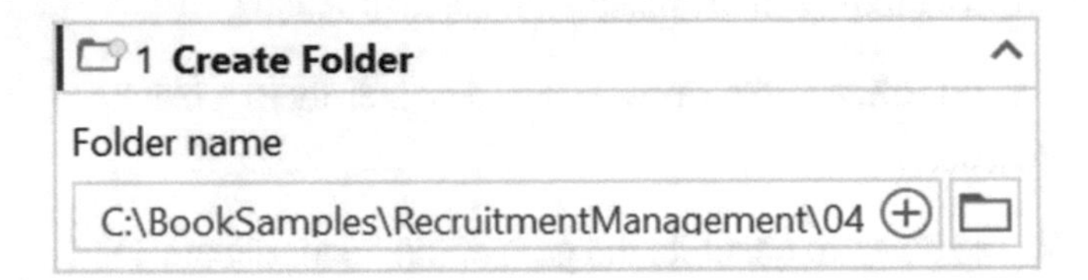

Figure 9-16. *Final configuration of Create Folder activity exercise*

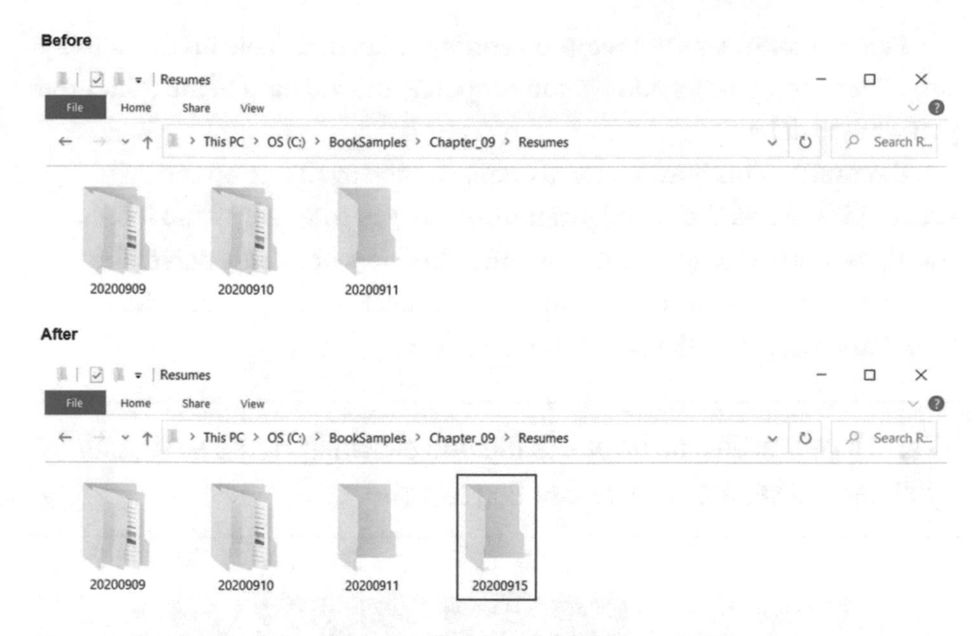

***Figure 9-17.** Result of Create Folder activity exercise*

Delete Folder

The **Delete Folder** activity allows you to delete a specified folder.

Configuration

This section provides instructions on how to configure a **Delete Folder** activity, shown in Figure 9-18.

***Figure 9-18.** Activity card for Delete Folder*

Folder name: This is a required configuration available on the activity card. This configuration allows you to specify the full path of the folder that you want to delete.

Recursive: This is an optional configuration available on the `Properties` panel. This configuration allows you to specify if the folder should be deleted with all content and subfolders or to only delete the folder if it is empty. By default, the value is checked, that is, the folder will be deleted with all of the contents inside.

Tip If you receive an error like the one shown in Figure 9-19, then you need to set the `Recursive` flag to `True`.

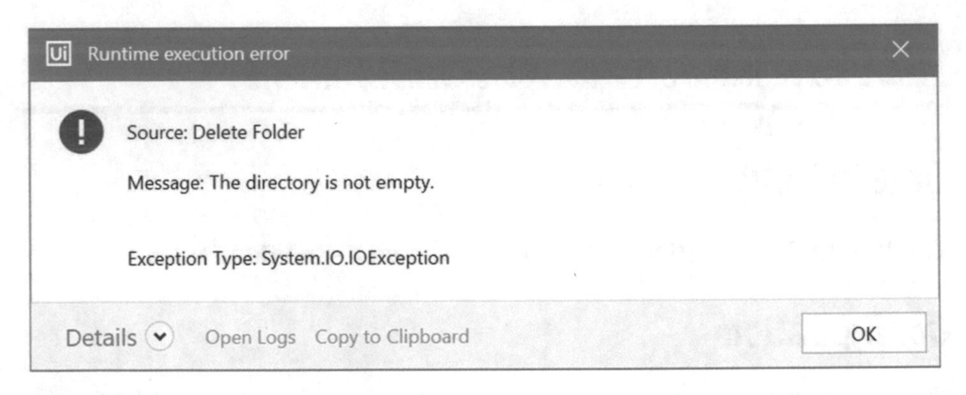

***Figure 9-19.** Error message while deleting a folder that has contents*

EXERCISE

Goal: Use the `Delete Folder` activity to delete `C:\BookSamples\Chapter_09\Archive\20200908` folder and all its contents.

Source Code: Chapter_9_FolderActivitiesExercise

Setup: Here are step-by-step implementation instructions:

1. In `StudioX`, add the `Delete Folder` activity to a blank process.
2. Click the `Folder` icon and select `C:\BookSamples\Chapter_09\Archive\20200908` folder.
3. Leave the `Recursive` property set to the default value of `True`. This will ensure all subfolders and files are also deleted.

Once you have completed the exercise, the final configuration of the **Delete Folder** activity should resemble Figure 9-20. Figure 9-21 shows the folder structure of the target location before and after this exercise.

Figure 9-20. *Final configuration of Delete Folder activity exercise*

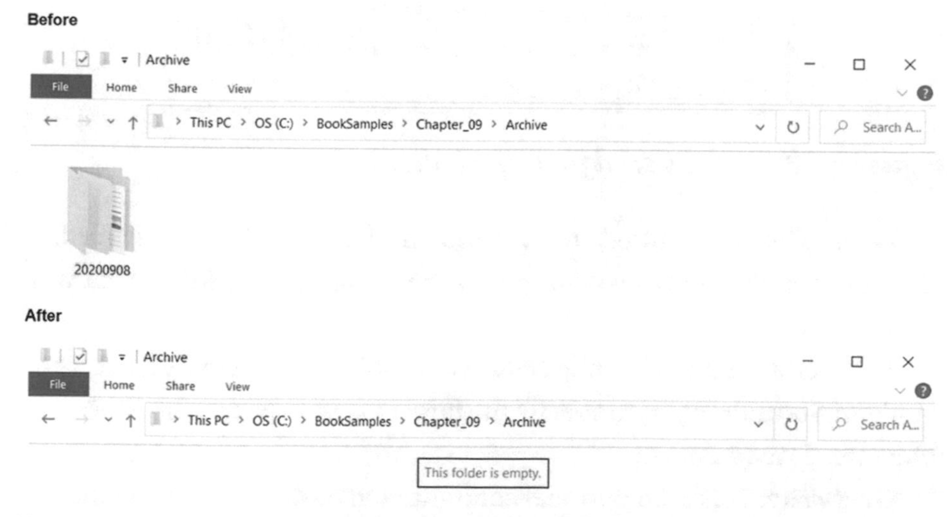

Figure 9-21. *Result of Delete Folder activity exercise*

Copy Folder

The **Copy Folder** activity allows you to copy a specified folder to another location.

Note This activity leaves the original folder(s) in place.

Configuration

This section provides instructions on how to configure a **Copy Folder** activity, shown in Figure 9-22.

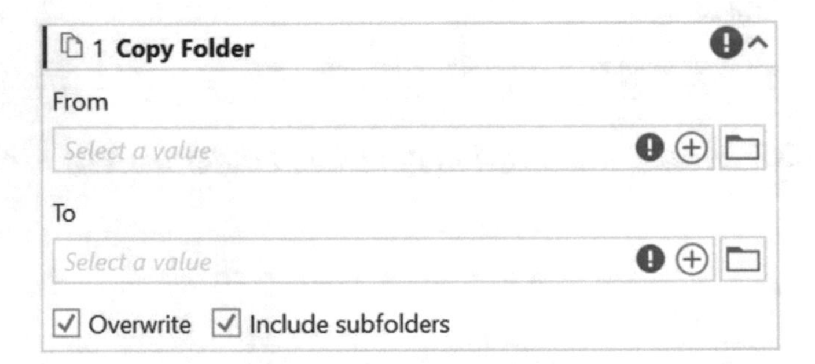

Figure 9-22. *Activity card for Copy Folder*

From: This is a required configuration available on the activity card. This configuration allows you to specify the full path of the folder that you want to copy.

To: This is a required configuration available on the activity card. This configuration allows you to specify the target location where the `From` folder needs to be copied.

Overwrite: This is an optional configuration available on the activity card. This configuration allows you to specify that in case a folder with the same name exists in the target location, should the automation overwrite the contents.

Include subfolders: This is an optional configuration available on the activity card. This configuration allows you to specify if all subfolders of the source folder need to be copied to the target location as well.

EXERCISE

Goal: Use the `Copy Folder` activity to copy all contents of `C:\BookSamples\Chapter_09\Templates` folder to `C:\BookSamples\Chapter_09\Resumes\20200911`. The target folder is empty before this exercise.

Source Code: Chapter_9_FolderActivitiesExercise

Setup: Here are step-by-step implementation instructions:

1. In `StudioX`, add the `Copy Folder` activity to a blank process.
2. In the `From` field, click the `Folder` icon and select `C:\BookSamples\Chapter_09\Templates\` folder.
3. In the `To` field, click the `Folder` icon and select `C:\BookSamples\Chapter_09\Resumes\20200911` folder.
4. The `From` folder in this exercise does not have any subfolders, so you can leave the `Overwrite` and `Include subfolders` checked as is.

Once you have completed the exercise, the final configuration of the **Copy Folder** activity should resemble Figure 9-23. Figure 9-24 shows the folder structure of the target location before and after this exercise.

1 Copy Folder

From

C:\BookSamples\Chapter_09\Templates\

To

C:\BookSamples\Chapter_09\Resumes\20200911

☑ Overwrite ☑ Include subfolders

***Figure 9-23.** Final configuration of Copy Folder activity exercise*

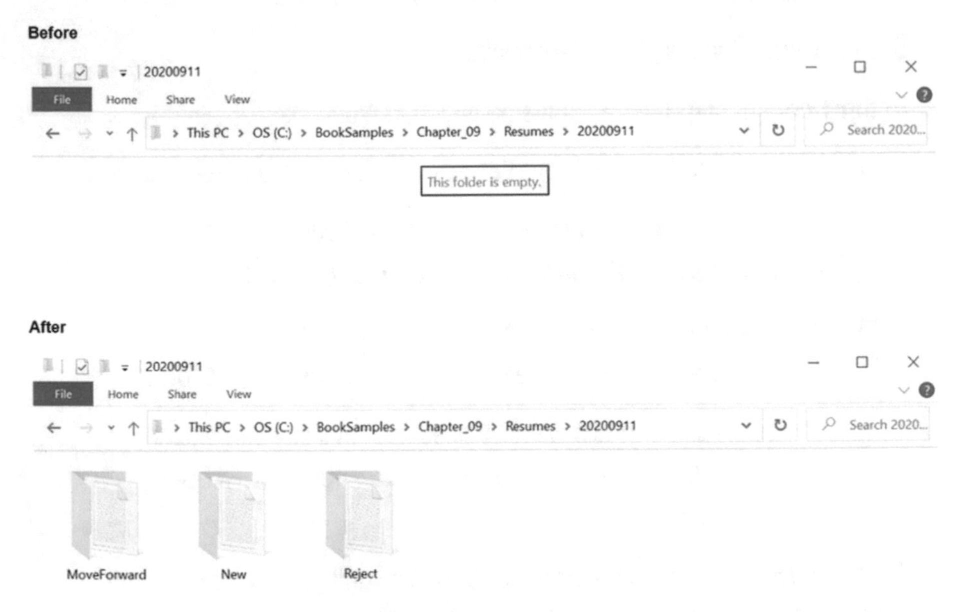

***Figure 9-24.** Result of Copy Folder activity exercise*

Move Folder

The **Move Folder** activity allows you to copy a specified folder to another location.

Note Unlike the Copy Folder activity that leaves the original folder as is, this activity moves the original folder.

Configuration

This section provides instructions on how to configure a **Move Folder** activity, shown in Figure 9-25.

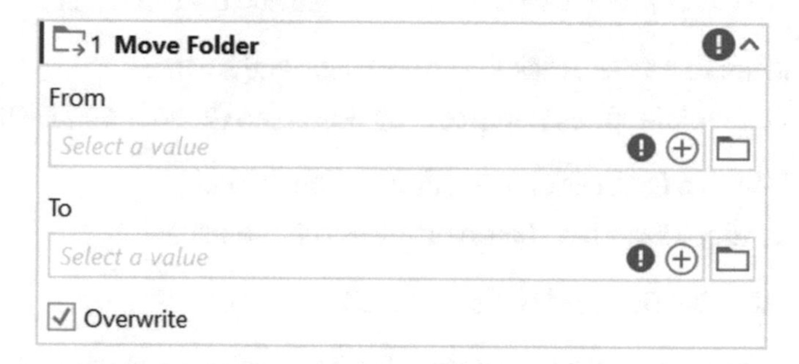

***Figure 9-25.** Activity card for Move Folder*

From: This is a required configuration available on the activity card. This configuration allows you to specify the full path of the folder that you want to move.

To: This is a required configuration available on the activity card. This configuration allows you to specify the target location where From folder needs to be moved.

Overwrite: This is an optional configuration available on the activity card. This configuration allows you to specify that in case a folder with the same name exists in the target location, should the automation overwrite the contents.

EXERCISE

Goal: Use the `Move Folder` activity to move all contents of `C:\BookSamples\Chapter_09\Resumes\20200909` folder to `C:\BookSamples\Chapter_09\Archive`.

Source Code: Chapter_9_FolderActivitiesExercise

Setup: Here are step-by-step implementation instructions:

1. In `StudioX`, add the `Move Folder` activity to a blank process.
2. In the `From` field, click the `Folder` icon and select `C:\BookSamples\Chapter_09\Resumes\20200909` folder.
3. In the `To` field, click the `Folder` icon and select `C:\BookSamples\Chapter_09\Archive` folder.
4. Leave the `Overwrite` flag checked.

Once you have completed the exercise, the final configuration of the **Move Folder** activity should resemble Figure 9-26. Figure 9-27 shows the folder structure of the target location before and after this exercise.

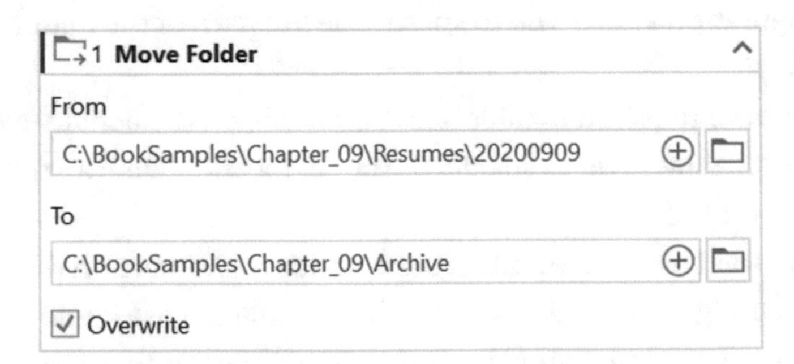

***Figure 9-26.** Final configuration of Move Folder activity exercise*

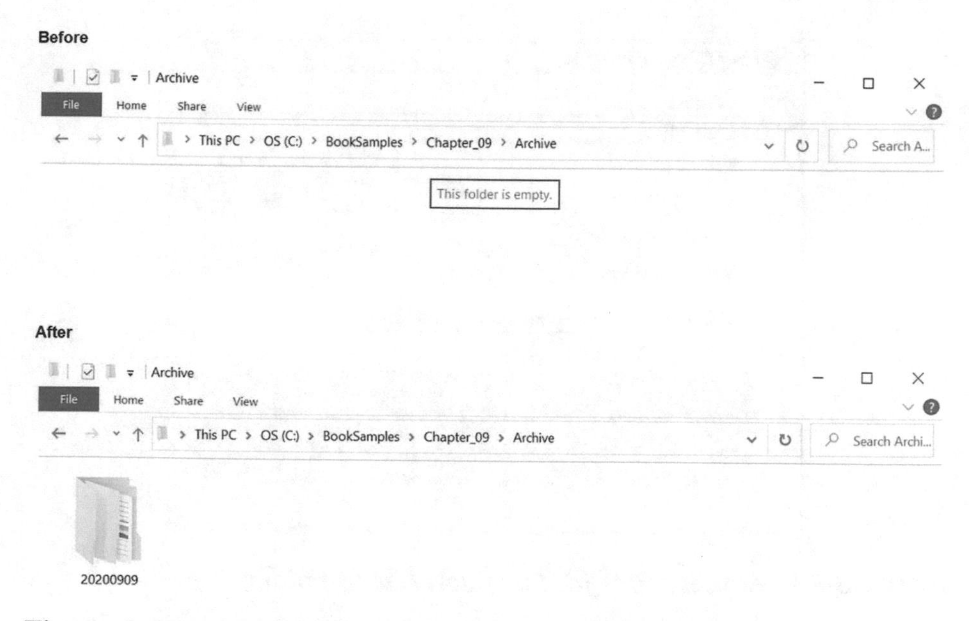

***Figure 9-27.** Result of the Move Folder activity exercise*

For Each File In Folder

The **For Each File In Folder** activity allows you to take actions on all files within a folder and its subfolders. You can add other activities in the body of **For Each File In Folder** activity, and all those activities will be executed for every file in the specified folder.

Configuration

This section provides instructions on how to configure a **For Each File In Folder** activity, shown in Figure 9-28.

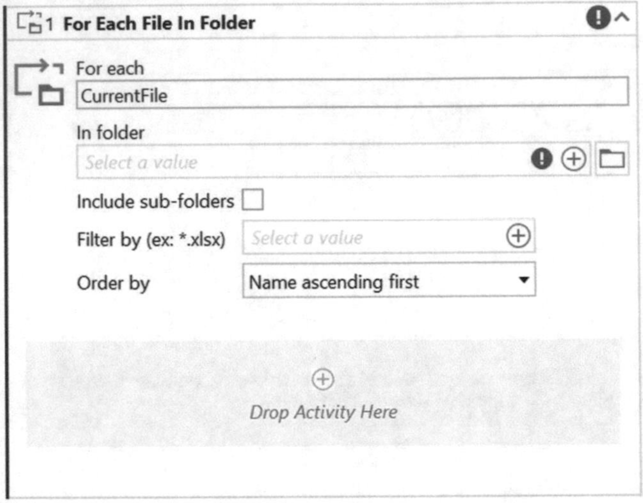

Figure 9-28. *Activity card for For Each File In Folder*

For each: This is a required configuration available on the activity card. This configuration allows you to specify how you want to reference the files in the nested activities. By default, the value is `CurrentFile`. Figure 9-29 shows how this is used by other activities in the body of `For Each File In Folder` activity.

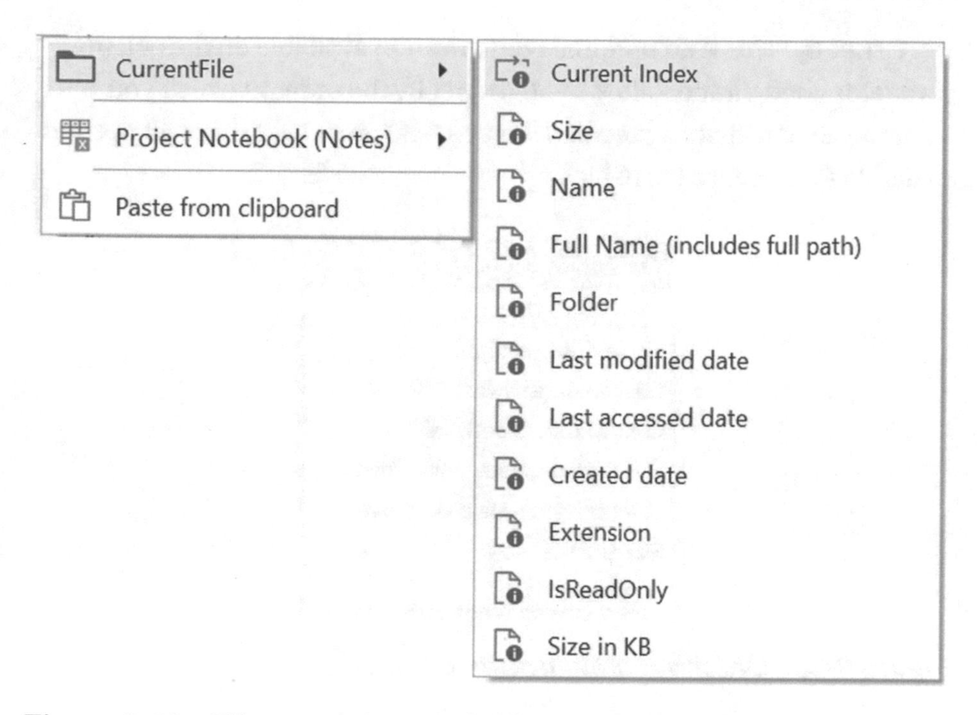

Figure 9-29. *File properties available to activities in body of For Each File In Folder activity*

In folder: This is a required configuration available on the activity card. This configuration allows you to specify the full path of the folder where the files you want to iterate over are located.

Include sub-folders: This is an optional configuration available on the activity card. This configuration allows you to specify if you want to execute the nested activities in the files from the subfolders as well. By default, this option is not selected.

Filter by: This is an optional configuration available on the activity card. This configuration allows you to pick specific types of files from the specified folder. For example, your folder might contain .txt, .docx, and .xlsx files, but you only want to look at .docx and ignore the rest of the files; you will use this property to specify *.docx.

Order by: This is an optional configuration available on the activity card. This configuration allows you to specify the order in which you want the child activities executed. Figure 9-30 provides a list of all options available for sorting a list of files.

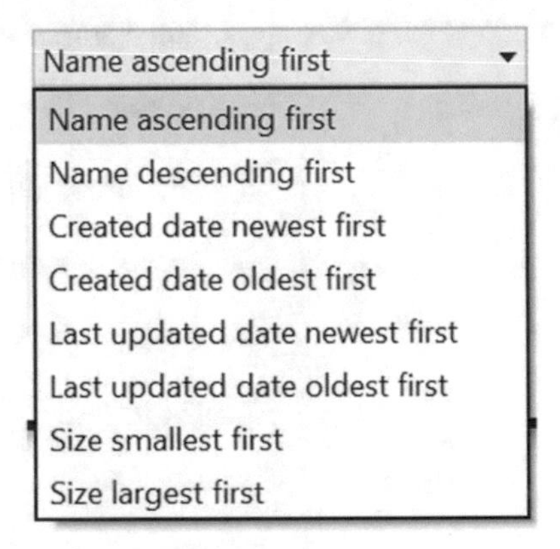

Figure 9-30. *Order by configuration options*

EXERCISE

Goal: Use the `For Each File In Folder` activity to print names of all Word documents (*.docx) in the `C:\BookSamples\Chapter_09\Resumes` folder (including subfolders).

Source Code: Chapter_9_FolderActivitiesExercise

Setup: Here are step-by-step implementation instructions:

1. In `StudioX`, add the `For Each File In Folder` activity to a blank process.
2. In the `In folder` field, click the `Folder` icon and select the `C:\BookSamples\Chapter_09\Resumes` folder.
3. Check the `Include sub-folders` flag.

4. In the `Filter by` field, click the `Plus` icon, select `Text`, and enter `*.docx`. This will ensure that only Word document files are returned.

5. Add the `Write Line` activity in the body of `For Each File In Folder` activity. In the `Text` field, click the `Plus` icon and hover over `Current File` to select `Full Name (includes full path)`.

Once you have completed the exercise, the final configuration of the **For Each File In Folder** activity should resemble Figure 9-31. Figure 9-32 shows the output of this exercise.

***Figure 9-31.** Final configuration of For Each File In Folder activity exercise*

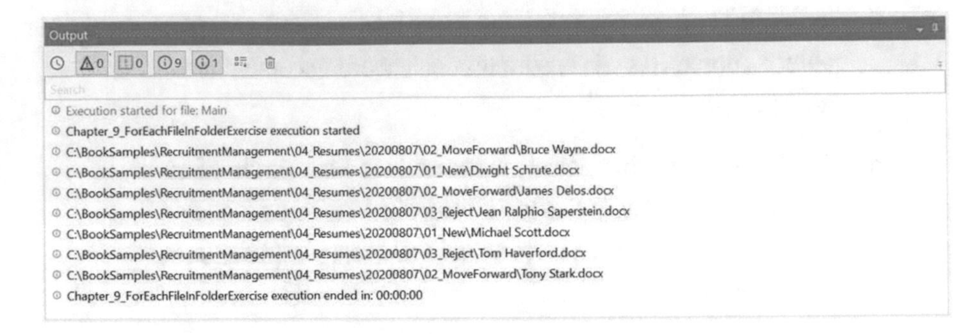

Figure 9-32. *Output of For Each File In Folder activity*

Compress/Zip Files

The **Compress/Zip Files** activity allows you to create a compressed file using multiple files or multiple folders.

Configuration

This section provides instructions on how to configure a **Compress/Zip Files** activity, shown in Figure 9-33.

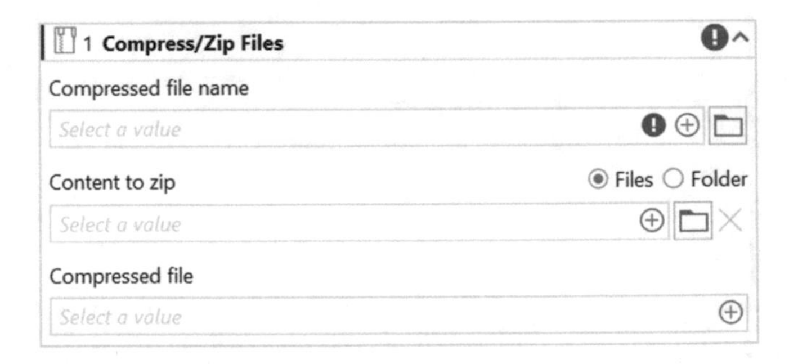

Figure 9-33. *Activity card for Compress/Zip Files*

Compressed file name: This is a required configuration available on the activity card. This configuration allows you to specify the name of the compressed file (.zip) that will be created.

Content to zip: This is an optional configuration available on the activity card. This configuration allows you to specify the target location where you want the file to be extracted.

Compressed file: This is an optional configuration available on the activity card. This configuration allows you to save information about the compressed file. This will create a variable that can be used later.

Compression level: This is an optional configuration available on the `Properties` panel. This configuration allows you to specify how small to compress the file. The more compression you want, the slower it will be. There are four options: None (default), `Fast`, `Normal`, and `Maximum`.

Overwrite existing file: This is an optional configuration available on the `Properties` panel. This configuration allows you to specify that in case a file with the same name exists in the target location, whether the automation should overwrite it. By default this is checked, that is, the automation should overwrite file with the same name.

Password: This is an optional configuration available on the `Properties` panel. This configuration allows you to set a password for the compressed file.

EXERCISE

Goal: Use the `Compress/Zip Files` activity to compress the contents of `C:\BookSamples\Chapter_09\Resumes\20200910\MoveForward` folder in `MoveForwardResumes_20200910.zip` file.

Source Code: Chapter_9_CompressedFilesExercise

Setup: Here are step-by-step implementation instructions:

1. In `StudioX`, add the `Compress/Zip Files` activity to a blank process.
2. In the `Compressed file name` field, click the Plus icon, select Text, and type `C:\BookSamples\Chapter_09\Resumes\20200910\MoveForward\MoveForwardResumes_20200910.zip`.

3. In the `Content to zip` field, select the `Folder` option. Click the `Folder` icon and select `C:\BookSamples\Chapter_09\Resumes\20200910\MoveForward` folder.

4. In the `Compressed file` field, click the `Plus` icon, select `Save for Later Use`, and name the value as `CompressedFileInfo`.

5. Add the `Write Line` activity right after the `Compress/Zip Files` activity. In the `Text` field, click the `Plus` icon, select `Use Saved Value`, and hover over `CompressedFileInfo` to select `Full Path`.

Once you have completed the exercise, the final configuration of the **Compress/Zip Files** activity should resemble Figure 9-34. Figure 9-35 shows the before and after folder structure of the target location.

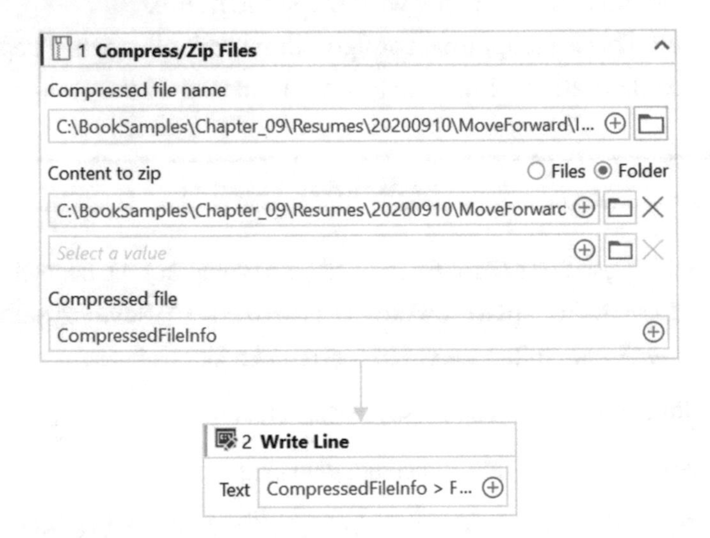

***Figure 9-34.** Final configuration of Compress/Zip Files activity*

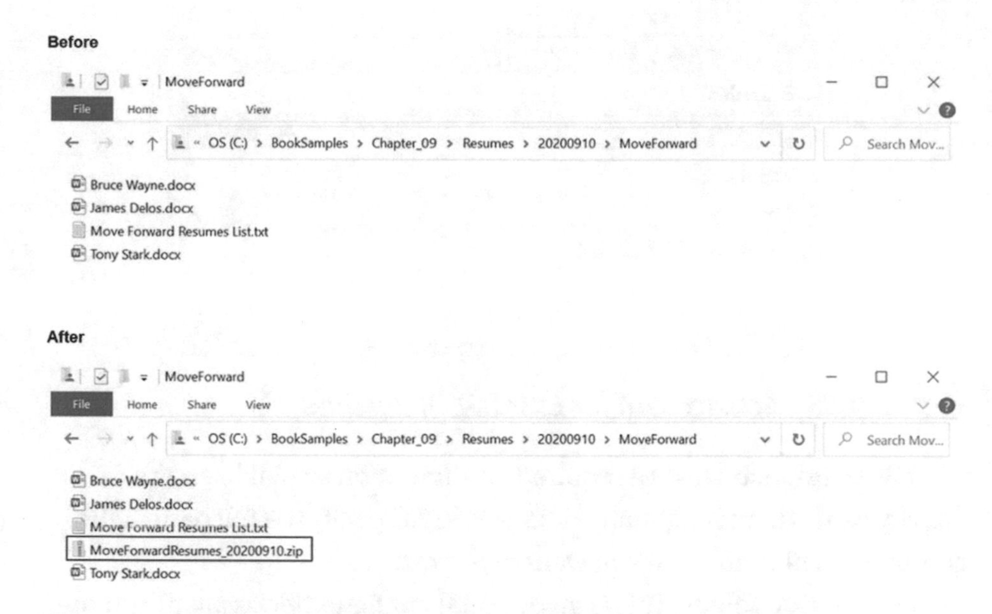

Figure 9-35. *Result of Compress/Zip Files activity exercise*

Extract/Unzip Files

The **Extract/Unzip Files** activity allows you to extract the contents of a compressed file to a specific folder.

Configuration

This section provides instructions on how to configure an **Extract/Unzip Files** activity, shown in Figure 9-36.

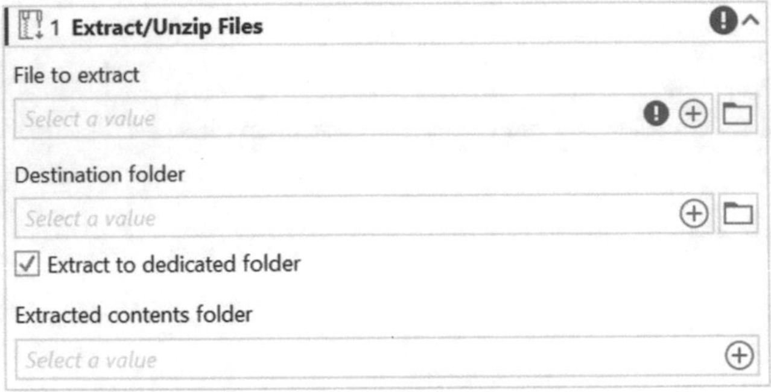

Figure 9-36. *Activity card for Extract/Unzip Files*

File to extract: This is a required configuration available on the activity card. This configuration allows you to specify the full path of the compressed file (.zip) that you want to extract.

Destination folder: This is an optional configuration available on the activity card. This configuration allows you to specify the target location where you want the file to be extracted.

Extract to dedicated folder: This is an optional configuration available on the activity card. This configuration allows you to specify if you want to extract files to a dedicated folder inside the `Destination folder`. The name of the new folder will be the same as the name of the compressed file. This can help ensure that files with the same name in the `Destination folder` are not overwritten.

Extracted contents folder: This is an optional configuration available on the activity card. This configuration allows you to save information about the extracted folder. If the `Extract to dedicated folder` flag was checked, then all properties of that folder, otherwise all properties of the `Destination folder` will be saved in `Data Manager`. By default, this is left unchecked.

Password: This is an optional configuration available on the `Properties` panel. This configuration allows you to specify a password in case the compressed file is password protected.

Skip unsupported files: This is an optional configuration available on the Properties panel. This configuration allows you to skip files that cannot be extracted. This helps avoid any errors that might be thrown if StudioX is unable to extract specified file. By default, this value is not set.

EXERCISE

Goal: Use the Extract/Unzip Files activity to extract the contents of C:\BookSamples\Chapter_09\Source\StaffingAgencyResumes_20200911.zip file to C:\BookSamples\Chapter_09\Resumes\20200911\New folder, and print information about the extracted folder.

Source Code: Chapter_9_CompressedFilesExercise

Setup: Here are step-by-step implementation instructions:

1. In StudioX, add the Extract/Unzip Files activity to a blank process.
2. In the Files to extract field, click the Folder icon and select C:\BookSamples\Chapter_09\Source\StaffingAgencyResumes_20200911.zip file.
3. In the Destination folder field, click the Folder icon and select C:\BookSamples\Chapter_09\Resumes\20200911\New folder.
4. Uncheck the Extract to dedicated folder flag as we want all files to be extracted in the specified folder.
5. In the Extracted contents folder field, click the Plus icon, select Save for Later Use, and name the value as DestinationFolderInfo.

6. Add the `Write Line` activity right after the `Extract/ Unzip Files` activity. In the `Text` field, click the `Plus` icon, select `Use Saved Value`, and hover over `DestinationFolderInfo` to select `Full Path`.

Once you have completed the exercise, the final configuration of the **Extract/ Unzip Files** activity should resemble Figure 9-37. Figure 9-38 shows the contents of the extracted file, while Figure 9-39 shows information about the extracted folder.

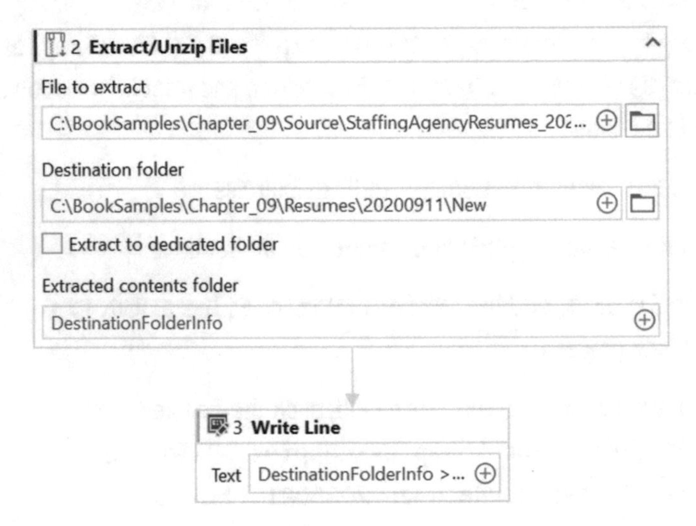

Figure 9-37. *Final configuration of Extract/Unzip Files activity*

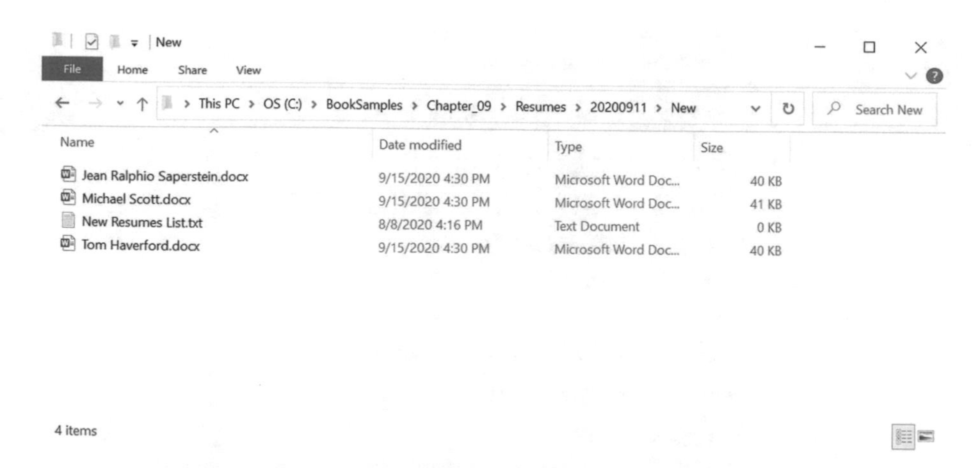

***Figure 9-38.** Contents of the extracted file*

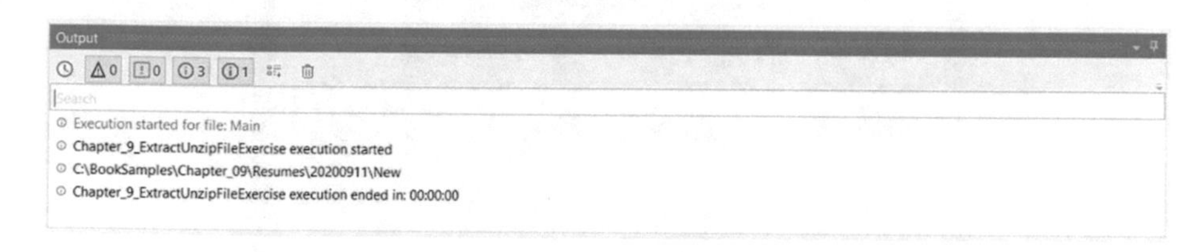

***Figure 9-39.** Output of the Extract/Unzip Files activity*

Get File Info

The **Get File Info** activity allows you to retrieve and save specific properties of a specified file. Figure 9-40 shows file properties that you can access using this activity.

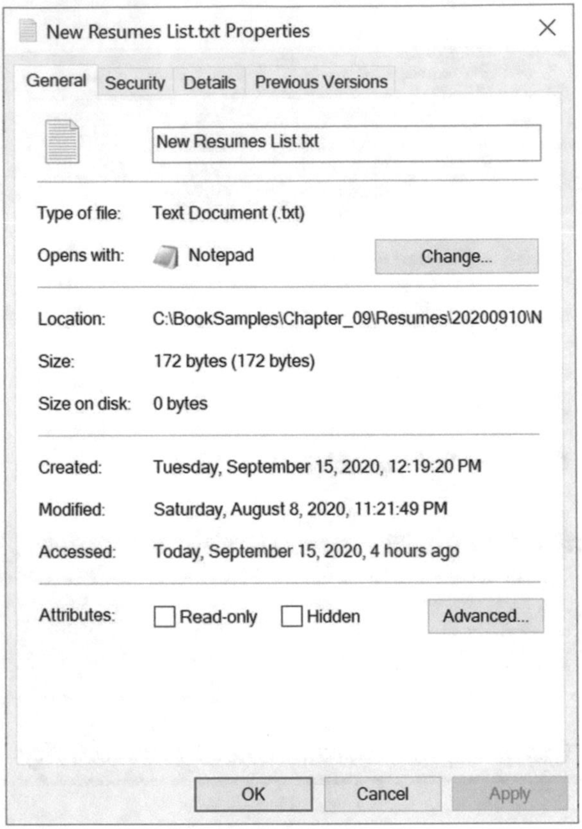

New Resumes List.txt Properties

General | Security | Details | Previous Versions

New Resumes List.txt

Type of file: Text Document (.txt)
Opens with: Notepad Change...

Location: C:\BookSamples\Chapter_09\Resumes\20200910\N
Size: 172 bytes (172 bytes)
Size on disk: 0 bytes

Created: Tuesday, September 15, 2020, 12:19:20 PM
Modified: Saturday, August 8, 2020, 11:21:49 PM
Accessed: Today, September 15, 2020, 4 hours ago

Attributes: Read-only Hidden Advanced...

OK Cancel Apply

Figure 9-40. *File properties window*

Configuration

This section provides instructions on how to configure a **Get File Info** activity, shown in Figure 9-41.

1 Get File Info
File path
Select a value
Output to
Select a value

Figure 9-41. *Activity card for Get File Info*

File path: This is a required configuration available on the activity card. This configuration allows you to specify the full path of the file that you want to retrieve the information from.

Output to: This is an optional configuration available on the activity card. This configuration allows you to save retrieved information in the `Data Manager` for use later.

Properties returned by this activity, shown in Figure 9-42, can later be used to decide if specific actions need to be performed. For example, if your automation frequently downloads data from a site, you can use the `Created date` **property** (see description in Table 9-2) of the previously downloaded file to check when was the last time data was downloaded. You can use this information to only download delta since the last download.

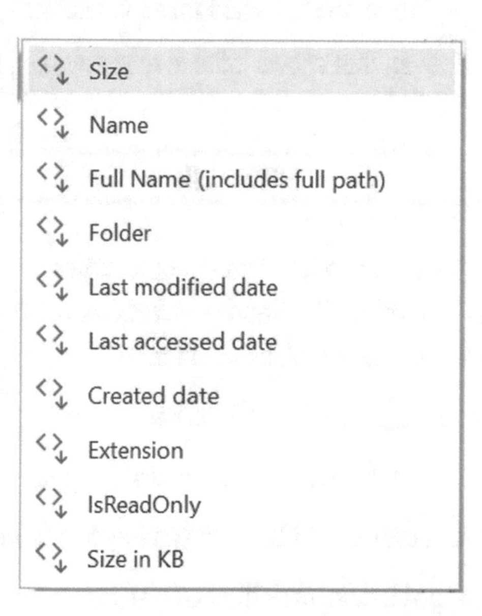

***Figure 9-42.** Properties returned by Get File Info activity*

Table 9-2 provides a brief description of all properties returned by this activity.

Table 9-2. Properties returned by Get File Info activity

Property	Description
Name	Name of the specified file (without folder path)
Full Name	Name of the specified file (with folder path)
Folder	Path of the folder that contains the file
Extension	Extension of the specified file (e.g., txt or docx)
Created date	Data and time when specified file was created
Last modified date	Data and time when specified file was last modified
Last accessed date	Data and time when specified file was last accessed
Size	Size of the file in bytes
Size in KB	Size of the specified folder in kilobytes (KB)
IsReadOnly	Flag that shows if file is read only, i.e., it cannot be edited

EXERCISE

Goal: Use the `Get File Info` activity to get properties of `C:\BookSamples\Chapter_09\Resumes\20200910\New\New Resumes List.txt` file, and print the last modified date.

Source Code: Chapter_9_FileActivitiesExercise

Setup: Here are step-by-step implementation instructions:

1. In StudioX, add the `Get File Info` activity to a blank process.
2. In the `File path` field, click the `Folder` icon and select `C:\BookSamples\Chapter_09\Resumes\20200910\New\New Resumes List.txt` file.
3. Click the `Plus` icon in `Output to` field, name the value as `FileInfo`, and click `Ok`.

4. Next, add a `Write Line` activity in the `Designer` panel after the `Get File Info` activity.
5. In the `Text` field, click the `Plus` icon, and select `Text`. In the `Text Builder`, type `File was last modified on:`.
6. While the `Text Builder` is still open, click the `Plus` icon, select `Use Saved Value` option, and hover over `FileInfo` to select `LastModifiedDate`. The complete text should read as `File was last modified on: FileInfo` ä `LastModifiedDate`. Click `Save`.

Once you have completed the exercise, the final configuration of the **Get File Info** activity should resemble Figure 9-43. Figure 9-44 shows the output of this exercise.

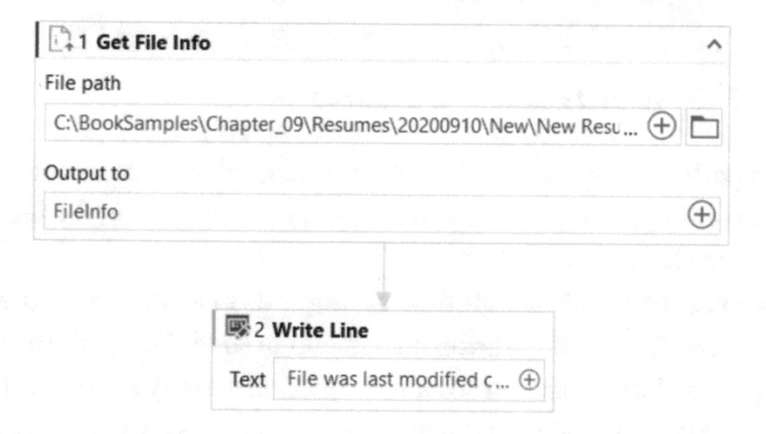

***Figure 9-43.** Final configuration of Get File Info activity exercise*

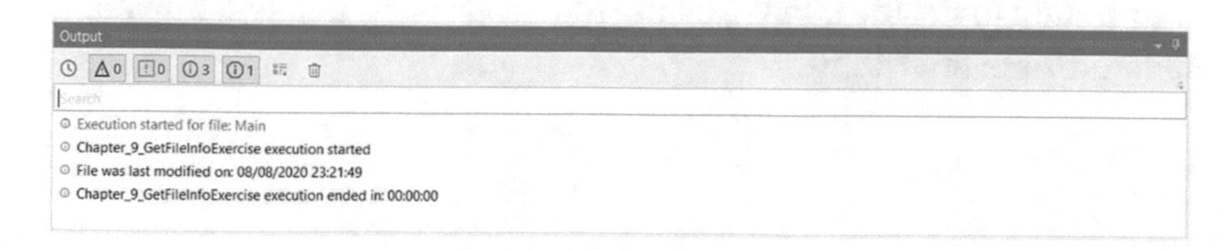

***Figure 9-44.** Output of Get File Info activity exercise*

File Exists

The **File Exists** activity allows you to check if a file already exists in the target location.

Configuration

This section provides instructions on how to configure a **File Exists** activity, shown in Figure 9-45.

Figure 9-45. *Activity card for File Exists*

File path: This is a required configuration available on the activity card. This configuration allows you to specify the full path of the file that you want to check exists or not.

Save result: This is an optional configuration available on the activity card. This configuration allows you to save the result from this activity in the `Data Manager` for later use. This activity returns a `True` if the file exists and `False` if the file does not exist. As an example, let us say this activity returns that a file already exists; you can use this information to decide if that existing file needs to be deleted, moved, updated, or overwritten.

EXERCISE

Goal: Use the `File Exists` activity to check if the `C:\BookSamples\Chapter_09\Resumes\20200911\New\New Resumes List.txt` already exists, and print the result.

Source Code: Chapter_9_FileActivitiesExercise

Setup: Here are step-by-step implementation instructions:

1. In `StudioX`, add the `File Exists` activity to a blank process.
2. In the `File path` field, click the `Plus` icon, select `Text` option, and type `C:\BookSamples\Chapter_09\Resumes\20200911\New\New Resumes List.txt`.
3. Click the `Plus` icon in the `Save result` field, name the value as `FileExists`, and click Ok.
4. Add the `Write Line` activity to the `Designer` panel after the `File Exists` activity.
5. In the `Text` field, click the `Plus` icon and select the `Text` option. In the `Text Builder`, type `File exists:`.
6. While the `Text Builder` is still open, click the `Plus` icon and hover over `Use Saved Value` to select `FileExists`. Click `Save`.

Once you have completed the exercise, the final configuration of the **File Exists** activity should resemble Figure 9-46. Figure 9-47 shows the output of this exercise.

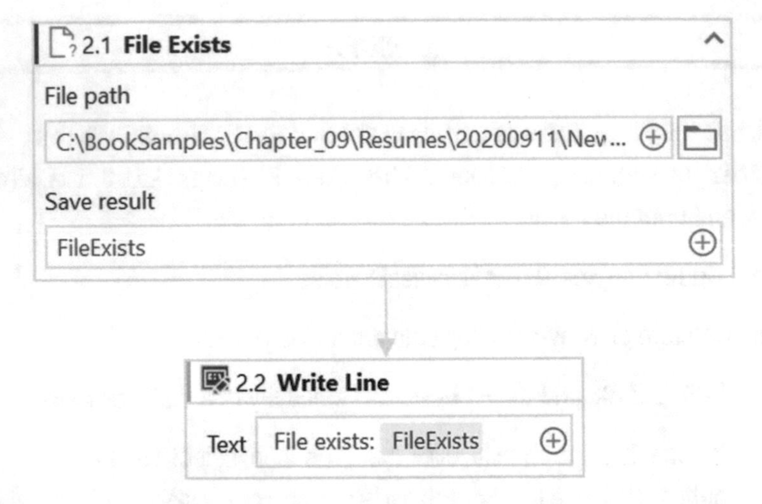

***Figure 9-46.** Final configuration of File Exists activity exercise*

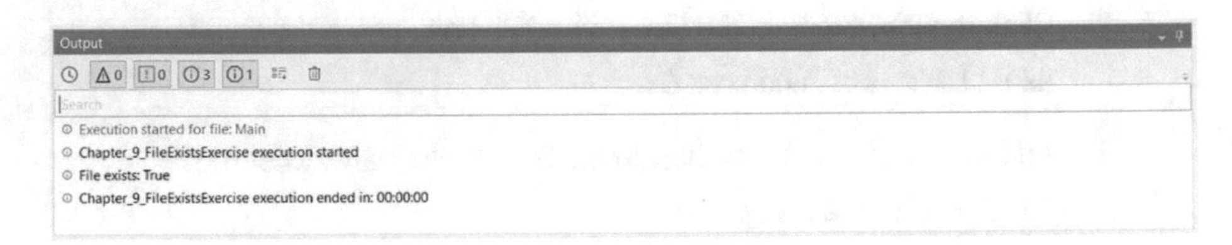

***Figure 9-47.** Output of File Exists activity exercise*

Create File

The **Create File** activity allows you to create a new file in the specified location.

Configuration

This section provides instructions on how to configure a **Create File** activity, shown in Figure 9-48.

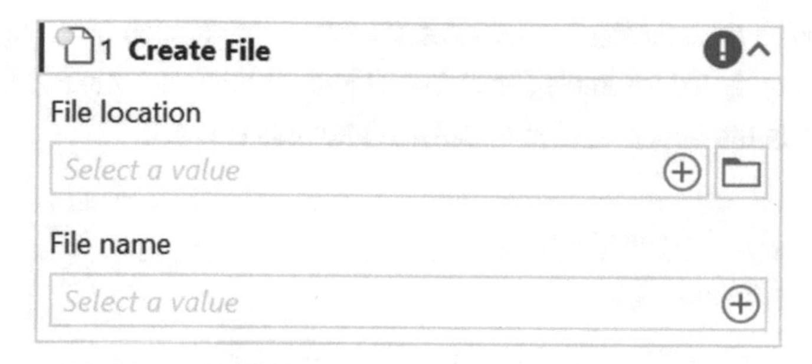

Figure 9-48. Activity card for Create File

File location: This is a required configuration available on the activity card. This configuration allows you to specify the full path of the folder where you want to create the new file.

File name: This is a required configuration available on the activity card. This configuration allows you to specify the name of the new file that you want to create.

EXERCISE

Goal: Use the `Create File` activity to create a new file `Pending Interview List.txt` in the `C:\BookSamples\Chapter_09\Interviews\Pending` location.

Source Code: Chapter_9_FileActivitiesExercise

Setup: Here are step-by-step implementation instructions:

1. In `StudioX`, add the `Create File` activity to a blank process.
2. In the `File location` field, click the `Folder` icon and select `C:\BookSamples\Chapter_09\Interviews\Pending`.
3. In the `File name` folder, click the `Plus` icon, select the `Text` option, and type `Pending Interview List.txt`. Click `Save`.

Once you have completed the exercise, the final configuration of the **Create File** activity should resemble Figure 9-49. Figure 9-50 shows the folder structure of the target location before and after this exercise.

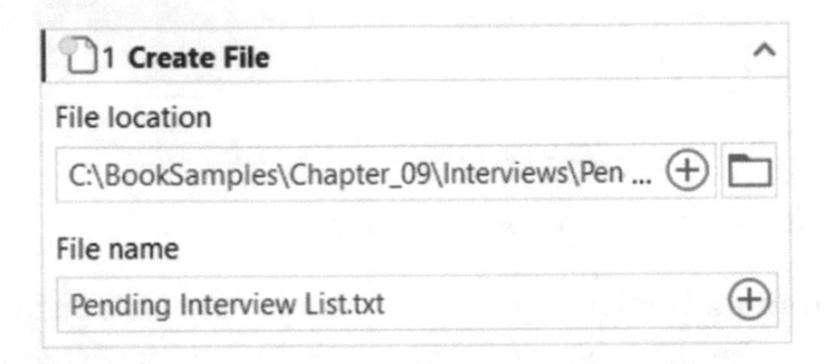

Figure 9-49. *Final configuration of Create Folder activity*

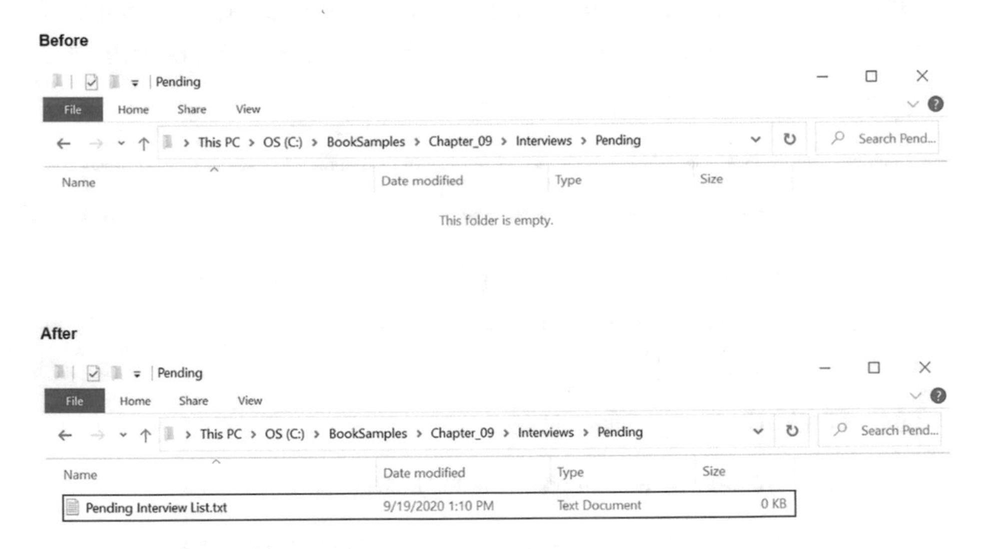

Figure 9-50. *Result of Create File activity exercise*

Delete File

The **Delete File** activity allows you to delete a specified file.

Configuration

This section provides instructions on how to configure a **Delete File** activity, shown in Figure 9-51.

***Figure 9-51.** Activity card for Delete File*

File name: This is a mandatory configuration available from the activity card. This configuration allows you to specify the full path of the file that you want to delete.

EXERCISE

Goal: Use the `Delete File` activity to delete the `C:\BookSamples\Chapter_09\Source\StaffingAgencyResumes_20200911.zip` file.

Source Code: Chapter_9_FileActivitiesExercise

Setup: Here are step-by-step implementation instructions:

1. In `StudioX`, add the `Delete File` activity to a blank process.
2. Click the `Folder` icon and select `C:\BookSamples\Chapter_09\Source\StaffingAgencyResumes_20200911.zip` file.

Once you have completed the exercise, the final configuration of the **Delete File** activity should resemble Figure 9-52. Figure 9-53 shows the before and after folder structure of the target location.

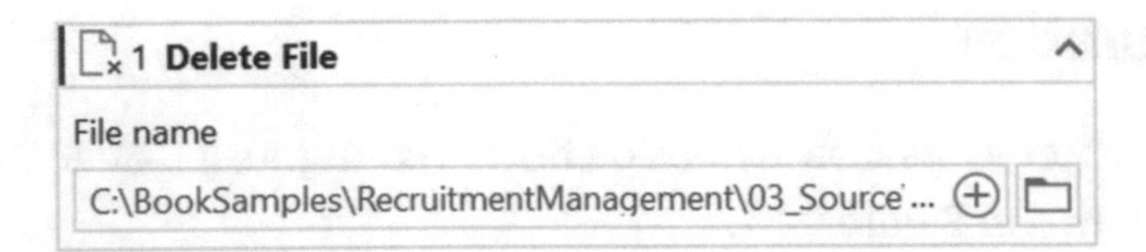

***Figure 9-52.** Final configuration of Delete File activity exercise*

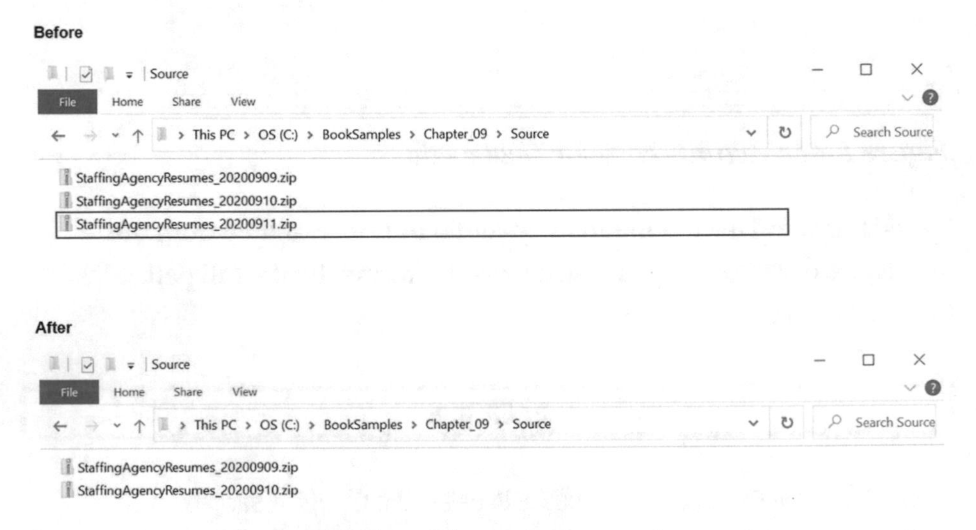

***Figure 9-53.** Result of Delete File activity exercise*

Copy File

The **Copy File** activity allows you to copy a specified file to another location.

Note This activity leaves the original file in place.

Configuration

This section provides instructions on how to configure a **Copy File** activity, shown in Figure 9-54.

Figure 9-54. Activity card for Copy File

From: This is a required configuration available on the activity card. This configuration allows you to specify the full path of the file that you want to copy.

To: This is a required configuration available on the activity card. This configuration allows you to specify the target location where `From` file needs to be copied. You will need to specify the file name – it can be the same as or different from the original.

Overwrite: This is an optional configuration available on the activity card. This configuration allows you to specify that in case a file with the same name exists in the target location, should the automation overwrite contents.

EXERCISE

Goal: Use the `Copy File` activity to copy `C:\BookSamples\Chapter_09\Source\StaffingAgencyResumes_20200909.zip` file to `C:\BookSamples\Chapter_09\Archive` location.

Source Code: Chapter_9_FileActivitiesExercise

Setup: Here are step-by-step implementation instructions:

1. In StudioX, add the Copy File activity to a blank process.
2. In the From field, click the Folder icon and select C:\BookSamples\Chapter_09\Source\StaffingAgencyResumes_20200909.zip file.
3. In the To field, click the Plus icon, select the Folder icon, and select C:\BookSamples\Chapter_09\Archive folder. You will need to specify the file name here as well, and type StaffingAgencyResumes_20200909.zip. Click Save.
4. Check the Overwrite flag.

Once you have completed the exercise, the final configuration of the **Copy File** activity should resemble Figure 9-55. Figure 9-56 shows the before and after folder structure of the target location. If you look at the original file, it would still be in the original folder.

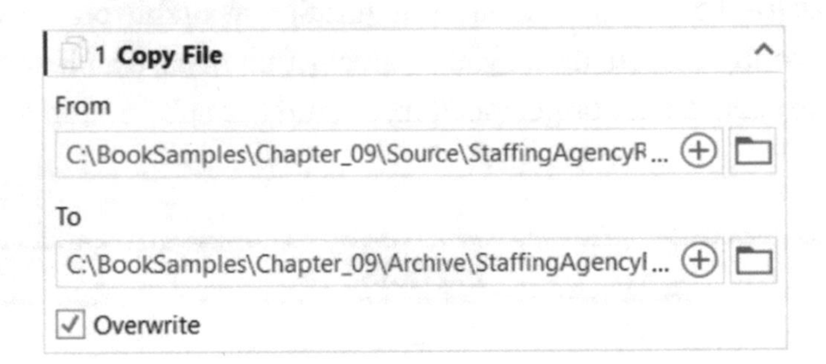

***Figure 9-55.** Final configuration of Copy File activity exercise*

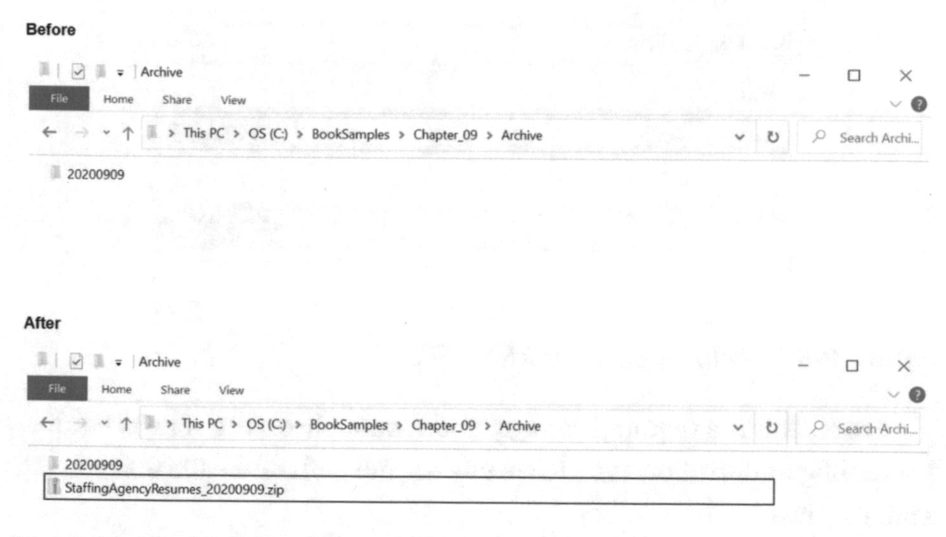

***Figure 9-56.** Result of Copy File activity exercise*

Move File

The **Move Folder** activity allows you to copy a specified folder to another location.

Note Unlike the `Copy File` activity, this activity does not leave the original file in place.

Configuration

This section provides instructions on how to configure a **Move File** activity, shown in Figure 9-57.

Figure 9-57. *Activity card for Move File*

From: This is a required configuration available on the activity card. This configuration allows you to specify the full path of the file that you want to move.

To: This is a required configuration available on the activity card. This configuration allows you to specify the target location where `From` file needs to be copied.

Overwrite: This is an optional configuration available on the activity card. This configuration allows you to specify that in case a file with the same name exists in the target location, should the automation overwrite contents.

EXERCISE

Goal: Use the `Move File` activity to move `C:\BookSamples\Chapter_09\Resumes\20200910\MoveForward\Bruce Wayne.docx` file to `C:\BookSamples\Chapter_09\Interviews\Pending` location.

Source Code: Chapter_9_FileActivitiesExercise

Setup: Here are step-by-step implementation instructions:

1. In StudioX, add the Move File activity to a blank process.
2. In the From field, click the Folder icon and select C:\BookSamples\Chapter_09\Resumes\20200910\MoveForward\Bruce Wayne.docx folder.
3. In the To field, click the Folder icon and select C:\BookSamples\Chapter_09\Interviews\Pending folder.
4. Leave the Overwrite field unchecked.

Once you have completed the exercise, the final configuration of the **Move File** activity should resemble Figure 9-58. Figure 9-59 shows the before and after folder structure of the source and target locations. If you look at the original file, it would still be in the original folder.

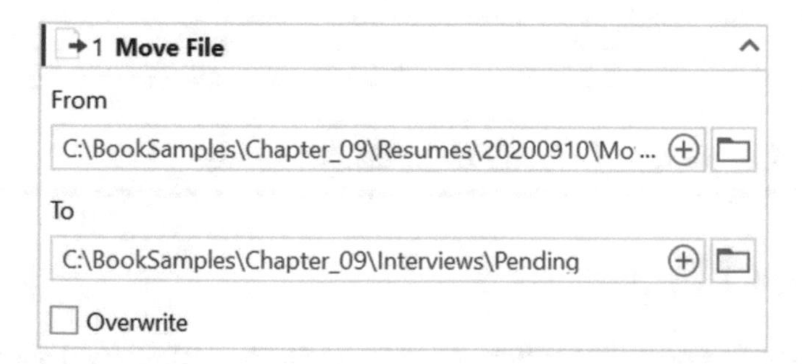

Figure 9-58. *Final configuration of Move File activity exercise*

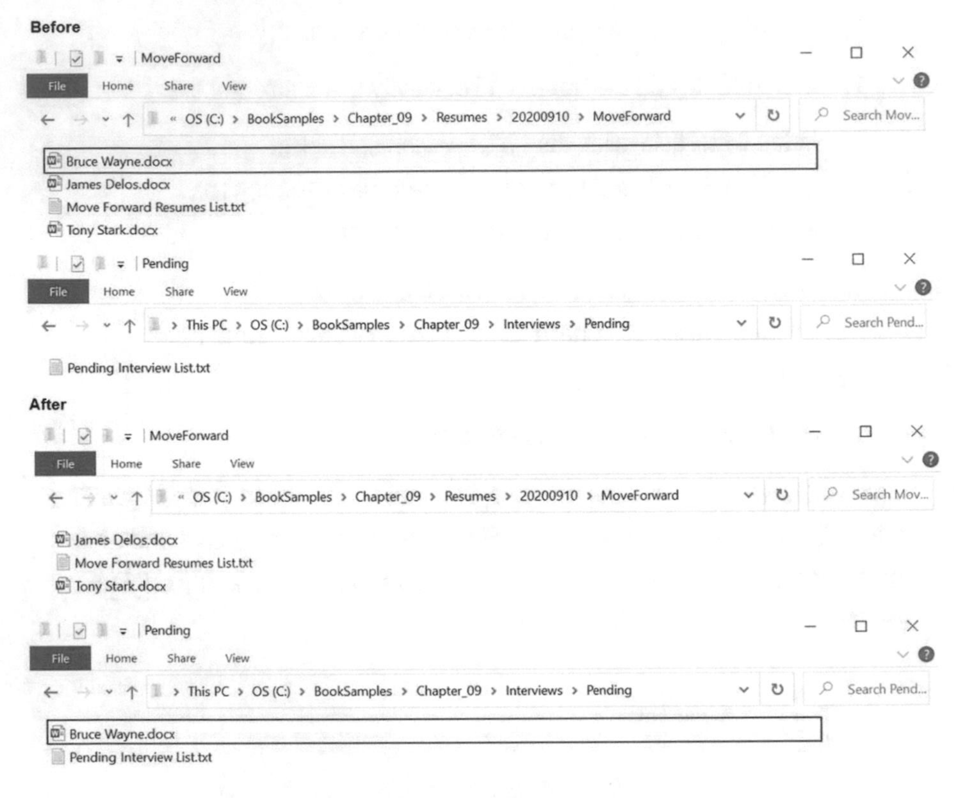

***Figure 9-59.** Result of Move File activity exercise*

Write Text File

The **Write Text File** activity allows you to write specified text to a text-based file.[1]

Note If the specified file does not exist, then this activity will first create the file and then write text to the file. Also, this activity will always overwrite any text already present in the specified file.

---

[1]Text-based files can be opened with any basic text editor, for example, TXT and CSV.

Configuration

This section provides instructions on how to configure a **Write Text File** activity, shown in Figure 9-60.

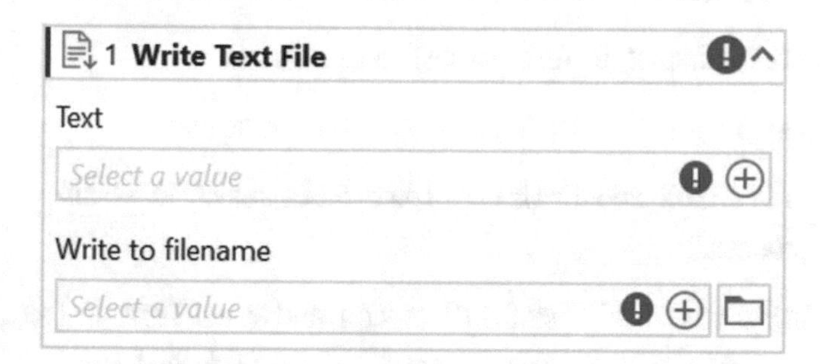

Figure 9-60. *Activity card for Write Text File*

Text: This is a required configuration available on the activity card. This configuration allows you to specify the text that you want to write in the target file. You can utilize the Text Builder to configure this.

Write to filename: This is a required configuration available on the activity card. This configuration allows you to specify the file where you want to write text.

Encoding: This is an optional configuration available on the `Properties` panel. This configuration allows you to specify the `Encoding`[2,3] in case the content contains any characters other than standard English.

---

[2]Learn more about character encoding from W3C website – `www.w3.org/International/questions/qa-what-is-encoding`

[3]Complete list of encoding types supported by UiPath – `https://docs.uipath.com/activities/docs/supported-character-encoding`

EXERCISE

Goal: Use the `Write Text File` activity to write text to `C:\BookSamples\Chapter_09\Resumes\20200911\New\New Resumes List.txt` file.

Source Code: Chapter_9_TextFileActivitiesExercise

Setup: Here are step-by-step implementation instructions:

1. In `StudioX`, add the `Write Text File` activity to a blank process.
2. In the `Text` field, click the `Plus` icon, and select `Text`. In `Text Builder`, enter `Resumes received on`, click the `Plus` icon, and hover over `Project Notebook (Notes)` to select `[Notes] Date!`YYYYMMDD.
3. In the `Write to filename` field, click `Folder` icon and select `C:\BookSamples\Chapter_09\Resumes\20200911\New\New Resumes List.txt` file.

Once you have completed the exercise, the final configuration of the **Write Text File** activity should resemble Figure 9-61. Figure 9-62 shows the text file with updated content.

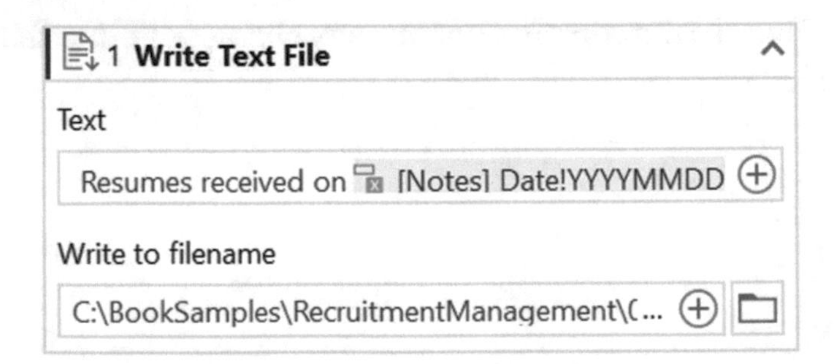

***Figure 9-61.** Final configuration of Write Text File activity exercise*

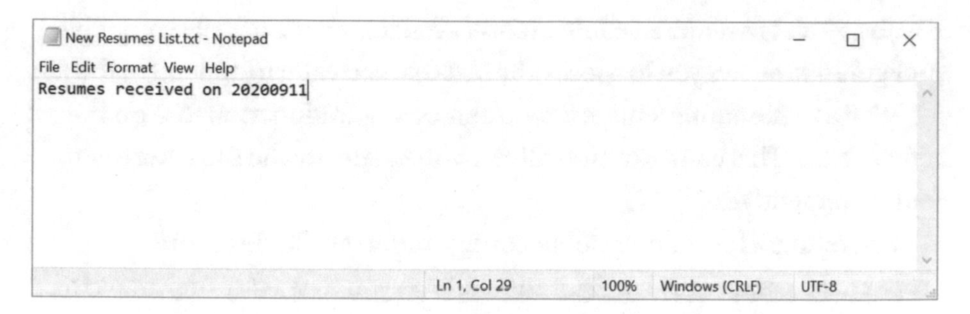

Figure 9-62. Result of Write Text File activity exercise

Append Line

The **Append Line** activity allows you to add a line of text to the end of a text-based file.

Note The Append Line activity will not overwrite or delete any text that exists in the file already.

Configuration

This section provides instructions on how to configure an **Append Line** activity, shown in Figure 9-63.

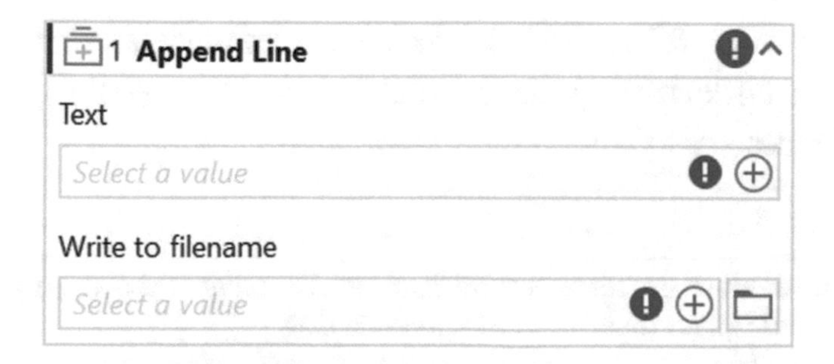

Figure 9-63. Activity card for Append Line

Text: This is a required configuration available on the activity card. This configuration allows you to specify the text that you want to append in the file.

Write to filename: This is a required configuration available on the activity card. This configuration allows you to specify the file where you want to append text.

Encoding: This is an optional configuration available on the `Properties` panel. This configuration allows you to specify the `Encoding`[4,5] in case the content contains any characters other than standard English.

Use default encoding: This is an optional configuration available on the `Properties` panel. This configuration allows you to specify if StudioX should be the default encoding of the file. By default, the value is set to False, that is, to not use the default encoding.

EXERCISE

Goal: Use the `Append Line` activity to append the name of each Word file in `C:\BookSamples\Chapter_09\Resumes\20200911\New` folder to `New Resumes List.txt` file (in the same folder).

Source Code: Chapter_9_TextFileActivitiesExercise

Setup: Here are step-by-step implementation instructions:

1. In `StudioX`, add the `For Each File In Folder` activity to a blank process.
2. In the `In folder` field, click the `Folder` icon and select the `C:\BookSamples\Chapter_09\Resumes\20200911\New` folder.

---

[4]Learn more about character encoding from W3C website – `www.w3.org/International/questions/qa-what-is-encoding`

[5]Complete list of encoding types supported by UiPath – `https://docs.uipath.com/activities/docs/supported-character-encoding`

3. In the Filter by field, click the Plus icon, select the Text option, and enter *.docx. This will ensure that only Word document files are returned.
4. Add the Append Line activity in the body of For Each File In Folder activity.
5. In the Text field, click the Plus icon, hover over CurrentFile, and select Name.
6. In the Write to filename field, click the Folder icon and select C:\BookSamples\Chapter_09\Resumes\20200911\New\New Resumes List.txt file.

Once you have completed the exercise, the final configuration of the **Append Line** activity should resemble Figure 9-64. Figure 9-65 shows the text file with updated content.

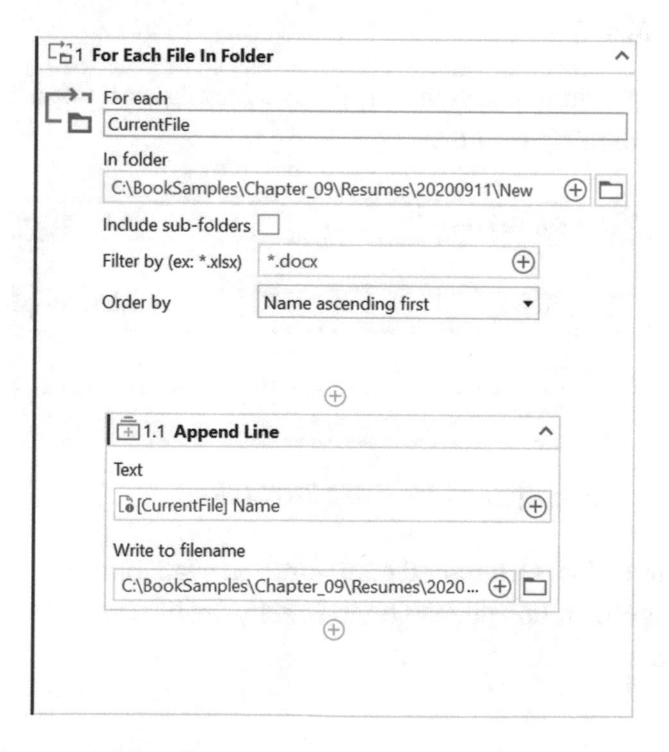

Figure 9-64. *Final configuration of Append Line activity exercise*

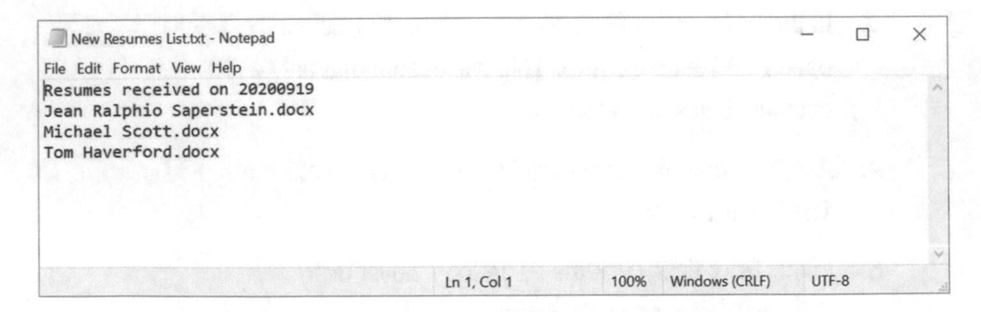

Figure 9-65. *Result of Append Line activity exercise*

Read Text File

The **Read Text File** activity allows you to read the contents (all characters) from a text file.

Configuration

This section provides instructions on how to configure a **Read Text File** activity, shown in Figure 9-66.

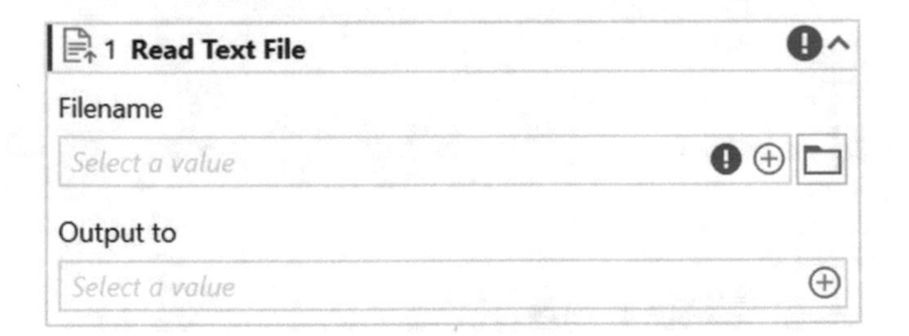

Figure 9-66. *Activity card for Read Text File*

Filename: This is a required configuration available on the activity card. This configuration allows you to specify the file from which you want to read text.

Output to: This is an optional configuration available on the activity card. This configuration allows you to save the contents of the text file for later use.

Encoding: This is an optional configuration available on the Properties panel. This configuration allows you to specify the `Encoding`[6,7] in case the content contains any characters other than standard English.

EXERCISE

Goal: Use the `Read Text File` activity to read contents of `C:\BookSamples\Chapter_09\Resumes\20200911\New\New Resumes List.txt` file.

Source Code: Chapter_9_TextFileActivitiesExercise

Setup: Here are step-by-step implementation instructions:

1. In `StudioX`, add the `Read Text File` activity to a blank process.
2. In the `Filename` field, click `Folder` icon and select the `C:\BookSamples\Chapter_09\Resumes\20200911\New\New Resumes List.txt` file.
3. In the `Output to` field, click the `Plus` icon and select `Copy to clipboard` option.
4. Add the `Write Line` activity right after the `Read Text File` activity, click the `Plus` icon, and select `Paste from clipboard` option.

---

[6]Learn more about character encoding from W3C website – `www.w3.org/International/questions/qa-what-is-encoding`

[7]Complete list of encoding types supported by UiPath – `https://docs.uipath.com/activities/docs/supported-character-encoding`

Once you have completed the exercise, the final configuration of the **Read Text File** activity should resemble Figure 9-67. Figure 9-68 shows the output of this exercise.

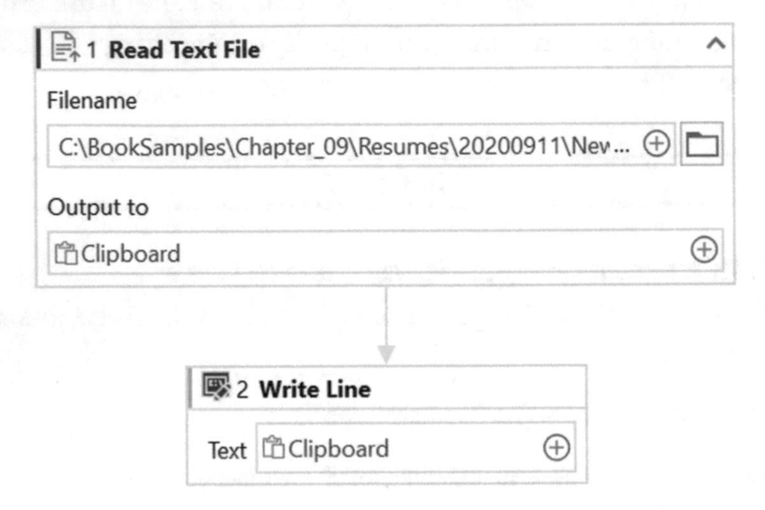

***Figure 9-67.** Final configuration of Read Text File activity exercise*

***Figure 9-68.** Output of Read Text File activity exercise*

CHAPTER 10

Presentation Automation

PowerPoint presentations are the most common and effective software used by organizations to display key information and engage audiences in various presentation settings from boardroom meetings and industry conferences to company training and project implementations. UiPath's StudioX Presentation Automation capabilities allow users to automate common tasks such as adding or replacing text, copy pasting slides, and adding data/images/files to slides in a PowerPoint.

Learning Objectives

At the end of this chapter, you will learn how to

- Add, delete, copy, and move slides in a presentation
- Add or replace text in a slide
- Add tables, image, video, and file to a slide
- Run macros in PowerPoint
- Save PowerPoint as PDF or another PowerPoint file type

A. Javed et al., *Robotic Process Automation using UiPath StudioX*,
https://doi.org/10.1007/978-1-4842-6794-3_10

Sample Overview

Throughout this chapter, we will be using a simplified version of a report generation process to showcase the usage of all Presentation Automation activities.

This section will familiarize you with the prerequisites for all exercises in this chapter.

File System Structure

Download the source code from the book's site, and make sure you move the entire BookSamples folder to your C:\ drive. All exercises in this chapter assume the folder paths will be **C:\BookSamples\Chapter_10**. Figure 10-1 shows the physical folder structure required for this sample. This folder structure comes with the source code.

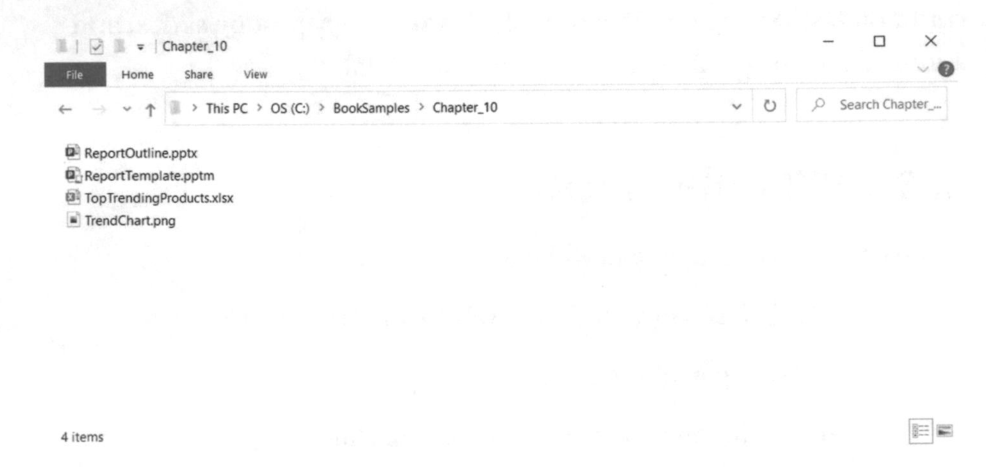

Figure 10-1. *Files used for the Report Generation exercise*

ReportTemplate.pptm: This is a macro-enabled PowerPoint that we will be using as a template to generate our reports. Figure 10-2 shows the initial state of the presentation.

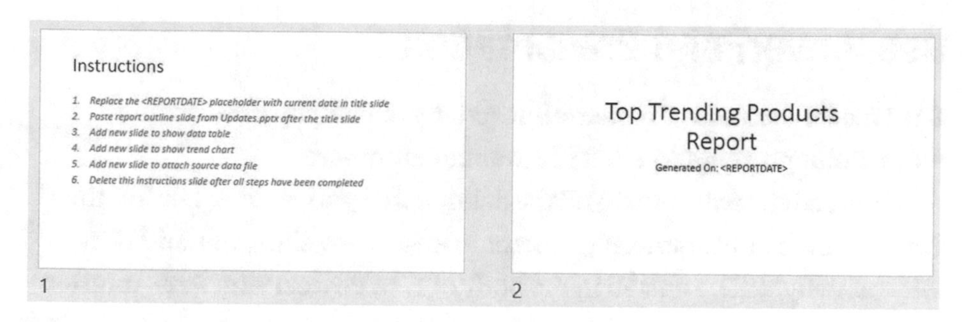

Figure 10-2. *Initial state of the presentation*

ReportOutline.pptx: This is a single-slide PowerPoint. We will be copying this slide to our report.

TopTrendingProducts.xlsx: This Excel spreadsheet contains a list of top trending products. We will be using the data from this spreadsheet in our report and attach this file in our report.

TrendChart.png: This is an image file. We will be displaying this image in our report.

Activities Reference

As shown in Figure 10-3, all Presentation Automation activities can be found under the Presentation category. The following sections will provide instructions on how to configure and use each activity.

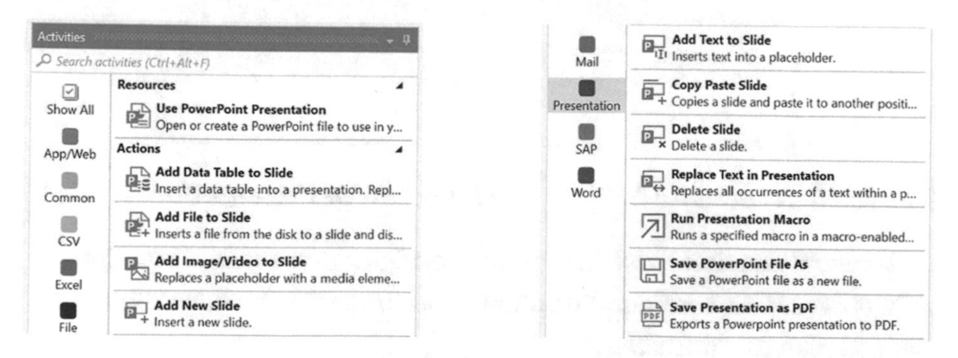

Figure 10-3. *Activities for Presentation Automation*

Use PowerPoint Presentation

The **Use PowerPoint Presentation** activity allows you to select the PowerPoint presentation that you want to automate.

This activity will contain all the actions that you want to take on the PowerPoint. For example, if you want to add a new slide, the **Add New Slide** activity will be nested in the body of **Use PowerPoint Presentation** activity.

Configuration

This section provides instructions on how to configure a **Use PowerPoint Presentation** activity, shown in Figure 10-4.

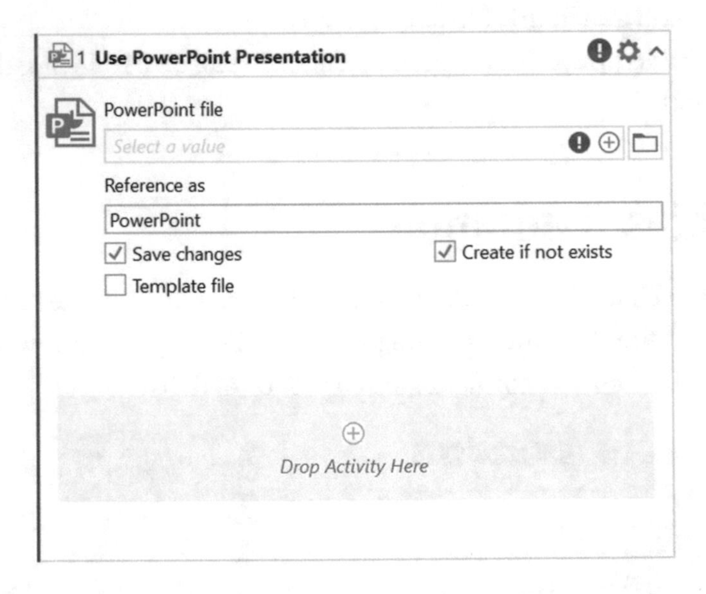

Figure 10-4. *Activity card for Use PowerPoint Presentation*

PowerPoint file: This is a required configuration available on the activity card. This configuration allows you to specify the PowerPoint file that you plan to automate.

Reference as: This is a required configuration available on the activity card. This configuration allows you to provide a name for your PowerPoint file. All the activities that need to use the selected PowerPoint will reference it by this name.

Save changes: This is an optional configuration available on the activity card. This configuration ensures that the PowerPoint is saved after each action is taken on an activity. By default, this option is checked.

Create if not exists: This is an optional configuration available on the activity card. This configuration ensures that a blank file is created if it does not exist in the target location. By default, this option is checked.

Template file: This is an optional configuration available on the activity card. This configuration is applicable when you are creating a new PowerPoint file. This configuration allows you to specify a template file, and StudioX uses the Slide Master of template file for the new PowerPoint file.

Edit password: This is an optional configuration available on the `Properties` panel. This field is used for editing a password-protected PowerPoint file. Enter the password in this field if necessary.

Password: This is an optional configuration available on the `Properties` panel. This field is used for opening a password-protected PowerPoint file. Enter the password in this field if necessary.

Read only: This is an optional configuration available on the `Properties` panel. If checked, the PowerPoint will open in read-only mode for automation. This option will allow the automation to extract data from a PowerPoint even if it is password protected. By default, this option is not checked.

Copy Paste Slide

The **Copy Paste Slide** activity allows you to copy or move a slide within the same PowerPoint presentation or from one PowerPoint presentation to another.

Configuration

This section provides instructions on how to configure a **Copy Paste Slide** activity, shown in Figure 10-5.

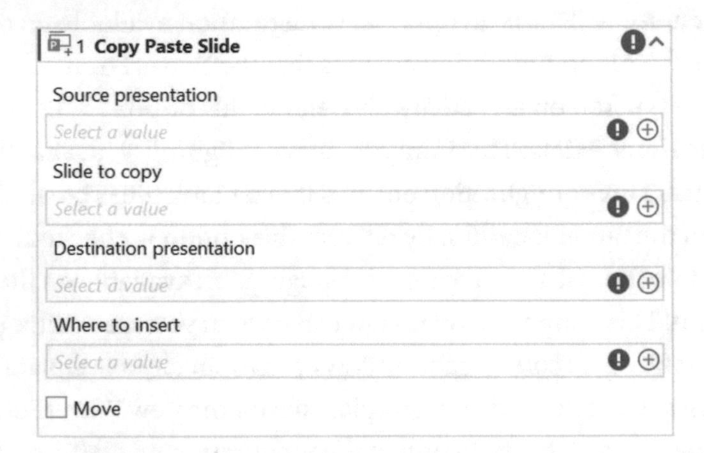

Figure 10-5. *Activity card for Copy Paste Slide*

Source presentation: This is a required configuration available on the activity card. This configuration specifies the PowerPoint presentation from which slide will be copied or moved.

Slide to copy: This is a required configuration available on the activity card. This configuration allows you to specify the slide from the source PowerPoint presentation that needs to be copied or moved.

Destination presentation: This is a required configuration available on the activity card. This configuration specifies the PowerPoint presentation to which the slide will be copied or moved.

Where to insert: This is a required configuration available on the activity card. This configuration allows you to specify where in the destination PowerPoint presentation the slide should be inserted.

Move: This is an optional configuration available on the activity card. This configuration specifies if the slide should be just copied or moved to the destination PowerPoint presentation. By default, this is not checked.

EXERCISE

Goal: Use the Copy Paste Slide activity to copy the Report Outline slide from ReportOutline.pptx and paste it after the Title slide in ReportTemplate.pptm. Figure 10-2 shows the state of the presentation before this exercise.

Source Code: Chapter_10-PresentationAutomationExercise

Setup: Here are step-by-step implementation instructions:

1. In StudioX, add the Use PowerPoint Presentation activity.
2. In the PowerPoint file field, click the Folder icon, and select C:\BookSamples\Chapter_10\ReportTemplate.pptm.
3. Set the value of the Reference as field to ReportTemplatePPT.
4. Uncheck Save changes and Create if not exists flags.
5. In the body of the Use PowerPoint Presentation activity, add another Use PowerPoint Presentation activity. The outer one will be used to reference the destination PowerPoint presentation, and the inner one will reference the source PowerPoint presentation.
6. In the PowerPoint file field, click the Folder icon, and select C:\BookSamples\Chapter_10\ReportOutline.pptx.
7. In the Reference as field, enter ReportOutlinePPT.
8. Uncheck Save changes and Create if not exists flags.
9. In the inner Use PowerPoint Presentation activity, add the Copy Paste Slide activity.

10. In the `Source presentation` field, select `ReportOutlinePPT`.

11. In the `Slide to copy` field, select `Slide 1`, or use the `Plus` icon to enter 1.

12. In the `Destination presentation` field, select `ReportTemplatePPT`.

13. In the `Where to insert` field, click the Plus icon, select the `Number` option, and type `3`.

Once you have completed the exercise, the final configuration of the **Copy Paste Slide** activity should resemble Figure 10-6.

Note We unchecked the Save changes option, so before you run the automation, open the ReportTemplate.pptm; this will allow you to view all the changes that the automation is making.

Once the automation runs, the Report Outline slide will be inserted at the end of the ReportTemplate.pptm, as shown in Figure 10-7.

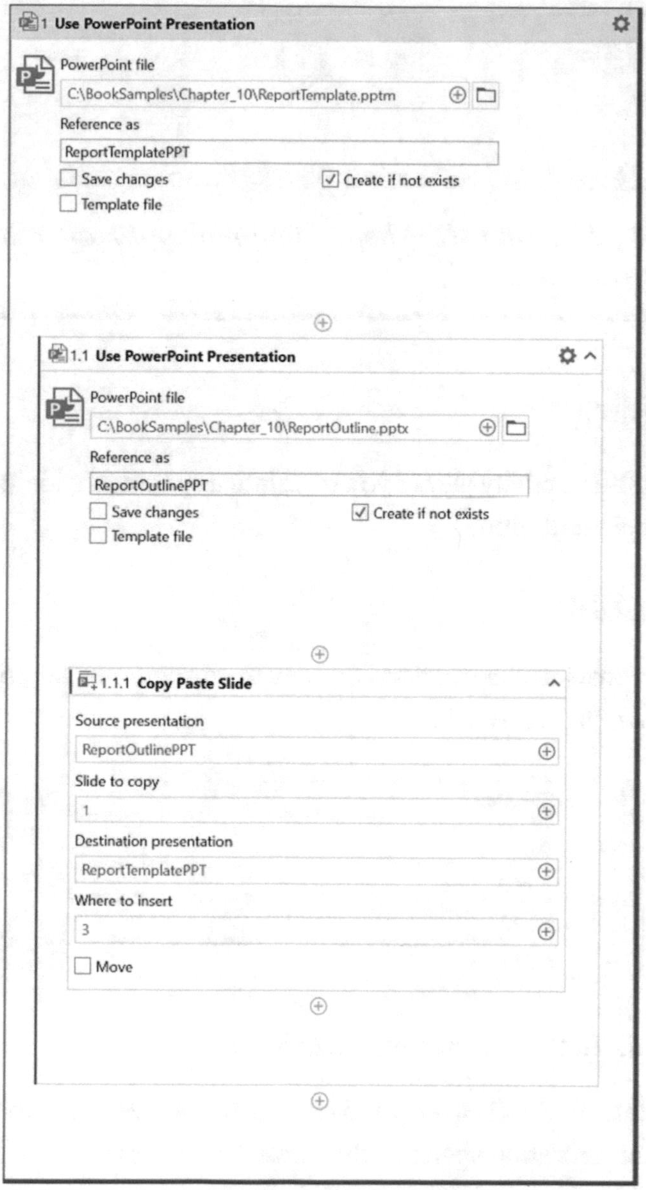

1 Use PowerPoint Presentation
PowerPoint file
C:\BookSamples\Chapter_10\ReportTemplate.pptm
Reference as
ReportTemplatePPT
Save changes
Create if not exists
Template file

1.1 Use PowerPoint Presentation
PowerPoint file
C:\BookSamples\Chapter_10\ReportOutline.pptx
Reference as
ReportOutlinePPT
Save changes
Create if not exists
Template file

1.1.1 Copy Paste Slide
Source presentation
ReportOutlinePPT
Slide to copy
1
Destination presentation
ReportTemplatePPT
Where to insert
3
Move

Figure 10-6. *Final configuration for Copy Paste Slide exercise*

***Figure 10-7.** The state of the ReportTemplate.pptm after the Copy Paste Slide exercise*

Delete Slide

The **Delete Slide** activity allows you to delete a specified slide from a PowerPoint presentation.

Configuration

This section provides instructions on how to configure a **Delete Slide** activity, shown in Figure 10-8.

***Figure 10-8.** Activity card for Delete Slide*

Presentation: This is a required configuration available on the activity card. This configuration specifies the PowerPoint presentation from which the slide needs to be deleted. This field is pre-populated with the PowerPoint presentation from the parent `Use PowerPoint Presentation` activity.

Slide number: This is a required configuration available on the activity card. This configuration specifies the slide number that needs to be deleted.

EXERCISE

Goal: Building on our previous exercise, use the `Delete Slide` activity to delete the Instructions slide from `C:\BookSamples\Chapter_10\ReportTemplate.pptm`. Figure 10-7 shows the current state of the presentation.

Source Code: Chapter_10-PresentationAutomationExercise

Setup: Here are step-by-step implementation instructions:

1. In `StudioX`, add the `Delete Slide` activity within the outer `Use PowerPoint Presentation` activity after the inner `Use PowerPoint Presentation` activity from the previous exercise.
2. The `Presentation` field will be pre-populated with the `ReportTemplatePPT` reference from the parent activity.
3. In the `Slide number` field, click the `Plus` icon, and select the `Number` option. The Instructions slide, as shown in Figure 10-7, is the first slide in the presentation, so type `1`.

Once you have completed the exercise, the final configuration of the **Delete Slide** activity should resemble Figure 10-9.

Note We unchecked the Save changes option, so before you run the automation, open the ReportTemplate.pptm; this will allow you to view all the changes that the automation is making.

Figure 10-10 shows the state of the presentation after this activity is run.

1 Use PowerPoint Presentation
PowerPoint file
C:\BookSamples\Chapter_10\ReportTemplate.pptm
Reference as
ReportTemplatePPT
Save changes
Create if not exists
Template file

1.1 Use PowerPoint Presentation
C:\BookSamples\Chapter_10\ReportOutline....
1 action(s) Double-click to view contents

1.2 Delete Slide
Presentation
ReportTemplatePPT
Slide number
1

***Figure 10-9.** Final configuration for Delete Slide activity exercise*

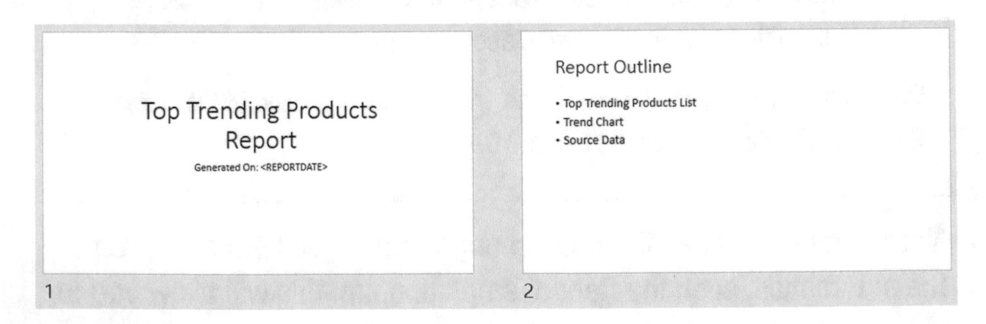

***Figure 10-10.** The state of the ReportTemplate.pptm after the Delete Slide exercise*

Add New Slide

The **Add New Slide** activity allows you to add a new slide to the PowerPoint presentation.

Configuration

This section provides instructions on how to configure an **Add New Slide** activity, shown in Figure 10-11.

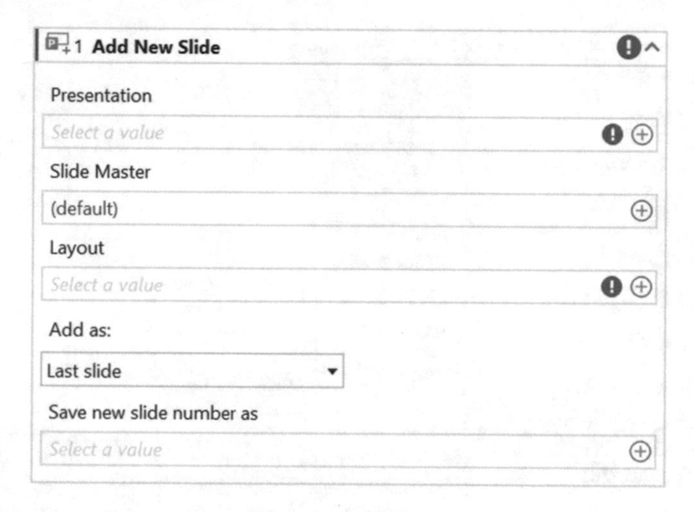

Figure 10-11. *Activity card for Add New Slide*

Presentation: This is a required configuration available on the activity card. This configuration specifies the PowerPoint presentation in which a new slide needs to be added. This field is pre-populated with the PowerPoint presentation from the parent `Use PowerPoint Presentation` activity.

Slide Master: This is a required configuration available on the activity card. In PowerPoint, you can update the Theme from the `Design` tab, shown in Figure 10-12. When this activity is placed within a `Use PowerPoint Presentation` activity, this lists the current Theme being used in the slides. This configuration can be used to specify what Theme the new slide should use. By default, the value of this configuration is `(default)`.

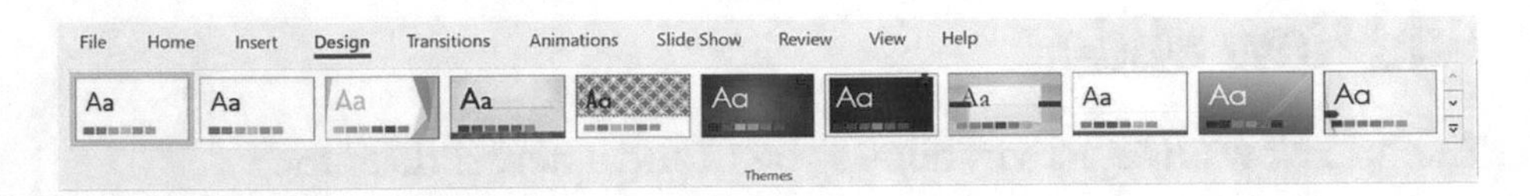

Figure 10-12. *Themes in PowerPoint*

Layout: This is a required configuration available on the activity card. In PowerPoint, you can view the layout options for a slide by right-clicking the slide. Figure 10-13 shows all the layout options available in the selected Theme. StudioX also allows you to specify the layout.

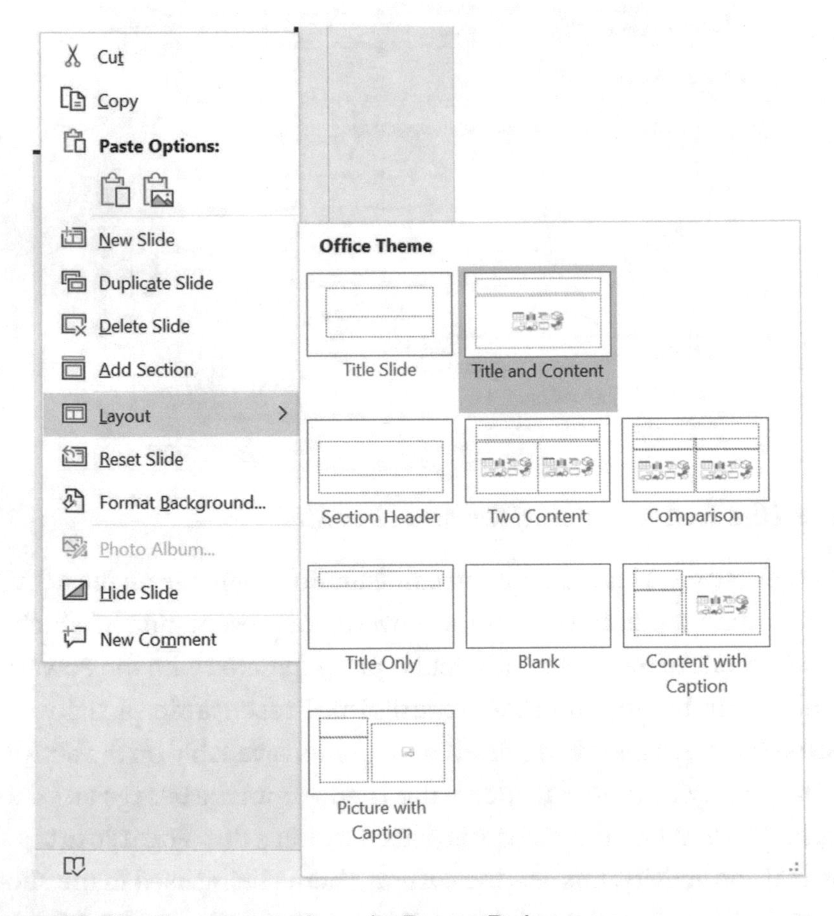

Figure 10-13. *Layout options in PowerPoint*

Add as: This is a required configuration available on the activity card. This configuration allows you to specify in what location this new slide should be added. You can select the `First Slide`, `Last Slide`, or `Slide number` as options. By default, `Last Slide` is selected.

Save new slide number as: This is an optional configuration available on the activity card. This configuration allows you to save the slide number of this newly created slide for later use.

EXERCISE

Goal: Building on our previous exercise, use the `Add New Slide` activity to add three new slides to the `C:\BookSamples\Chapter_10\ReportTemplate.pptm`. All three slides should use the Title and Content layout. Figure 10-10 shows the state of PowerPoint presentation before this exercise.

Source Code: Chapter_10-PresentationAutomationExercise

Setup: Here are step-by-step implementation instructions:

1. In `StudioX`, add the `Add New Slide` activity to the body of the outer `Use PowerPoint Presentation` activity after the `Delete Slide` activity from the previous exercise. This slide will be used later for adding a data table containing all top trending products.

2. The `Presentation` field will be pre-populated with the `ReportTemplatePPT` reference from the parent activity.

3. In the `Layout` field, select the `Title and Content` option.

4. In the `Add as` field, leave the default value of `Last slide` in the dropdown.

5. In the `Save new slide number as` field, click the `Plus` icon, select `Save for Later Use` option, and type `DataTableSlideNumber` as value.

6. Repeat steps 1–4. This second new slide will be used for showing an image of trending products chart. In the `Save new slide number as` field, click the `Plus` icon, select `Save for Later Use` option, and type `TrendChartSlideNumber` as the value.

7. Repeat steps 1–4. This third new slide will be used for attaching source data spreadsheet. In the `Save new slide number as` field, click the `Plus` icon, select `Save for Later Use` option, and type `SourceDataSlideNumber` as the value.

Once you have completed the exercise, the final configuration of the **Add New Slide** activity should resemble Figure 10-14.

Note We unchecked the Save changes option, so before you run the automation, open the ReportTemplate.pptm; this will allow you to view all the changes that the automation is making.

Figure 10-15 shows the state of the presentation after this activity is run.

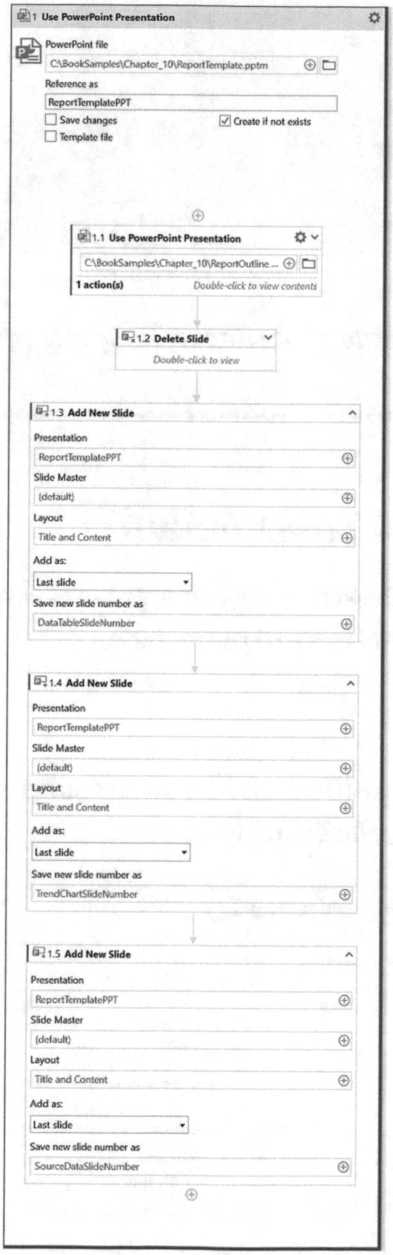

Figure 10-14. *Final configuration of the Add New Slide activity exercise*

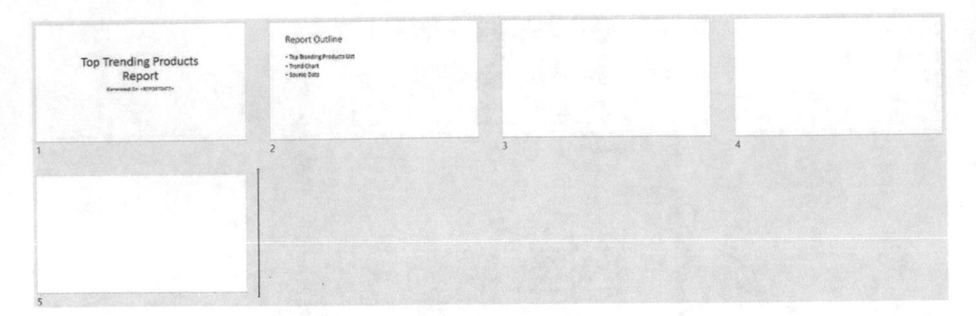

Figure 10-15. *The state of the ReportTemplate.pptm after the Add New Slide exercise*

Replace Text in Presentation

The **Replace Text in Presentation** activity allows you to replace all occurrences of specified text in a presentation.

Configuration

This section provides instructions on how to configure a **Replace Text in Presentation** activity, shown in Figure 10-16.

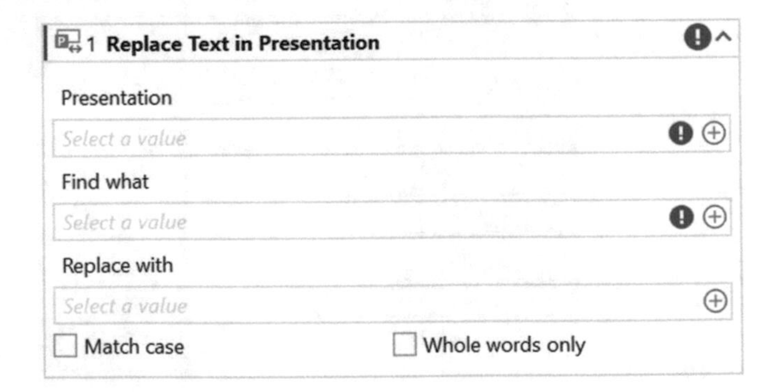

Figure 10-16. *Activity card for Replace Text in Presentation*

Presentation: This is a required configuration available on the activity card. This configuration specifies the PowerPoint presentation in which a new slide needs to be added. This field is pre-populated with the PowerPoint presentation from the parent `Use PowerPoint Presentation` activity.

Find what: This is a required configuration available on the activity card. You can use this configuration to specify the text that needs to be searched and replaced.

Replace with: This is an optional configuration available on the activity card. This configuration allows you to specify the new text that will replace the text specified in `Find what` configuration.

Match case: This is an optional configuration available on the activity card. This configuration allows you to specify if the search should be case-sensitive or not. If this option is checked, then the search will be case-sensitive.

Whole words only: This is an optional configuration available on the activity card. This configuration allows you to specify if StudioX should only replace whole words and not text that is part of larger word.

Number of replacements: This is an optional configuration available on the `Properties` panel. This configuration outputs the number of replacements that were made by the activity.

EXERCISE

Goal: Building on our previous exercise, use the `Replace Text in Presentation` activity to replace `<REPORTDATE>` with actual date on the Title slide. Figure 10-15 reflects the current state of presentation.

Source Code: Chapter_10-PresentationAutomationExercise

Setup: Here are step-by-step implementation instructions:

1. In StudioX, add the Replace Text in Presentation activity to the body of outer Use PowerPoint Presentation activity after the Add New Slide activity from the previous exercise.
2. The Presentation field will be pre-populated with the ReportTemplatePPT reference from the parent activity.
3. In the Find what field, click the Plus icon, select the Text option, and type <REPORTDATE>.
4. In the Replace with field, click the Plus icon, and hover the Notebook menu to select Date [Sheet]  Today [Cell] value. This will return the current date.
5. Leave the rest of the configurations as is.

Once you have completed the exercise, the final configuration of the **Replace Text in Presentation** activity should resemble Figure 10-17.

Note We unchecked the Save changes option, so before you run the automation, open the ReportTemplate.pptm; this will allow you to view all the changes that the automation is making.

Figure 10-18 shows the state of the title slide after this activity is run.

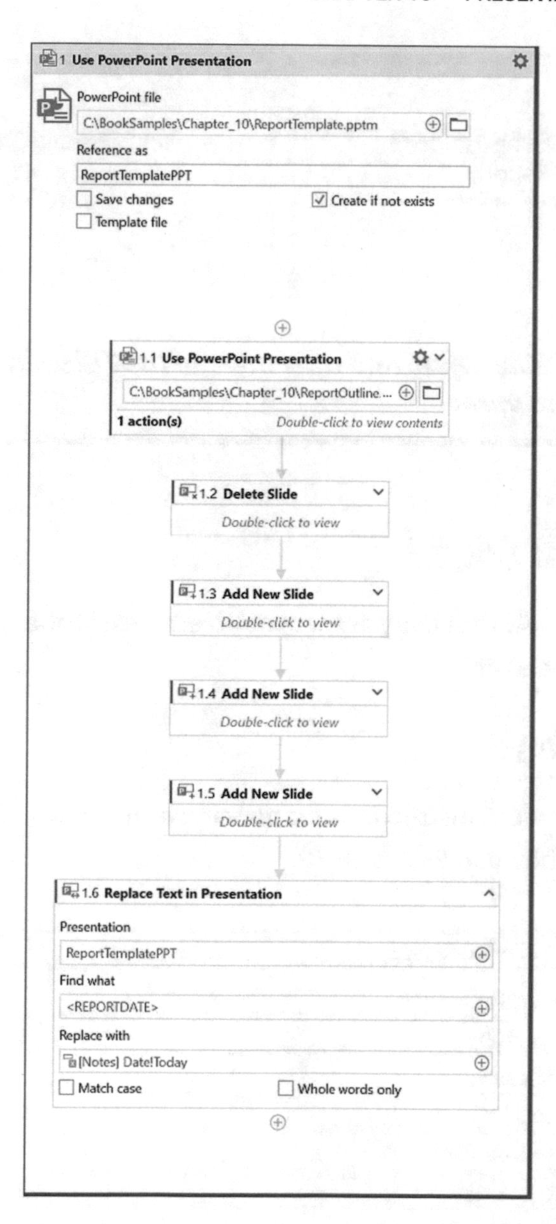

***Figure 10-17.** Final configuration of Replace Text in Presentation activity exercise*

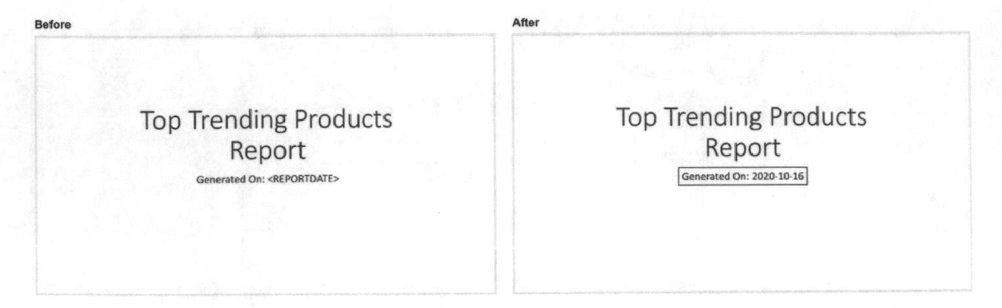

***Figure 10-18.** The before and after view of the Title slide in ReportTemplate.pptm*

Add Text to Slide

The **Add Text to Slide** activity allows you to add text to the specified placeholder on a slide.

Configuration

This section provides instructions on how to configure an **Add Text to Slide** activity, shown in Figure 10-19.

***Figure 10-19.** Activity card for Add Text to Slide*

Presentation: This is a required configuration available on the activity card. This configuration specifies the PowerPoint presentation where you want to add the text. This field is pre-populated with the PowerPoint presentation from the parent `Use PowerPoint Presentation` activity.

Slide number: This is a required configuration available on the activity card. This configuration specifies the slide number where you want to add the text.

Content placeholder: This is a required configuration available on the activity card. This configuration allows you to specify the name of the placeholder where you want the text to be added.

Tip To find the name of a placeholder, in PowerPoint, go to Home tab, and in the Editing group, click Select and chose Selection Pane, shown in Figure 10-20.

- Slide 3 title should be Top Trending Products List.
- Slide 4 title should be Trend Chart.
- Slide 5 title should be Source Data.

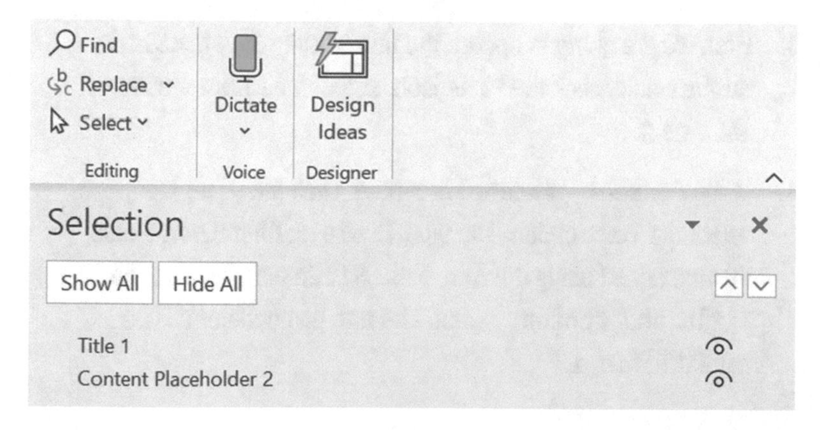

***Figure 10-20.** Selection Pane contains names of placeholders in PowerPoint*

Text to add: This is a required configuration available on the activity card. This configuration allows you to specify the actual text that will be added to the specified placeholder.

EXERCISE

Goal: Building on our previous exercise, use the `Add Text to Slide` activity to add titles to the three new slides (slides 3, 4, and 5) that were created in the `Add New Slide` exercise.

Figure 10-15 shows the state of PowerPoint presentation before this exercise.

Source Code: Chapter_10-PresentationAutomationExercise

Setup: Here are step-by-step implementation instructions:

1. In `StudioX`, add the `Add Text to Slide` activity to the body of outer `Use PowerPoint Presentation` activity after the `Replace Text in Presentation` activity from the previous exercise.
2. The `Presentation` field will be pre-populated with the `ReportTemplatePPT` reference from the parent activity.
3. First, we are going to update the title of slide 3. In the `Slide number` field, click the `Plus` icon, select the `Number` option, and type `3`.
4. In the `Content placeholder` field, click the `Plus` icon, select the `Text` option, and type `Title 1`. All the new slides that we added using the `Add New Slide` activity used the `Title and Content` layout. The title placeholder will be named `Title 1`.

5. Next, in the `Text to add` field, click the `Plus` icon, select the `Text` option, and type `Top Trending Products List`.

6. Repeat steps 1–5 for the other two slides, that is, slides 4 and 5. Update the `Slide number` and `Text to add` fields according to the information provided in the Goal section.

Once you have completed the exercise, the final configuration of the **Add Text to Slide** activity should resemble Figure 10-21.

Note We unchecked the Save changes option, so before you run the automation, open the ReportTemplate.pptm; this will allow you to view all the changes that the automation is making.

Figure 10-22 shows the state of the presentation after this activity is run.

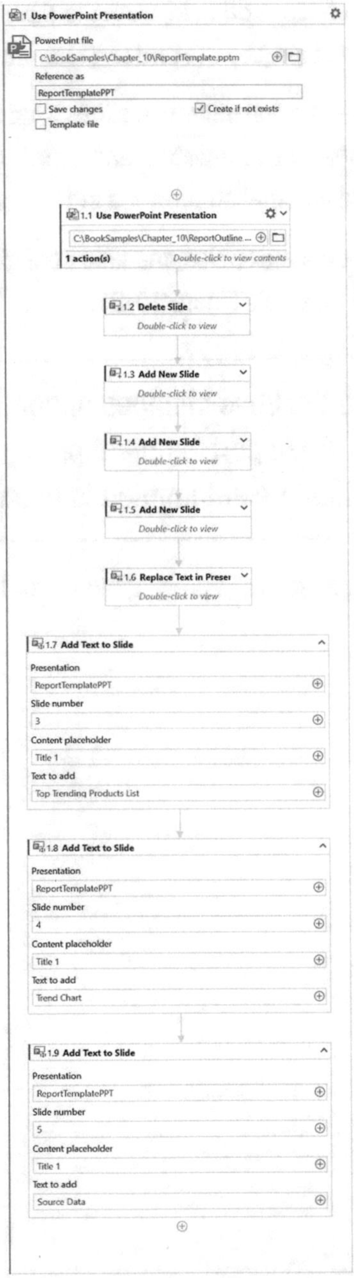

1 Use PowerPoint Presentation
PowerPoint file
C:\BookSamples\Chapter_10\ReportTemplate.pptm
Reference as
ReportTemplatePPT
Save changes
Create if not exists
Template file

1.1 Use PowerPoint Presentation
C:\BookSamples\Chapter_10\ReportOutline ...
1 action(s) Double-click to view contents

1.2 Delete Slide
Double-click to view

1.3 Add New Slide
Double-click to view

1.4 Add New Slide
Double-click to view

1.5 Add New Slide
Double-click to view

1.6 Replace Text in Preser
Double-click to view

1.7 Add Text to Slide
Presentation
ReportTemplatePPT
Slide number
3
Content placeholder
Title 1
Text to add
Top Trending Products List

1.8 Add Text to Slide
Presentation
ReportTemplatePPT
Slide number
4
Content placeholder
Title 1
Text to add
Trend Chart

1.9 Add Text to Slide
Presentation
ReportTemplatePPT
Slide number
5
Content placeholder
Title 1
Text to add
Source Data

***Figure 10-21.** Final configuration of Add Text to Slide exercise*

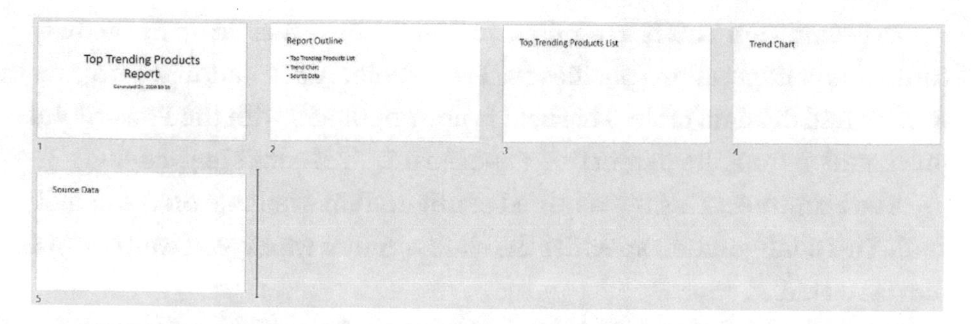

***Figure 10-22.** The state of the ReportTemplate.pptm after the Add Text to Slide exercise*

Add Data Table to Slide

The **Add Data Table to Slide** activity allows you to add a data table to the specified placeholder on a slide.

Configuration

This section provides instructions on how to configure an **Add Data Table to Slide** activity, shown in Figure 10-23.

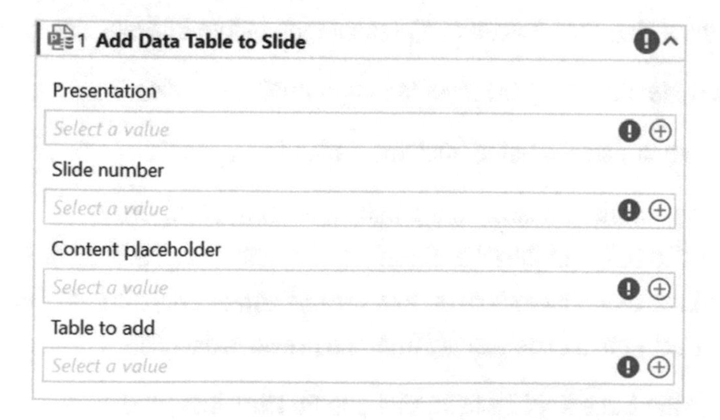

***Figure 10-23.** Activity card for Add Data Table to Slide*

Presentation: This is a required configuration available on the activity card. This configuration specifies the PowerPoint presentation where you want to add the data table. This field is pre-populated with the PowerPoint presentation from the parent `Use PowerPoint Presentation` activity.

Slide number: This is a required configuration available on the activity card. This configuration specifies the slide number where you want to add the data table.

Content placeholder: This is a required configuration available on the activity card. This configuration allows you to specify the name of the placeholder where you want the data table to be added. See `Add Text to Slide` activity to learn how to find the placeholder name in PowerPoint.

Table to add: This is a required configuration available on the activity card. This configuration allows you to specify the actual data table that will be added to the specified placeholder.

EXERCISE

Goal: Building on our previous exercise, use the `Add Data table to Slide` activity to add the data table from `C:\BookSamples\Chapter_10\TopTrendingProducts.xlsx` file to slide 3 of the presentation. Figure 10-22 shows the state of the PowerPoint presentation before this exercise.

Source Code: Chapter_10-PresentationAutomationExercise

Setup: Here are step-by-step implementation instructions:

1. We first need access to the data table from Excel. So, in `StudioX`, add the `Use Excel File` activity to the body of the outer `Use PowerPoint Presentation` activity after the `Add Text to Slide` activity from the previous exercise.

2. In the `Excel file` field, click the `Folder` icon, and select `C:\BookSamples\Chapter_10\TopTrendingProducts.xlsx`.

3. In the Reference as field, leave the default value as Excel.
4. Uncheck the Save changes and Create if not exists options.
5. Add the Add Data Table to Slide activity to the body of the Use Excel File activity.
6. The Presentation field will be pre-populated with the ReportTemplatePPT reference from the parent activity.
7. We are going to add the data table to slide 3. In the Slide number field, click the Plus icon, select the Number option, and type 3.
8. In the Content placeholder field, click the Plus icon, select the Text option, and type Content Placeholder 2. All the new slides that we added using the Add New Slide activity used the Title and Content layout. The title placeholder will be named Content Placeholder 2.
9. Next, in the Table to add field, click the Plus icon, hover over Excel, and select the Top Trending Products [Sheet].

Once you have completed the exercise, the final configuration of the **Add Data Table to Slide** activity should resemble Figure 10-24.

Note We unchecked the Save changes option, so before you run the automation, open the ReportTemplate.pptm; this will allow you to view all the changes that the automation is making.

Figure 10-25 shows the state of title slide after this activity is run.

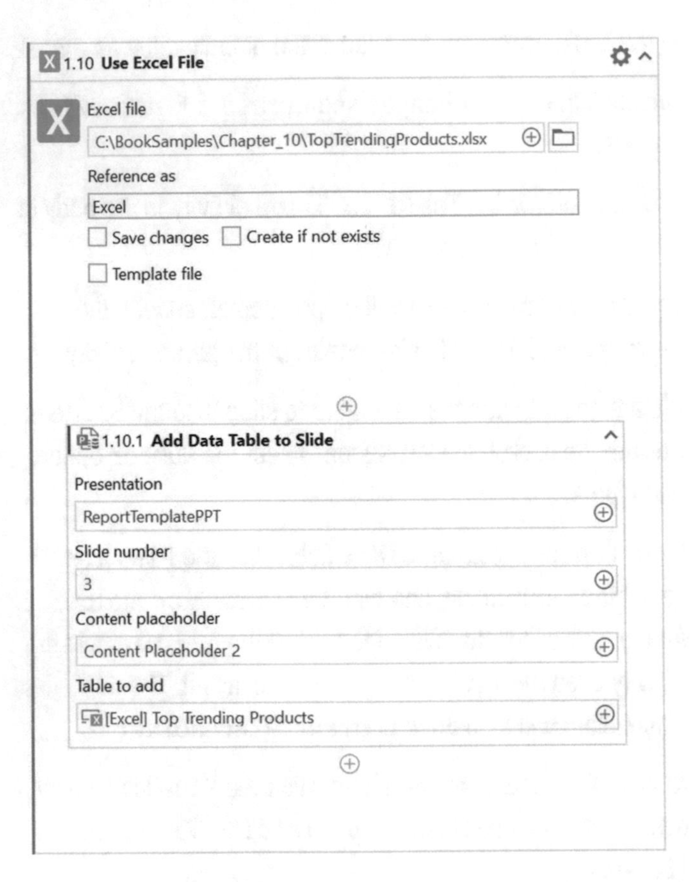

Figure 10-24. *Final configuration of Add Data Table to Slide exercise*

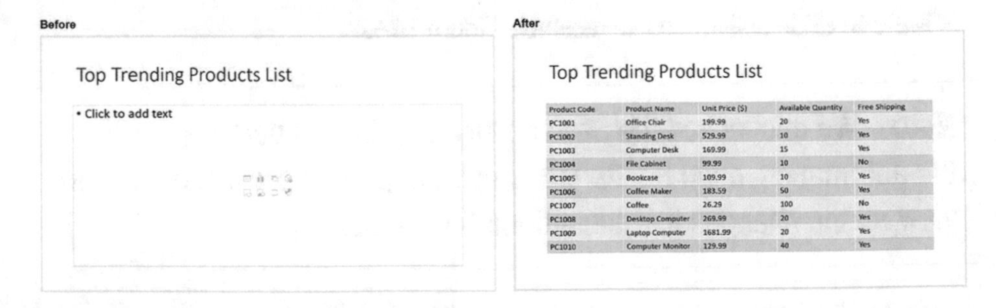

Product Code	Product Name	Unit Price ($)	Available Quantity	Free Shipping
PC1001	Office Chair	199.99	20	Yes
PC1002	Standing Desk	529.99	10	Yes
PC1003	Computer Desk	169.99	15	Yes
PC1004	File Cabinet	99.99	10	No
PC1005	Bookcase	109.99	10	Yes
PC1006	Coffee Maker	183.59	50	Yes
PC1007	Coffee	26.29	100	No
PC1008	Desktop Computer	269.99	20	Yes
PC1009	Laptop Computer	1681.99	20	Yes
PC1010	Computer Monitor	129.99	40	Yes

Figure 10-25. *The before and after view of the Data Table Slide in ReportTemplate.pptm*

Add Image/Video to Slide

The **Add Image/Video to Slide** activity allows you to add an image or a video to the specified placeholder on a slide.

Configuration

This section provides instructions on how to configure an **Add Image/Video to Slide** activity, shown in Figure 10-26.

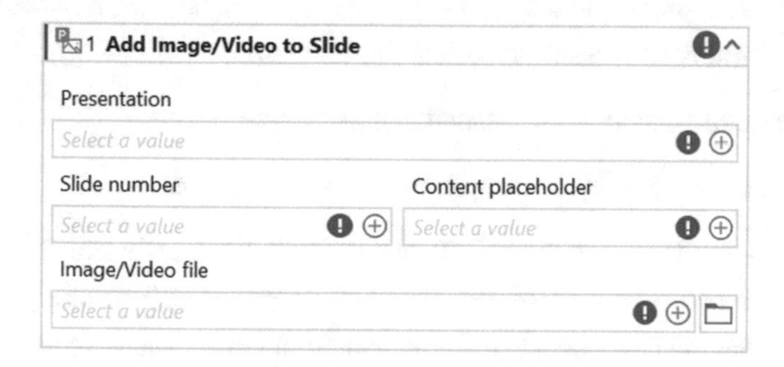

Figure 10-26. *Activity card for Add Image/Video to Slide*

Presentation: This is a required configuration available on the activity card. This configuration specifies the PowerPoint presentation where you want to add the image or video. This field is pre-populated with the PowerPoint presentation from the parent `Use PowerPoint Presentation` activity.

Slide number: This is a required configuration available on the activity card. This configuration specifies the slide number where you want to add the image or video.

Content placeholder: This is a required configuration available on the activity card. This configuration allows you to specify the name of the placeholder where you want the image or video to be added. See `Add Text to Slide` activity to learn how to find the placeholder name in PowerPoint.

Image/Video file: This is a required configuration available on the activity card. This configuration allows you to specify the actual image or video file that will be added to the specified placeholder.

EXERCISE

Goal: Building on our previous exercise, use the `Add Image/Video to Slide` activity to add the Trend Chart image from `C:\BookSamples\Chapter_10\TrendChart.png` file to slide 4 of the presentation. Figure 10-22 shows the state of PowerPoint presentation before this exercise.

Source Code: Chapter_10-PresentationAutomationExercise

Setup: Here are step-by-step implementation instructions:

1. In `StudioX`, add the `Add Image/Video to Slide` activity to the body of outer `Use PowerPoint Presentation` activity after the `Use Excel File` activity from the previous exercise.
2. The `Presentation` field will be pre-populated with the `ReportTemplatePPT` reference from the parent activity.
3. We are going to add the image to slide 4. In the `Slide number` field, click the `Plus` icon, select the `Number` option, and type `4`.
4. In the `Content placeholder` field, click the `Plus` icon, select the `Text` option, and type `Content Placeholder 2`. All the new slides that we added using the `Add New Slide` activity used the `Title and Content` layout. The title placeholder will be named `Content Placeholder 2`.
5. Next, in the `Image/Video file` field, click the `Folder` icon, and select `C:\BookSamples\Chapter_10\TrendChart.png` file.

Once you have completed the exercise, the final configuration of the **Add Image/Video to Slide** activity should resemble Figure 10-27.

Note We unchecked the Save changes option, so before you run the automation, open the ReportTemplate.pptm; this will allow you to view all the changes that the automation is making.

Figure 10-28 shows the state of the title slide after this activity is run.

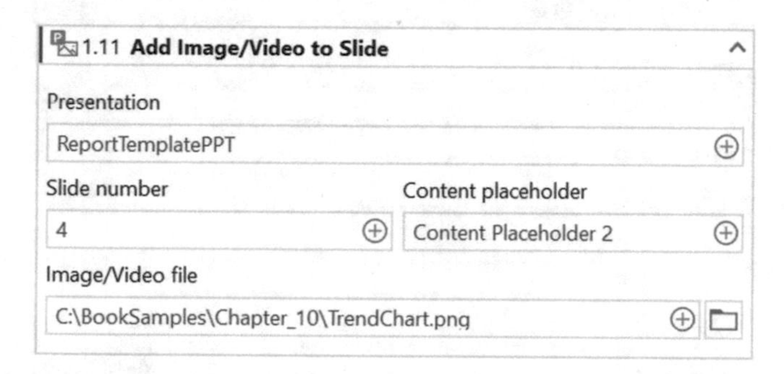

***Figure 10-27.** Final configuration of Add Image/Video to Slide exercise*

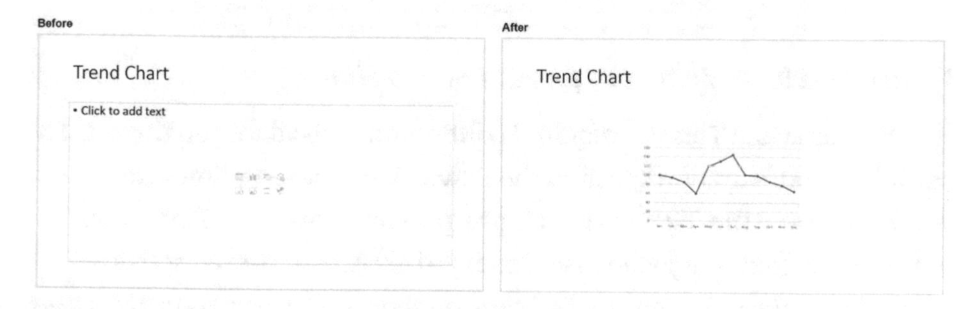

***Figure 10-28.** The before and after view of the Trend Chart slide*

Add File to Slide

The **Add File to Slide** activity allows you to add a file as an attachment to a slide in presentation.

Configuration

This section provides instructions on how to configure an **Add File to Slide** activity, shown in Figure 10-29.

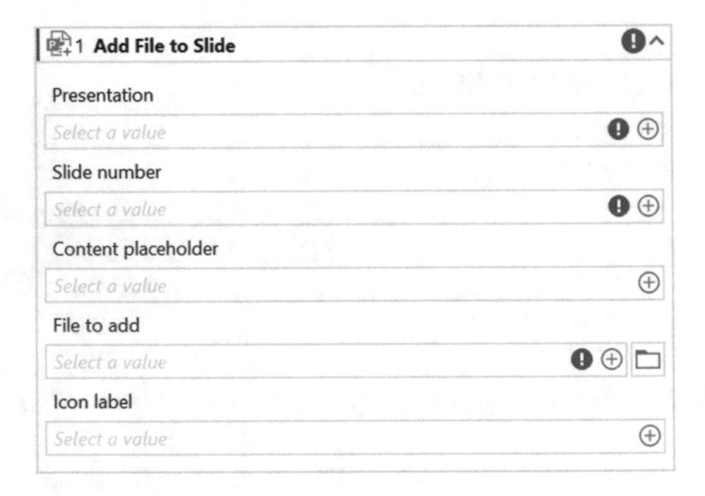

Figure 10-29. *Activity card for Add File to Slide*

Presentation: This is a required configuration available on the activity card. This configuration specifies the PowerPoint presentation where you want to add the file. This field is pre-populated with the PowerPoint presentation from the parent `Use PowerPoint Presentation` activity.

Slide number: This is a required configuration available on the activity card. This configuration specifies the slide number where you want to add the file.

Content placeholder: This is an optional configuration available on the activity card. This configuration allows you to specify the name of the placeholder where you want the file to be added. See `Add Text to Slide` activity to learn how to find the placeholder name in PowerPoint.

File to add: This is a required configuration available on the activity card. This configuration allows you to specify the actual file that will be added to the specified placeholder.

Icon label: This is an optional configuration available on the activity card. This configuration allows you to specify the label of the icon that will be displayed on the slide.

EXERCISE

Goal: Building on our previous exercise, use the `Add File to Slide` activity to add the source data spreadsheet `C:\BookSamples\Chapter_10\TopTrendingProducts.xlsx` to slide 5 of the presentation. Figure 10-22 shows the state of PowerPoint presentation before this exercise.

Source Code: Chapter_10-PresentationAutomationExercise

Setup: Here are step-by-step implementation instructions:

1. In `StudioX`, add the `Add File to Slide` activity to the body of the outer `Use PowerPoint Presentation` activity after the `Add Image/Video to Slide` activity from the previous exercise.
2. The `Presentation` field will be pre-populated with the `ReportTemplatePPT` reference from the parent activity.
3. We are going to add the file to slide 5. In the `Slide number` field, click the `Plus` icon, select the `Number` option, and type `5`.
4. In the `Content placeholder` field, click the `Plus` icon, select the `Text` option, and type `Content Placeholder 2`. All the new slides that we added using the `Add New Slide` activity used the `Title and Content` layout. The title placeholder will be named `Content Placeholder 2`.

5. Next, in the `File to add` field, click the `Folder` icon, and select `C:\BookSamples\Chapter_10\TopTrendingProducts.xlsx` file.

6. In the `Icon label` field, click the `Plus` icon, select the `Text` option, and type `Source Data`.

Once you have completed the exercise, the final configuration of the **Add File to Slide** activity should resemble Figure 10-30.

Note We unchecked the Save changes option, so before you run the automation, open the ReportTemplate.pptm; this will allow you to view all the changes that the automation is making.

Figure 10-31 shows the state of title slide after this activity is run.

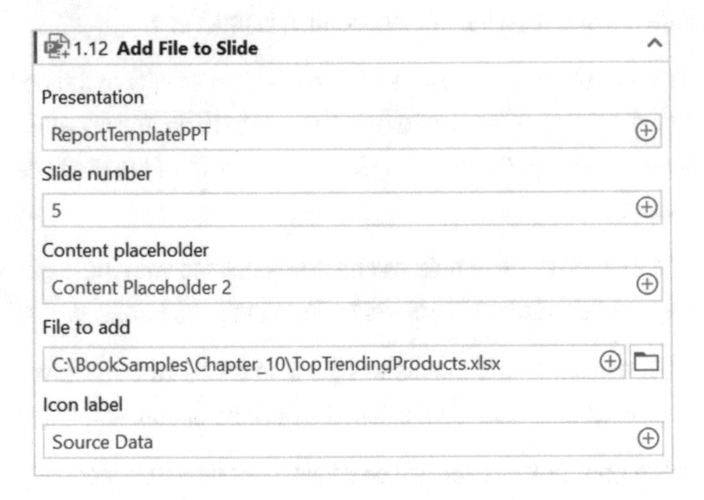

Figure 10-30. *Final configuration of Add File to Slide exercise*

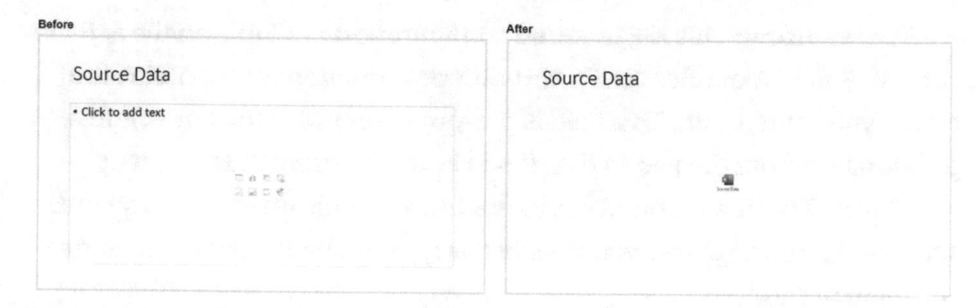

Figure 10-31. The before and after view of the Source Data File Slide in ReportTemplate.pptm

Run Presentation Macro

The **Run Presentation Macro** activity allows you to run the specified macro in PowerPoint presentation.

Configuration

This section provides instructions on how to configure a **Run Presentation Macro** activity, shown in Figure 10-32.

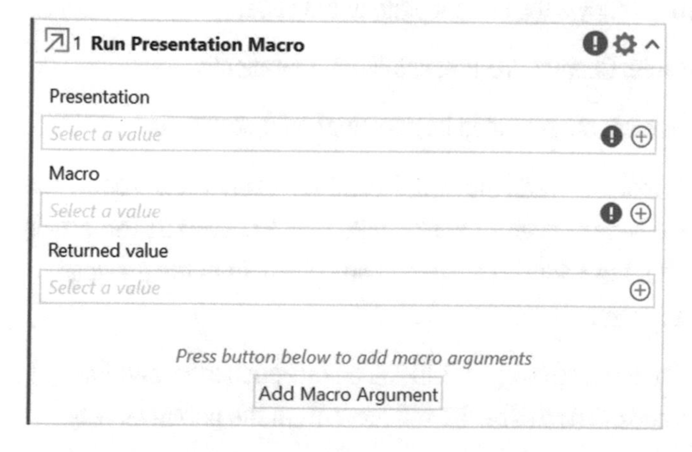

Figure 10-32. Activity card for Run Presentation Macro

Presentation: This is a required configuration available on the activity card. This field identifies the PowerPoint presentation that contains the macro you want to run. This field is pre-populated with the PowerPoint presentation from the parent `Use PowerPoint Presentation` activity.

Macro: This is a required configuration available on the activity card. This configuration allows you to select or type the name of the macro that you want to run.

Returned value: This is an optional configuration available on the activity card. This configuration allows you to save any value returned by the macro.

Add Macro Argument: This is an optional configuration available on the activity card. This configuration allows you to pass one or more arguments (data) to the macro.

EXERCISE

Goal: Building on our previous exercise, use `Run Presentation Macro` activity to run the `ResizeImageToCoverFullSlide` macro. This macro will resize the Trend Chart image on slide 4 so that it occupies the full slide. Figure 10-28 shows the current state of the slide.

Source Code: Chapter_10-PresentationAutomationExercise

Setup: Here are step-by-step implementation instructions:

1. In `StudioX`, add the `Run Presentation Macro` activity to the body of outer `Use PowerPoint Presentation` activity after the `Add File to Slide` activity from the previous exercise.

2. The `Presentation` field will be pre-populated with the `ReportTemplatePPT` reference from the parent activity.

3. In the `Macro` field, click the `Plus` icon, select Text option, and type `ResizeImage`.

Once you have completed the exercise, the final configuration of the **Run Presentation Macro** activity should resemble Figure 10-33.

Note We unchecked the Save changes option, so before you run the automation, open the ReportTemplate.pptm; this will allow you to view all the changes that the automation is making.

Figure 10-34 shows the state of title slide after this activity is run.

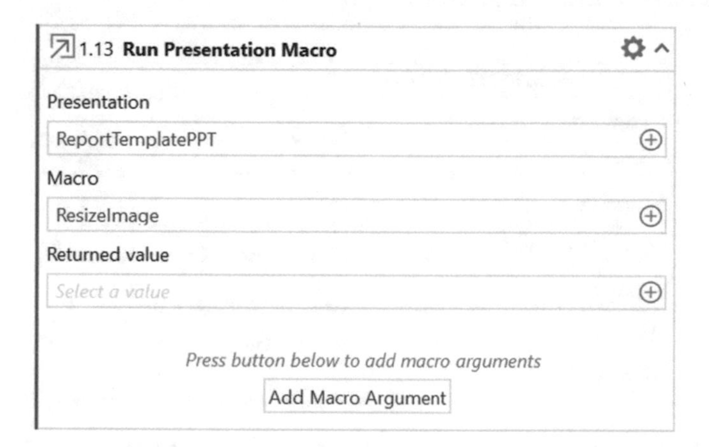

***Figure 10-33.** Final configuration for Run Presentation Macro exercise*

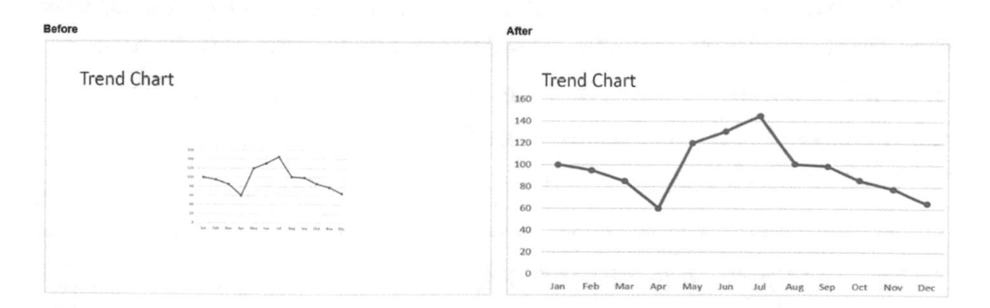

***Figure 10-34.** The before and after view of the Trend Chart slide in ReportTemplate.pptm*

Save PowerPoint File As

The **Save PowerPoint File As** activity allows you to save the referenced PowerPoint file as a different PowerPoint file type including .pptx, .pptm, and .ppt.

Configuration

This section provides instructions on how to configure a **Save PowerPoint File As** activity, shown in Figure 10-35.

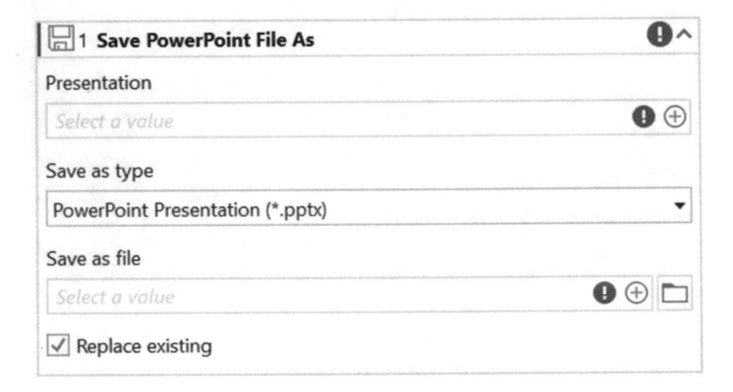

Figure 10-35. *Activity card for Save PowerPoint File As*

Presentation: This is a required configuration available on the activity card. This field identifies the PowerPoint presentation that needs to be saved with a different PowerPoint file type. This field is pre-populated with the PowerPoint presentation from the parent `Use PowerPoint Presentation` activity.

Note The `Save PowerPoint File As` activity only saves a PowerPoint file to .pptx, .ppt, and .pptm types.

Save as type: This is a required configuration available on the activity card. This field identifies the new PowerPoint format that the presentation needs to be saved as through the activity. The following are the dropdown options available:

- PowerPoint Presentation (*.pptx)
- PowerPoint Macro Enabled Presentation (*.pptm)
- PowerPoint 97-2003 Presentation (*.ppt)

Save as file: This is a required configuration available on the activity card. This field provides the name of the PowerPoint file that will be saved as a new PowerPoint format.

Replace existing: This is an optional configuration available on the activity card. If checked, this will replace the existing file of the same name in the target location. By default, this field is checked.

EXERCISE

Goal: Building on our previous exercise, use the `Save PowerPoint File As` activity to save the `ReportTemplate.pptm` file as `Report_YYYYMMDD.pptx` file. The `YYYYMMDD` will be replaced with actual date the report was generated.

Setup: Here are step-by-step implementation instructions:

1. In `StudioX`, add the `Save PowerPoint File As` activity to the body of outer `Use PowerPoint Presentation` activity after the `Run Presentation Macro` activity from the previous exercise.

2. The `Presentation` field will be pre-populated with the `ReportTemplatePPT` reference from the parent activity.

3. Next, click the `Save as type` dropdown to select `PowerPoint Presentation (*.pptx)` option.

4. Next, click the Plus icon in the Save as file field, select the Text option, and type in the file path C:\BookSamples\Chapter_10\Report_.
5. Next, from within the Text Builder, click the Plus icon, and hover over Notebook to select Date [Sheet]  YYYYMMDD [Cell].
6. Add .pptx at the end and click Save.
7. Leave the Replace existing option checked.

Once you have completed the exercise, the final configuration of the **Save PowerPoint File As** activity should resemble Figure 10-36.

Once you run this activity, a new file will be generated and saved in the Destination folder.

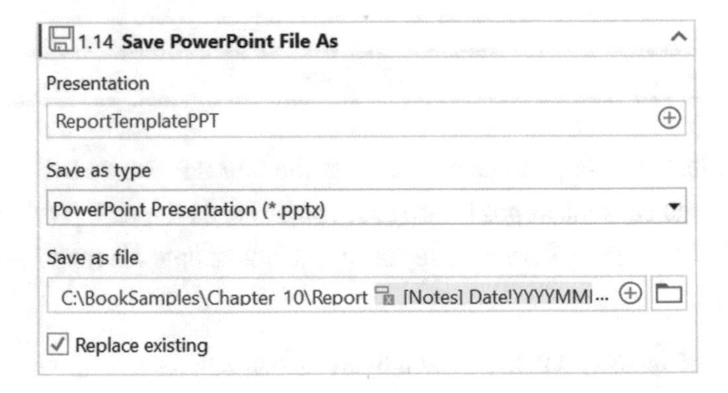

Figure 10-36. *Final configuration for Save PowerPoint File As exercise*

Save Presentation as PDF

The **Save Presentation as PDF** activity allows you to save the referenced PowerPoint presentation as a PDF file.

Configuration

This section provides instructions on how to configure a **Save Presentation as PDF** activity, shown in Figure 10-37.

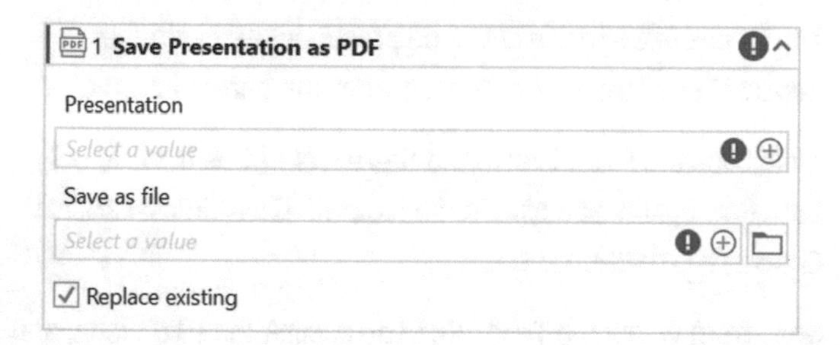

Figure 10-37. *Activity card for Save Presentation as PDF*

Presentation: This is a required configuration available on the activity card. This field identifies the PowerPoint presentation that needs to be saved as a PDF. This field is pre-populated with the PowerPoint from the parent `Use PowerPoint Presentation` activity.

Save as file: This is a required configuration available on the activity card. This field provides the file path to save the new PDF file created from the activity.

Replace existing: This is an optional configuration available on the activity card. If checked, this will replace the existing file of the same name in the target location. By default, this field is checked.

EXERCISE

Goal: Building on our previous exercise, use the `Save Presentation as PDF` activity to save the `ReportTemplate.pptm` file as `Report_YYYYMMDD.pdf` file. The `YYYYMMDD` will be replaced with actual date the report was generated.

Setup: Here are step-by-step implementation instructions:

1. In StudioX, add the Save Presentation as PDF activity to the body of outer Use PowerPoint Presentation activity after the Save PowerPoint File As activity from the previous exercise.

2. The Presentation field will be pre-populated with the ReportTemplatePPT reference from the parent activity.

3. Next, click the Plus icon in the Save as file field, select the Text option, and type in the file path C:\BookSamples\Chapter_10\Report_.

4. Next, from within the Text Builder, click the Plus icon, and hover over Notebook to select Date [Sheet]  YYYYMMDD [Cell].

5. Add .pdf at the end and click Save.

6. Leave the Replace existing option checked.

Once you have completed the exercise, the final configuration of the **Save Presentation as PDF** activity should resemble Figure 10-38.

Once you run this activity, a new file will be generated and saved in the Destination folder.

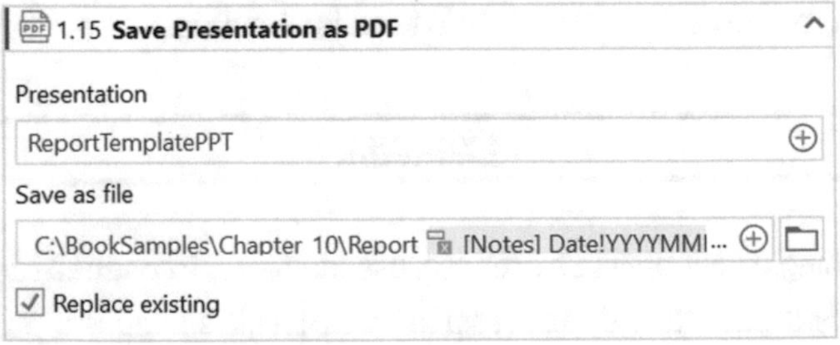

Figure 10-38. *Final configuration for Save Presentation as PDF exercise*

PART III

Prototypes

CHAPTER 11

Product Data Entry Automation

In this chapter, we are going to automate a daily task to enter product data as part of an inventory management process.

Learning Objectives

This automation will help you understand the following topics:

- Creating folder structures, copying, and moving files
- Fetching a filtered list of emails, saving email attachments, moving emails, and sending emails
- Reading, filtering, and saving data in Excel
- Using Excel to keep track of status
- Entering data in a web application

Manual Task Overview

The task starts when an inventory manager receives a new inventory email from the procurement team. Figure 11-1 shows a high-level overview of the product data entry task.

A. Javed et al., *Robotic Process Automation using UiPath StudioX*,
https://doi.org/10.1007/978-1-4842-6794-3_11

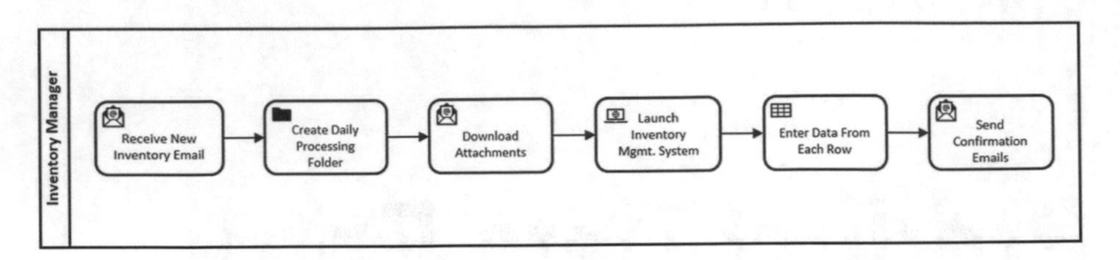

***Figure 11-1.** A high-level overview of the product data entry task*

The inventory manager on a frequent basis checks the Outlook Inbox for unread new inventory emails. Figure 11-2 shows a sample of the email.

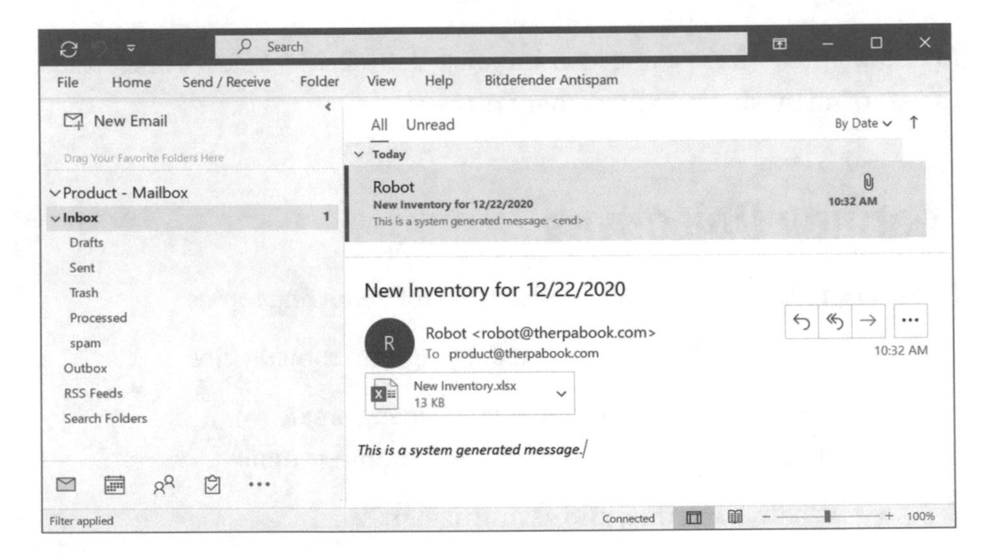

***Figure 11-2.** Sample of new inventory email*

If there is an unread new inventory email with products list as an attachment, the inventory manager downloads the attachment to a folder for further processing. To ensure that the same email is not processed multiple times, the inventory manager moves the email to the Processed folder in Outlook.

If this is the first email of the day, then before downloading the attachments, the inventory manager creates a new folder structure to manage processing. This new folder is created for the current date in YYYYMMDD format under the DailyProcessing folder. Under the current date folder, the inventory manager also creates two subfolders, Pending and Processed.

All attachments are by default downloaded to the Pending folder, and once all new products have been entered from a file in the inventory management system, files are moved to the Processed folder. Figure 11-3 shows the folder structure used for managing product data entry tasks.

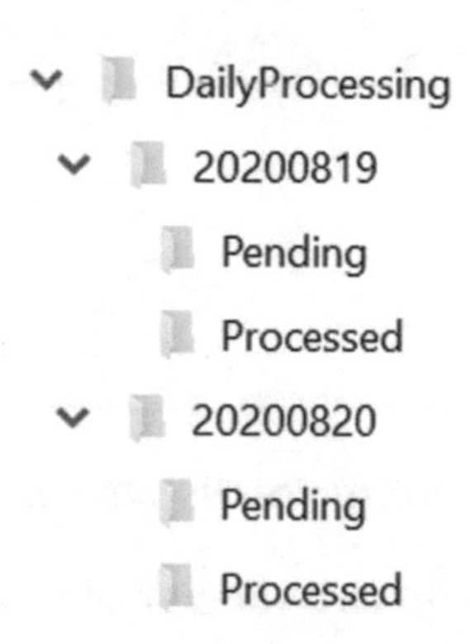

Figure 11-3. *Folder structure for managing product data entry tasks*

The products that need to be entered into the inventory are included in the email as an Excel attachment. This Excel file is in a predefined format, shown in Figure 11-4.

	A	B	C	D	E
1	**Product Code**	**Product Name**	**Unit Price ($)**	**Available Quantity**	**Free Shipping**
2	PC1001	Office Chair	199.99	20	Yes
3	PC1002	Standing Desk	529.99	10	Yes
4	PC1003	Computer Desk	169.99	15	Yes
5	PC1004	File Cabinet	99.99	10	No

List

Figure 11-4. *Sample of products list Excel attachment*

The inventory manager launches the inventory management system in a web browser, shown in Figure 11-5. In parallel, the inventory manager opens the products list in Excel.

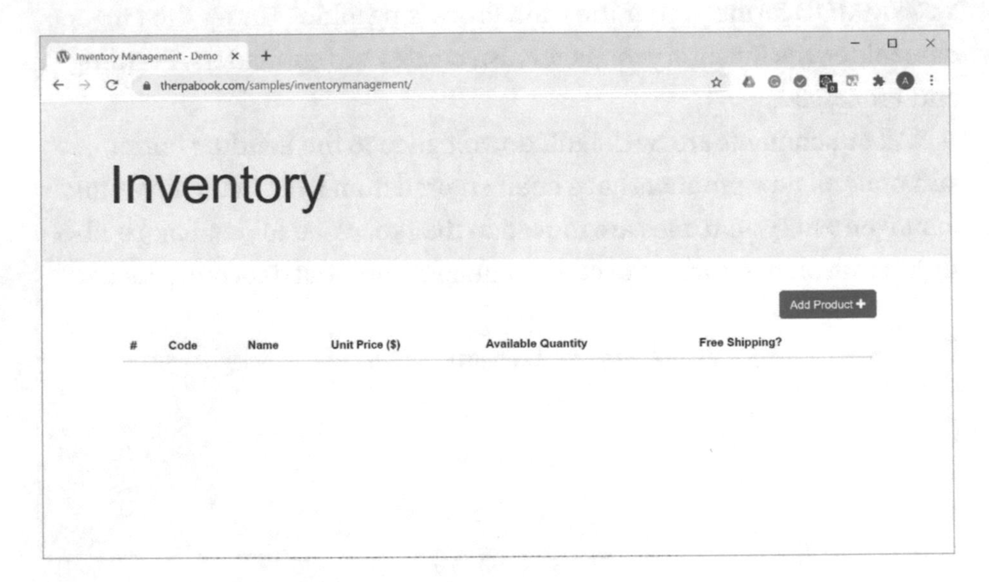

***Figure 11-5.** Inventory management system*

The inventory manager goes through each row of the Excel file, copies data from each column, and pastes it in the Product Details dialog of the inventory management system, shown in Figure 11-6.

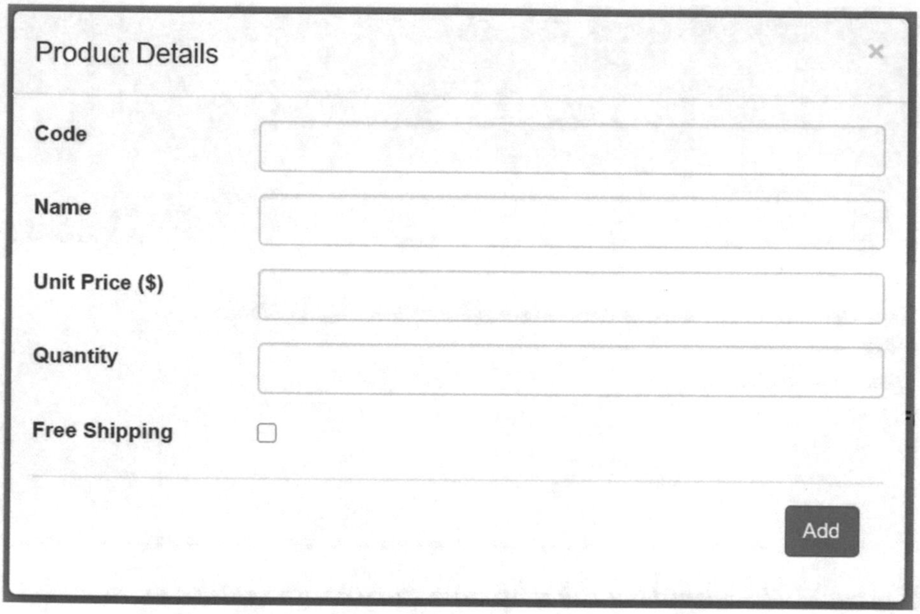

***Figure 11-6.** Product Details dialog of the inventory management system*

Once all products have been entered into the inventory management system, the inventory manager sends an email to the sender, confirming that the processing has been completed.

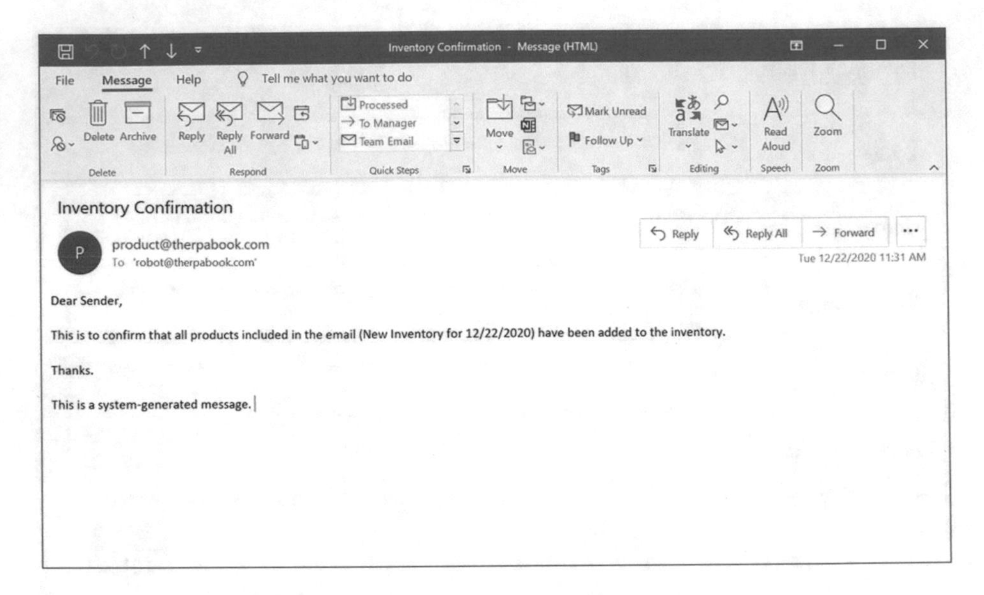

Figure 11-7. *Sample of the new inventory confirmation email*

Solution Design

This section provides an overview of how we are going to design the automation. Figure 11-8 shows the process flow diagram for automating the product data entry task.

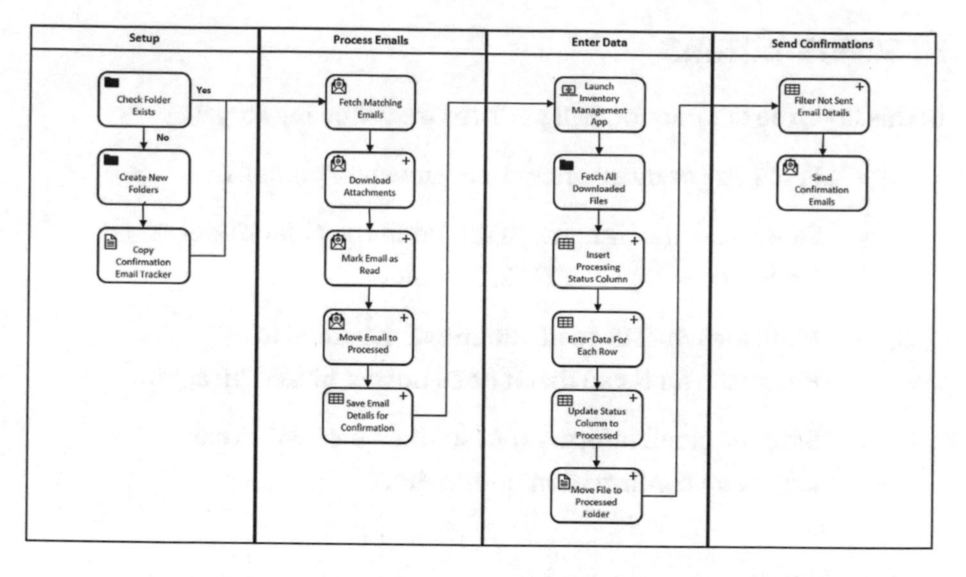

***Figure 11-8.** Solution design of the product data entry task automation*

Dividing an automation into smaller and logical flows helps simplify the implementation and debugging process. As shown in Figure 11-8, we are going to divide the automation into four logical flows.

Initialize

In this flow, the automation will perform the following steps:

- The automation will start by checking if the required folder structure exists for the current date.
- If the folder does not exist, then the automation will create the required folder structure. This will also include copying the Confirmation Email Tracker file to the current date folder.

Process Emails

In this flow, the automation will perform the following steps:

- Fetch one or more unread new inventory emails.
- Download attachments to the Pending folder under DailyProcessing\YYYYMMDD folder.
- Mark the email as read and move the email to Processed folder so that it does not get picked up again.
- Save the email information, so that automation can later send confirmation notification.

Enter Data

In this flow, the automation will perform the following steps:

- Launch the inventory management system in a browser.
- Fetch all new inventory Excel files that are in the Pending folder.
- Insert a Status column in each Excel file, so that processing status can be tracked at individual row level.
- The automation will copy data from all rows and enter the data into the inventory management web application.
- Once data from a single row has been entered, the automation will update the Status to Processed.
- Once the complete file has been processed, the automation will move the file to the Processed folder. This will ensure the file does not get reprocessed.

Send Confirmation

In this flow, the automation will perform the following steps:

- Automation will fetch a list of emails from the Confirmation Email Tracker that have a status of Not Sent.
- Automation will send a confirmation email to all senders.

Implementation

Now that we have a good understanding of how we will design the automation, we can get started with the implementation.

Step 1: Setup

The first step of this implementation is to ensure we have downloaded the demo application and associated templates from the book's GitHub repository.

Note This chapter assumes that the automation folder `ProductDataEntryAutomation` is placed inside the `C:\BookSamples\` folder.

The `ProductDataEntryAutomation` folder contains three subfolders, as shown in Figure 11-9.

Figure 11-9. *The folder structure of product data entry automation*

App

This subfolder contains the inventory management web application. You can launch the web application by clicking `home.html`, and it will open it in your default web browser.

You can also access the inventory management web application in your browser at `https://therpabook.com/samples/inventorymanagement/`.

Templates

This subfolder contains two Excel templates used by the automation.

The `ProductsList.xlsx` template helps UiPath understand the structure of the Excel attachments received via email. Figure 11-10 shows the structure of this Excel template.

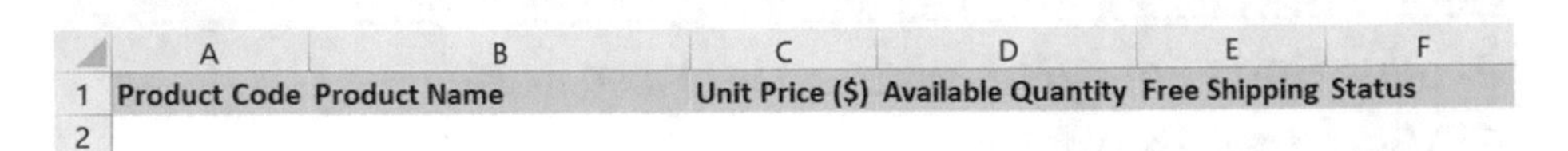

	A	B	C	D	E	F
1	Product Code	Product Name	Unit Price ($)	Available Quantity	Free Shipping	Status
2						

Figure 11-10. *Excel template for products list received as an email attachment*

A copy of `ConfirmationEmailTracker.xlsx` template will be made in the `DailyProcessing\YYYYMMDD` folder. This Excel file will be used by the automation to store email details so that the automation can send confirmation at the end of each run. Figure 11-11 shows the structure of this Excel template.

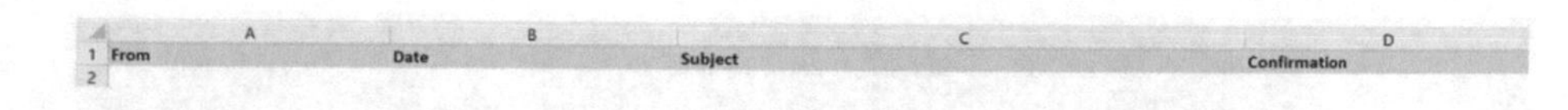

Figure 11-11. *Excel template for the confirmation email tracker*

DailyProcessing

This subfolder is empty and will be used by the automation to create new folders and to save and process attachments.

Samples

This subfolder contains a sample inventory file that can be used for testing.

Step 2: Create Project

Next, we are going to create a new project in UiPath StudioX. Here are the steps to creating a new project. You do not have to use the exact same name and description.

1. Launch `UiPath StudioX`.
2. From the `Home` screen, click `Blank Task` to create a new project from scratch.
3. In the `Process name:` field, enter `Chapter_11_ProductDataEntryPrototype`.
4. In the `Description:` field, enter `Automation for product data entry prototype`.
5. Click the `Create` button to create the project.

Figure 11-12 shows the final configuration of this step.

Blank Task

Blank Task
Start with a blank project to design a new task automation

Process name:
Chapter_11_ProductDataEntryPrototype

Location:
C:\Users\adeel\Documents\UiPath

Description:
Automation for product data entry prototype

Show advanced options

Create Cancel

Figure 11-12. *Create new project*

Step 3: Setup Project Notebook

This automation is going to use custom formulas for generating folders and file paths.

We are going to configure a custom notebook to manage all our custom formulas. Figure 11-13 shows the custom notebook with all formulas used by the automation.

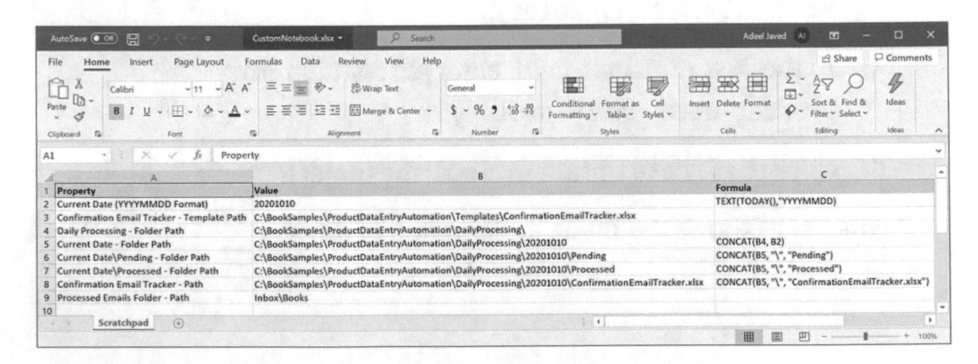

Property	Value	Formula
Current Date (YYYYMMDD Format)	20201010	TEXT(TODAY(),"YYYYMMDD)
Confirmation Email Tracker - Template Path	C:\BookSamples\ProductDataEntryAutomation\Templates\ConfirmationEmailTracker.xlsx	
Daily Processing - Folder Path	C:\BookSamples\ProductDataEntryAutomation\DailyProcessing\	
Current Date - Folder Path	C:\BookSamples\ProductDataEntryAutomation\DailyProcessing\20201010	CONCAT(B4, B2)
Current Date\Pending - Folder Path	C:\BookSamples\ProductDataEntryAutomation\DailyProcessing\20201010\Pending	CONCAT(B5, "\", "Pending")
Current Date\Processed - Folder Path	C:\BookSamples\ProductDataEntryAutomation\DailyProcessing\20201010\Processed	CONCAT(B5, "\", "Processed")
Confirmation Email Tracker - Path	C:\BookSamples\ProductDataEntryAutomation\DailyProcessing\20201010\ConfirmationEmailTracker.xlsx	CONCAT(B5, "\", "ConfirmationEmailTracker.xlsx")
Processed Emails Folder - Path	Inbox\Books	

Figure 11-13. *Custom formulas in the Project Notebook*

Table 11-1 describes how each value shown in Figure 11-13 is used.

Table 11-1. *Description of each value in Scratchpad*

Property	Description
Current Date	The formula in cell B2 will generate the current date in YYYYMMDD format that will be used by all paths in subsequent rows.
Confirmation Email Tracker – Template Path	Static path of the ConfirmationEmailTracker.xlsx template.
Daily Processing – Folder Path	Static path of the folder used by the automation to store folders and files for daily processing.
Current Date – Folder Path	Dynamic path of the folder used by the automation to manage processing for the current day.
Current Date\Pending – Folder Path	Dynamic path of the folder used by the automation to store attachments that are pending processing.
Current Date\Processed – Folder Path	Dynamic path of the folder used by the automation to store attachments that have been processed.
Confirmation Email Tracker – Path	Dynamic path where the copy of ConfirmationEmailTracker.xlsx for current day is placed.
Processed Emails Folder – Path	Static path of the Outlook folder where all emails are moved after processing.

Here are the step-by-step instructions to configure a custom notebook:

1. In StudioX, click the Notebook ➤ Configure Notebook menu from the ribbon on top.
2. In the Notebook file field, click the ... and select C:\BookSamples\ProductDataEntryAutomation\Templates\CustomNotebook.xlsx file, as shown in Figure 11-14.
3. Click the OK button.

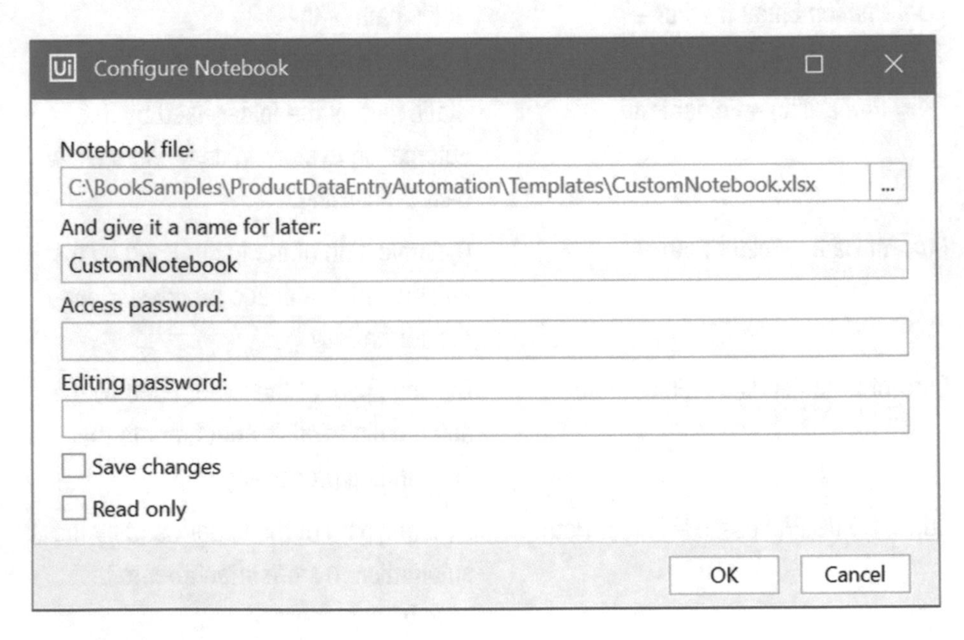

***Figure 11-14.** Configure custom notebook*

Step 4: Check Folder Structure Exists

The automation can be run multiple times a day. The first run of the day will create the required folder structure, so the subsequent runs do not need to repeat this step. To accomplish this, the automation needs to check if a folder for the current date (YYYYMMDD format) already exists under the `DailyProcessing` folder:

1. In `StudioX`, add the `Group` activity to a blank process.
2. Update `Display Name` to `Initialize`.
3. Add the `Folder Exists` activity to the body of `Initialize` group activity.
4. Update `Display Name` to `Check Folder Structure Exists`.
5. In the `Folder path` field, select the `Notebook` ➤ `Indicate in Excel` option. Select cell `B5` from the `Scratchpad` worksheet. As mentioned in Table 11-1, this cell will contain the dynamic path of the current date folder.
6. In the `Save result` field, use `Save for Later Use` option and name your saved value as `CurrentDateFolderExists`.

In this step, we are not going to specify what needs to happen in case the folder exists. We will do that in the next step. Figure 11-15 shows the configuration of this step.

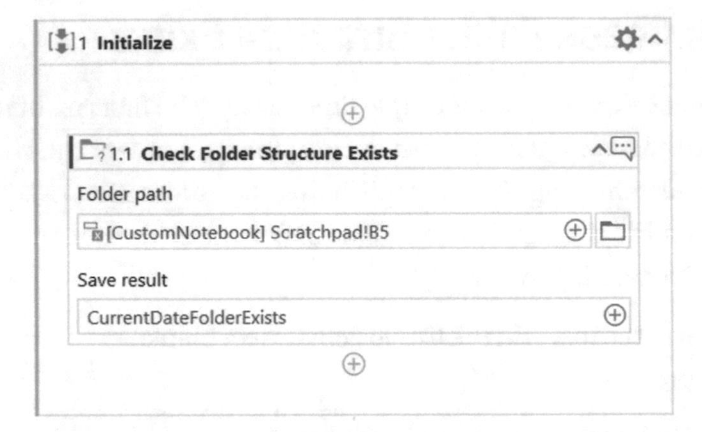

Figure 11-15. *Configuration of Check Folder Structure Exists step*

Step 5: Create New Folders

Next, we are going to use the result `CurrentDateFolderExists` from `Check Folder Structure Exists` activity and specify what actions the automation needs to perform.

If `CurrentDateFolderExists` is true, then that means the folder structure has already been created by a prior run. On the other hand, if `CurrentDateFolderExists` is false, then the automation needs to create the necessary folder structure before it can proceed.

Once the folder structure has been created, the final step in this flow will be to copy the `ConfirmationEmailTracker.xlsx` to the current date folder. This Excel will be used by the automation to save email details so that at the end of the run, it can send completion confirmation emails to senders:

1. In `StudioX`, add the `If` activity to the body of `Initialize` group activity after the `Check Folder Structure Exists` activity.
2. Update `Display Name` to `Create Folder Structure`.

3. Using the Condition Builder, create a condition CurrentDateFolderExists is false.

4. Add a Create Folder activity in the Then block of Create Folder Structure activity.

5. Update Display Name to Create Current Date Folder.

6. In the Folder name field, select the Notebook ➤ Indicate in Excel option. Select cell B5 from the Scratchpad worksheet.

7. Add another Create Folder activity in the Then block of the Create Folder Structure activity after the Create Current Date Folder activity.

8. Update Display Name to Create Pending Sub-Folder.

9. In the Folder name field, select the Notebook ➤ Indicate in Excel option. Select cell B6 from Scratchpad worksheet. As mentioned in Table 11-1, this cell will contain the dynamic path of the Pending folder for the current date.

10. Add another Create Folder activity in the Then block of the Create Folder Structure activity after the Create Pending Sub-Folder activity.

11. Update Display Name to Create Processed Sub-Folder.

12. In the Folder name field, select the Notebook ➤ Indicate in Excel option. Select cell B7 from Scratchpad worksheet. As mentioned in Table 11-1, this cell will contain the dynamic path of the Processed folder for the current date.

13. Add a Copy File activity in the Then block of the Create Folder Structure activity after the Create Processed Sub-Folder activity.

14. Update Display Name to Copy Confirmation Email Tracker.

15. In the From field, select the Notebook ➤ Indicate in Excel option. Select cell B3 from the Scratchpad worksheet. As mentioned in Table 11-1, this cell will contain the static path of the ConfirmationEmailTracker.xlsx template file.

16. In the To field, select the Notebook ➤ Indicate in Excel option. Select cell B8 from the Scratchpad worksheet. As mentioned in Table 11-1, this cell will contain the dynamic path where the tracker will be placed.

Figure 11-16 shows the configuration of this step.

Figure 11-16. *Configuration of Create Folder Structure step*

Step 6: Process Emails

Next, the automation needs to connect with an Outlook account and fetch relevant emails. To narrow down the list of emails that the automation needs to fetch and process, we will also add a filter. For each email, the automation will need to download attachments, mark the email as read, and move the email to the Processed folder:

1. In `StudioX`, add the `Use Desktop Outlook App` activity after the `Initialize` group activity.
2. Update the `Display Name` to `Process Emails`.
3. If you have multiple accounts set up in Outlook, then select the appropriate one from `Account` dropdown. Otherwise, you can use the `Default Email Account`. You can use the same account for sending and receiving emails for this automation.
4. Leave the value in `Reference as` field as `Outlook`.
5. Add the `For Each Email` activity in the body of `Process Emails` activity. This will allow you to fetch all matching emails and go through them one by one.
6. Update `Display Name` to `Fetch Matching Emails`.
7. In the `For each` field, leave the value as `CurrentMail`. All activities in the body of `Fetch Matching Emails` activity will be able to access the current email attributes by referencing `CurrentMail`.
8. In the `Emails from` field, select the folder where emails are received; this is typically the `Inbox` folder.

9. Update `Limit emails to first` field to 10. Since this is a prototype and we do not expect too many emails, so 10 is okay, but this will vary based on your actual use case.

10. The automation is only interested in Unread emails with attachments, so check the `Unread mail` and `With attachments only` checkboxes.

11. To further limit emails that the automation fetches, add an `Additional Filter` on the email subject, as shown in Figure 11-17.

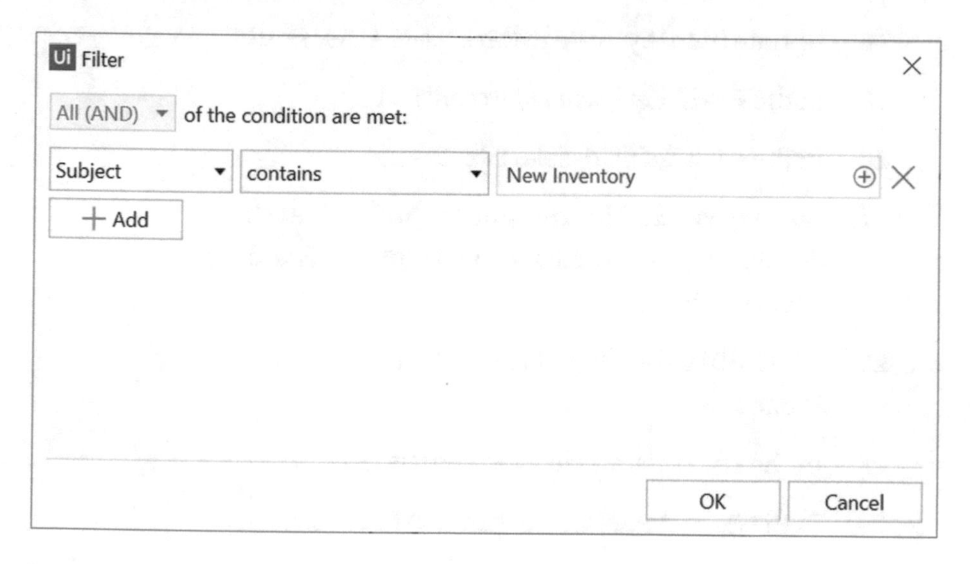

***Figure 11-17.** Additional filters dialog for fetching Outlook emails*

12. Add a `Save Email Attachments` activity in the body of `Fetch Matching Emails` activity.

13. Update `Display Name` to `Download Attachments`.

14. In the `Email` field, select `CurrentMail`.

15. The email will only contain Excel attachments, but to ensure that the automation only downloads Excel attachments from the email, enter `*.xls*` in the `Filter by file name` field.

16. In the `Save to folder` field, select the `Notebook` ➤ `Indicate in Excel` option. Select cell B6 from `Scratchpad` worksheet.

17. Add a `Mark Email As Read/Unread` activity in the body of `Fetch Matching Emails` activity after the `Download Attachments` activity.

18. Update `Display Name` to `Mark Email As Read`.

19. In the `Email` field, select `CurrentMail`.

20. In the `Mark as` field, select `Read`.

21. Add a `Move Email` activity in the body of `Fetch Matching Emails` activity after the `Mark Email As Read` activity.

22. Update `Display Name` to `Move Email to Processed`.

23. In the `Email` field, select `CurrentMail`.

24. In the `Move to` field, select `Inbox\Processed`.

Figure 11-18 shows the configuration of the Process Emails step.

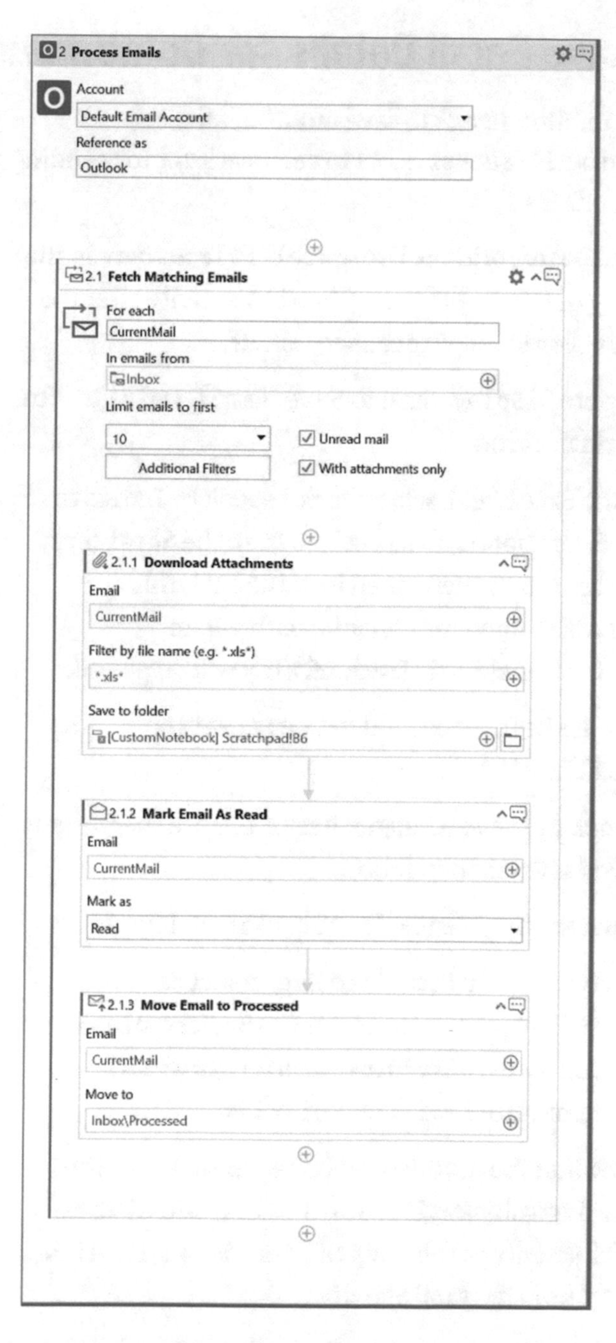

Figure 11-18. *Configuration of Process Emails step*

Step 7: Save Email Details for Confirmation

Next, the automation needs to save email details in `ConfirmationEmailTracker.xlsx` to retrieve later for sending confirmation emails:

1. In `StudioX`, add the `Use Excel File` activity in the body of `Fetch Matching Emails` activity after the `Move Email to Processed` activity.
2. Update `Display Name` to `Save Email Details for Confirmation`.
3. In the `Excel` field, select the `Notebook` ➤ `Indicate in Excel` option. Select cell `B8` from the `Scratchpad` worksheet. As mentioned in Table 11-1, this cell will contain the dynamic path where `ConfirmationEmailTracker.xlsx` will be placed.
4. Update `Reference` field to `ConfirmationEmailTracker`.
5. Check the `Save changes` flag to ensure the file gets saved after each iteration.
6. Uncheck the `Create if not exists` flag.
7. Check the `Template file` flag, select `Browse for file` option, and select `C:\BookSamples\ProductDataEntryAutomation\Templates\ConfirmationEmailTracker.xlsx`.
8. Now that the automation has a reference to the `ConfirmationEmailTracker.xlsx`, add a `Write Cell` activity in the body of `Save Email Details for Confirmation` activity.
9. Update `Display Name` to `Save From Email Address`.

10. In the What to write field, select [Current Email] From; this contains the from email address of the current email.

11. In the Where to write field, you need to provide a cell where data needs to be entered. Select ConfirmationEmailTracker ➤ Indicate in Excel option, and select cell A2. You just need to provide the starting cell number.

12. Check the Auto increment row option. As mentioned in the previous step, you only need to provide the starting cell because this auto-increment configuration will make sure that the automation keeps moving to the next row with each iteration of the For Each Email loop.

13. Copy and paste the Write Cell activity. Update Display name to Save Date Sent. Update Where to write field to [Current Mail] Date. Update Where to write field to cell B2.

14. Copy and paste the Write Cell activity. Update Display name to Save Email Subject. Update Where to write field to [CurrentMail] Subject. Update Where to write field to cell C2.

15. Copy and paste the Write Cell activity. Update Display name to Save Default Confirmation Status. In the What to write field, type Not Sent. Update Where to write field to cell D2. The automation will update the status to Sent later in the process. The reason for adding this default Not Sent status is so that if emails need to be sent out multiple times a day, the automation has a way to filter Sent and Not Sent rows in Excel.

Figure 11-19 shows the configuration of this step.

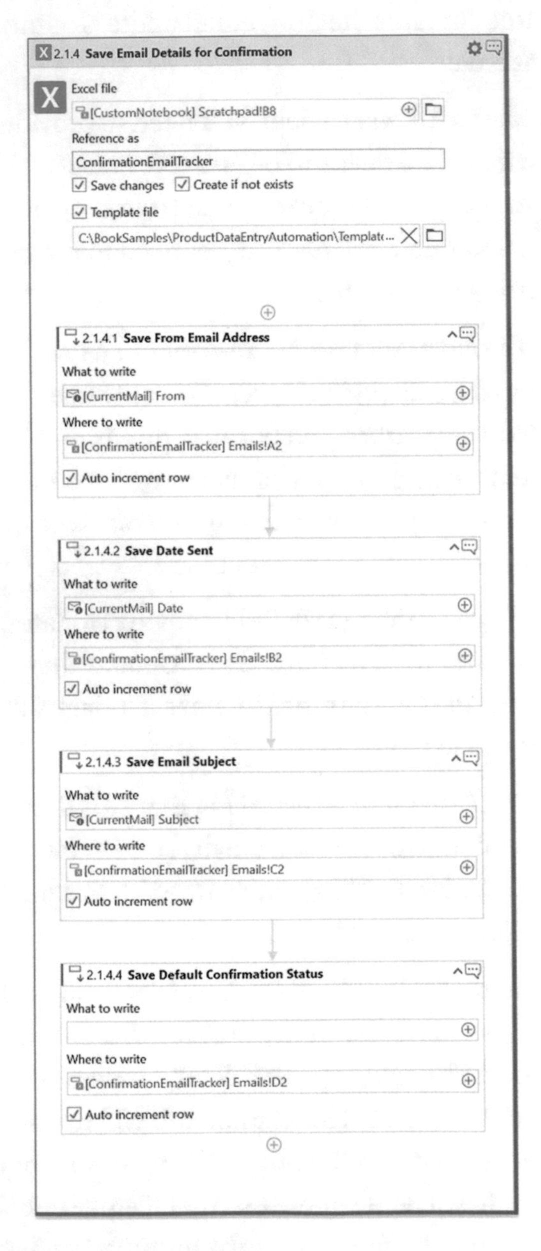

***Figure 11-19.** Configuration of Save Email Details for Confirmation activity*

Step 8: Launch Inventory Management App

Next, we will configure the automation to launch a new instance of the inventory management system:

1. Open your favorite browser, and open `https://therpabook.com/samples/inventorymanagement/` web application.
2. In `StudioX`, add the `Use Application/Browser` activity after the `Process Emails` activity.
3. Update `Display Name` to `Enter Data`.
4. Click the `Indicate application to automation (I)` link and select your browser that has the web application currently open. UiPath will auto-detect that you are trying to interact with a browser, and it will also populate the `URL`.
5. From the `Properties` panel, select `Always` option for `Options` ➤ `Open` property. This will ensure a new browser is opened each time the automation runs.
6. Add `Get Active Window` activity in the body of `Enter Data` activity.
7. Open the `Properties` panel, and in the `Output` ➤ `ApplicationWindow` property, click the `Plus` icon. Use the `Save for Later Use` option and name your saved value as `InventoryManagementWindow`.
8. Add `Maximize Window` activity in the body of `Get Active Window` activity.

9. Open the `Properties` panel, and in the `Output` ➤ `ApplicationWindow` property, click the `Plus` icon. Hover over the `Use Saved Value` option to select `InventoryManagementWindow`.

Figure 11-20 shows the configuration of this step.

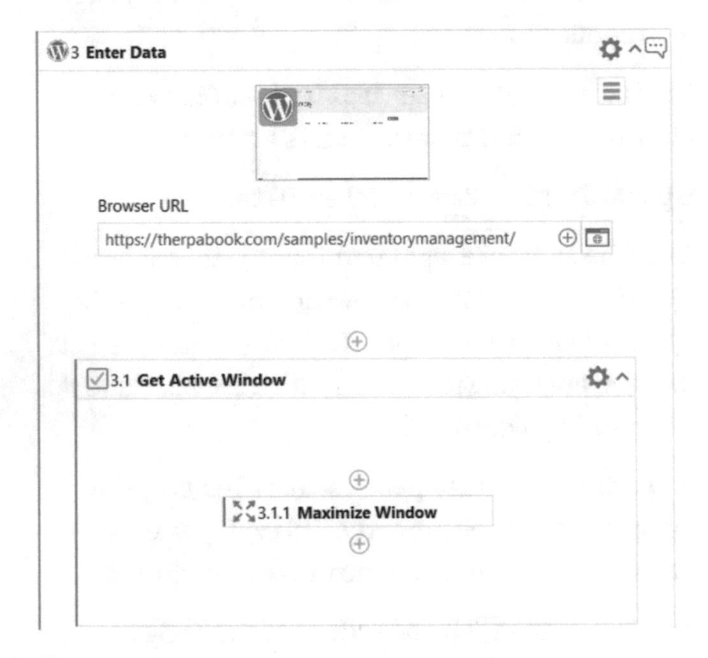

***Figure 11-20.** Configuration of Enter Data activity*

Step 9: Fetch All Downloaded Files

As mentioned in the solution design and implemented in the Process Emails flow, we can receive one or more new inventory emails. Each email will contain an attachment, so the Pending folder can contain multiple files (equal to the number of emails received). In this step, we will enable the automation to fetch all files from the Pending folder:

1. In `StudioX`, add the `For Each File In Folder` activity in the body of `Enter Data` activity after the `Refresh Web Page` activity.
2. Update `Display Name` to `Fetch All Downloaded Files`.
3. In the `For each` field, leave the reference as `CurrentFile`.
4. In the `In folder` field, select the `Notebook` ➤ `Indicate in Excel` option. Select cell `B6` from the `Scratchpad` worksheet.
5. Uncheck the `Include subfolders` option since there is no subfolder.
6. In the `Filter by` field, type `*.xlsx`. This will ensure that only Excel files are picked up from the folder.
7. In the `Order by` field, enter `Created date oldest first`. This will return files in older to newer order.

Figure 11-21 shows the configuration of this step.

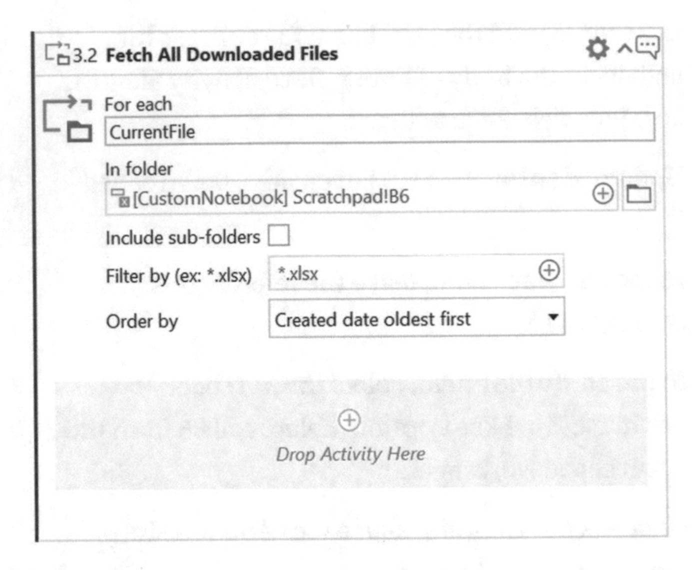

Figure 11-21. *Configuration of Fetch All Downloaded Files activity*

Step 10: Insert Processing Status Column

As the automation processes each row of each file, we also want to keep track of the rows that were processed. There is no status column in the new inventory file, so first, we will insert a new column in each Excel file:

1. In StudioX, add the `Use Excel File` activity in the body of `Fetch All Downloaded Files` activity.
2. Update `Display Name` to `Process Products List Excel`.
3. In the `Excel file` field, use `[CurrentFile] Full Path`. This will reference the current Excel file being processed by `For Each File In Folder` loop.
4. Update `Reference as` to `Products`.
5. Check `Save changes` option.
6. Uncheck `Create if not exists` option.

7. Check Template option, and use Browse for file option to select C:\BookSamples\ProductDataEntryAutomation\Templates\ProductsList.xlsx.

8. Add Insert Column activity in the body of Process Products List Excel activity.

9. Update Display Name to Insert Processing Status Column.

10. In the Range field, select [Products] List option. Since we specified a template, UiPath can extract metadata about the file that it will receive at runtime.

11. Check the Has headers option.

12. From Where dropdown, select After option. In Relative to column field, select Range ➤ Free Shipping column. This will ensure that the new column is created after the Free Shipping column in Excel.

13. In the Add Header field, type Status.

14. Set Format of the column to Text using the Category list.

Figure 11-22 shows the final configuration of this step.

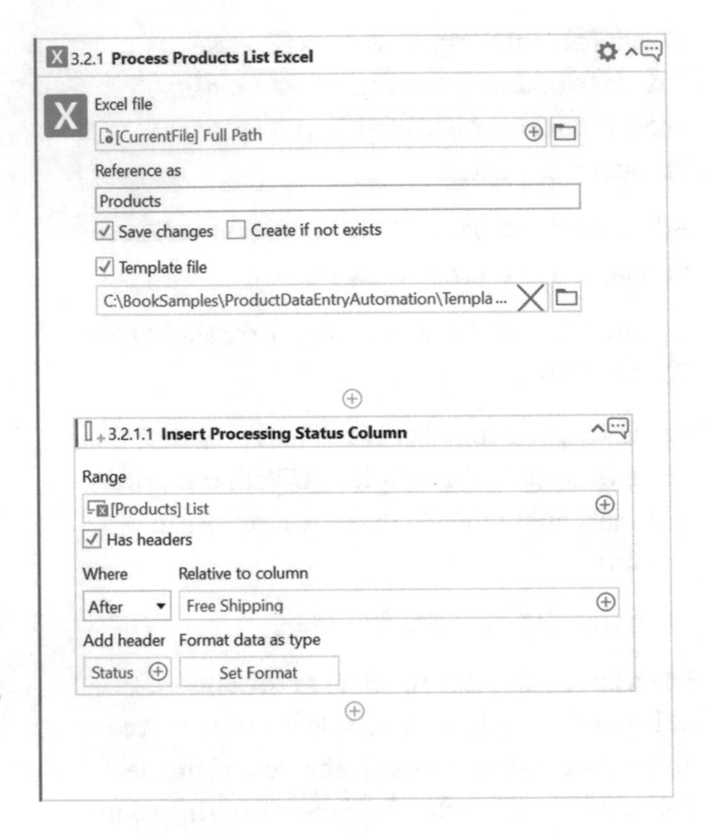

Figure 11-22. *Configuration of Insert Processing Status Column step*

Step 11: Enter Data & Update Status For Each Row

Next, we will configure the automation so that it enters data from each Excel row in the inventory management system:

1. In `StudioX`, add the `For Each Excel Row` activity in the body of `Process Products List Excel` activity after the `Insert Processing Status Column` activity.

2. Update Display Name to Enter Data & Update Status For Each Row.

3. In the For each field, enter CurrentProduct as the reference name of the current row.

4. In the In Range field, select [Products] List.

5. Check both the Has headers and Save after each row options. Now we are ready to start entering data in the web application. At this point, configuration for this activity should resemble Figure 11-23.

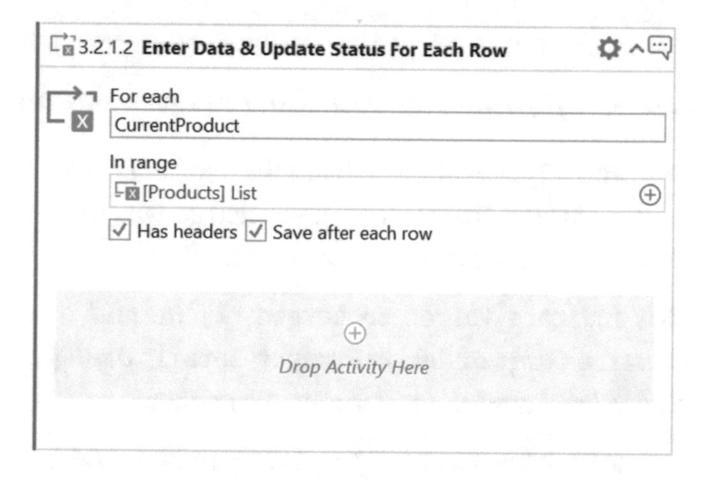

***Figure 11-23.** Configuration of the For Each Excel Row activity*

6. The web application is already open in the browser; now, we need to click Add Product + button to open the Product Details dialog. Add a Click activity in the body of Enter Data & Update Status For Each Row.

7. Click Indicate target on screen (I) link and select the Add Product + button. Use the screen header as an anchor.

8. Update Display Name to Click Add Product + Button. Figure 11-24 shows how the activity should look like at this point.

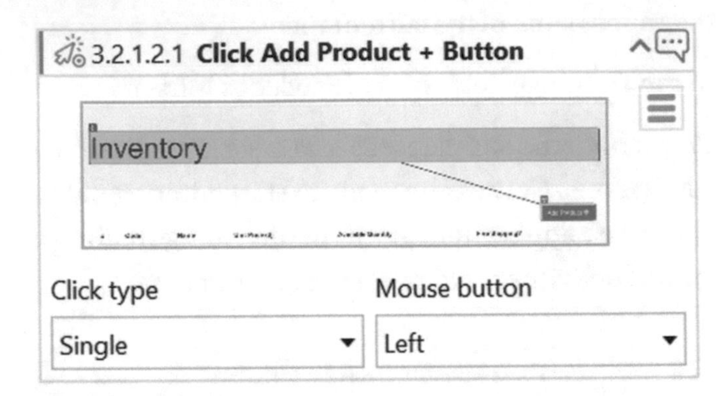

***Figure 11-24.** Configuration of Click Add Product + Button activity*

9. Next, add a Type Into activity in the body of Enter Data & Update Status For Each Row activity after the Click Add Product + Button activity.

10. Click Indicate target on screen (I) link and select the Code text field in Product Details dialog. StudioX will auto-detect label as the anchor.

11. Update Display Name to Type Into Product Code.

12. In the Type this field, click the Plus icon, hover over CurrentProduct, and select Product Code. We are specifying to type value from the current cell of the Product Code column of the new inventory Excel file. Figure 11-25 shows how the activity should look like at this point.

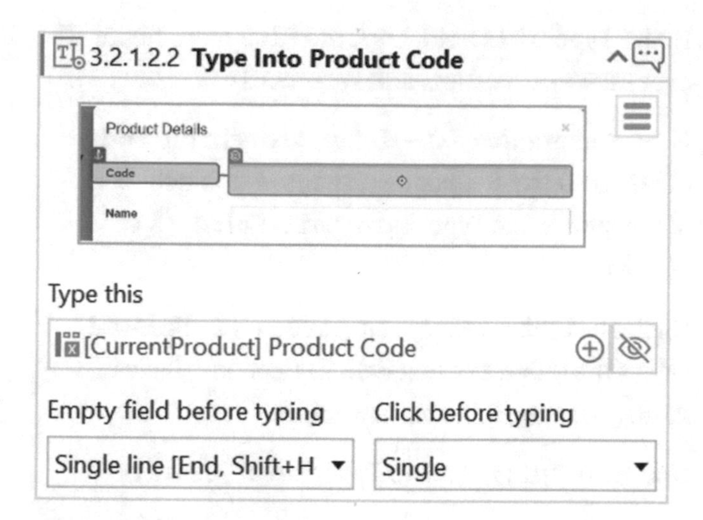

Figure 11-25. *Configuration of Type Into Product Code activity*

13. Next, add another `Type Into` activity in the body of `Enter Data & Update Status For Each Row` activity after the `Type Into Product Code` activity.

14. Click `Indicate target on screen (I)` link and select the `Name` text field in `Product Details` dialog. Use the label as an anchor.

15. Update `Display Name` to `Type Into Product Name`.

16. In the `Type this` field, click the `Plus` icon, hover over `CurrentProduct`, and select `Product Name`.

17. Next, add another `Type Into` activity in the body of `Enter Data & Update Status For Each Row` activity after the `Type Into Product Name` activity.

18. Click `Indicate target on screen (I)` link and select the `Unit Price ($)` text field in `Product Details` dialog. Use the label as an anchor.

19. Update `Display Name` to `Type Into Unit Price ($)`.

20. In the Type this field, click the Plus icon, hover over CurrentProduct, and select Unit Price ($).

21. Next, add another Type Into activity in the body of Enter Data & Update Status For Each Row activity after the Type Into Unit Price ($) activity.

22. Click Indicate target on screen (I) link and select the Quantity text field in Product Details dialog. Use the label as an anchor.

23. Update Display Name to Type Into Quantity.

24. In the Type this field, click the Plus icon, hover over CurrentProduct, and select Available Quantity.

25. Next, we are going to add logic to check the Free Shipping flag. In the source Excel file, this information is stored as Yes or No. Yes will indicate that the Free Shipping flag needs to be checked; No will indicate the flag should be left unchecked. To accomplish this, we first need to add an If activity in the body of Enter Data & Update Status For Each Row activity after the Type Into Quantity activity.

26. Update Display Name to Enter Free Shipping Data.

27. Use Condition Builder to create a new condition [CurrentProduct] Free Shipping equal to Yes. Click Save.

28. Since we need to perform an action only if the condition is met, we will add Check/Uncheck activity in the body of Then block in the Enter Free Shipping Data activity.

29. Click Indicate target on screen (I) link and select the Free Shipping checkbox in the Product Details dialog. Use the label as an anchor.

30. Update Display Name to Check Free Shipping Flag. Figure 11-26 shows the final state of Check/Uncheck activity after these configurations.

Figure 11-26. *Configuration of Enter Free Shipping Data activity*

31. Add a Click activity in the body of Enter Data & Update Status For Each Row after Enter Free Shipping Data activity.

32. Click `Indicate target on screen (I)` link and select the Add button. Use the dialog footer as an anchor.

33. Update `Display Name` to `Click Add Button`.

34. Now that we have entered data from a single Excel row in the inventory management application, we need to mark the `Status` as `Processed`. To do so, add a `Write Cell` activity in the body of `Enter Data & Update Status For Each Row` activity after the `Click Add Button` activity.

35. Update `Display Name` to `Update Status to Processed`.

36. In the `What to write` field, enter `Processed`.

37. In `Where to write`, you need to provide a cell where data needs to be entered. Click the `Plus` icon, hover over `CurrentProduct`, and select `Status`.

38. Check the `Auto increment row` option. Figure 11-27 shows the final state of `Write Cell` activity after these configurations.

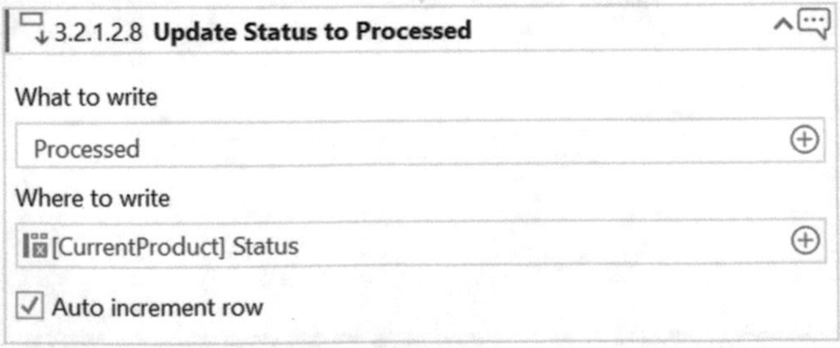

***Figure 11-27.** Configuration of Update Status to Processed activity*

Figure 11-28 shows the sequence of all activities from this step.

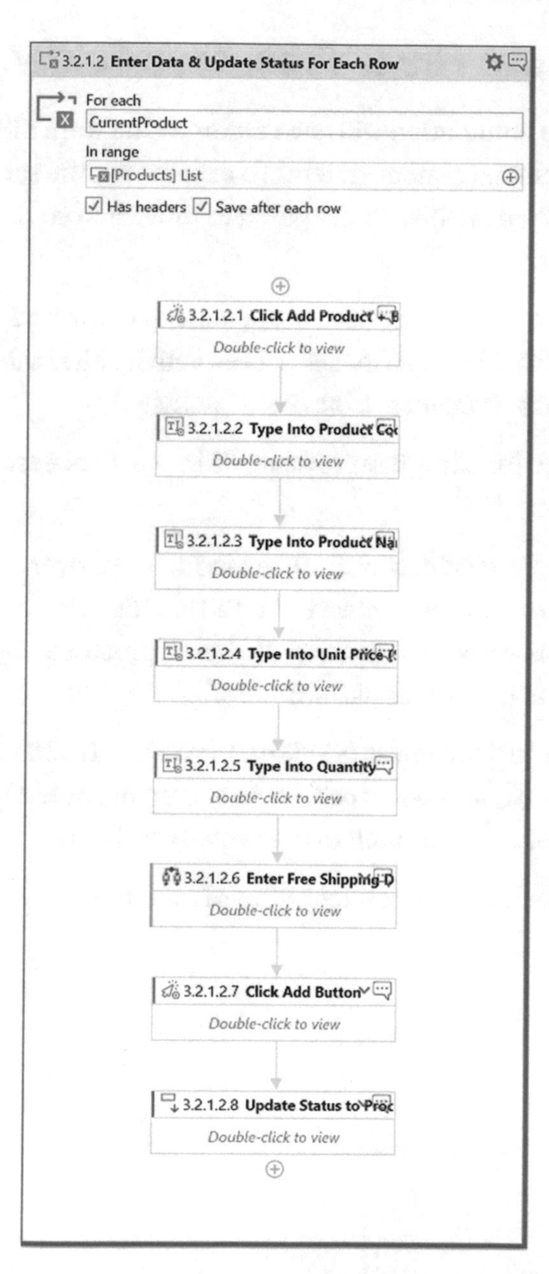

***Figure 11-28.** Configuration of Enter Data & Update Status For Each Row activity*

Step 12: Move File to Processed Folder

At this point, the automation will have entered data from all rows of Excel in the inventory management system. To ensure that the same Excel file is not processed repeatedly, we are going to move it from the Pending to Processed folder:

1. In `StudioX`, add the `Move File` activity in the body of `Fetch All Downloaded Files` activity after the `Process Products List Excel` activity.
2. Update `Display Name` to `Move File to Processed Folder`.
3. In the `From` field, click the `Plus` icon, hover over `CurrentFile`, and select `Full Path`. This will reference the current Excel file being processed by `For Each File In Folder` loop.
4. In the `To` field, hover over `Notebook`, click `Indicate in Excel`, and select cell `B7`. As shown in Table 11-1, this cell contains path to the Processed folder.

Figure 11-29 shows the configuration of this step.

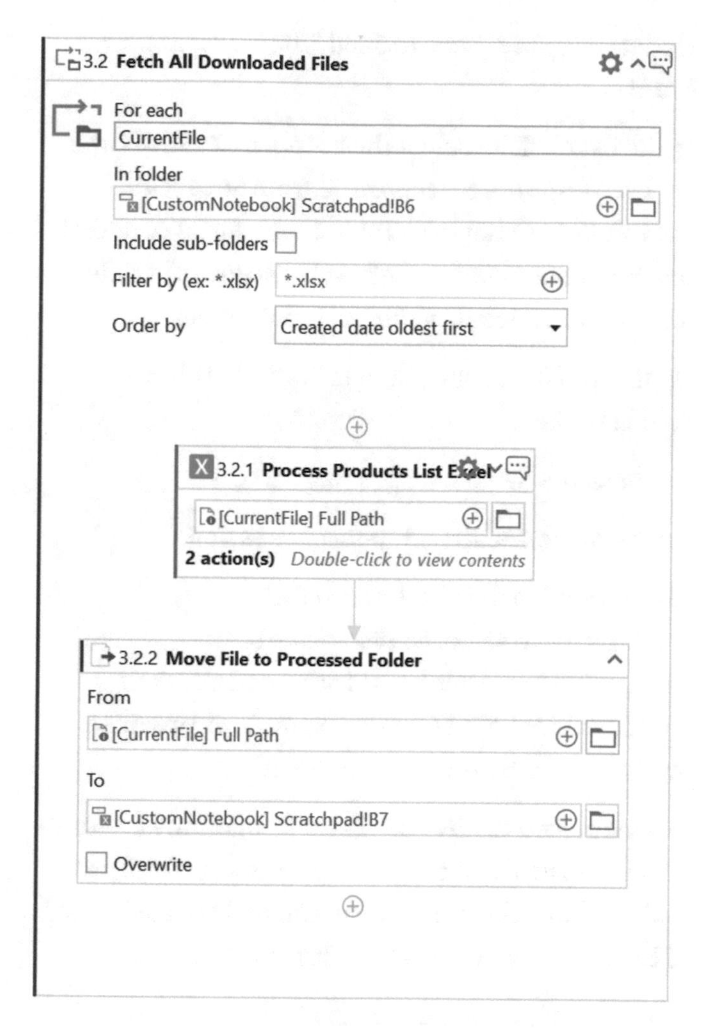

Figure 11-29. *Configuration of Move File to Processed Folder activity*

Step 13: Send Confirmation Emails

The automation's final step is to send confirmation emails to the procurement team that sent new inventory emails:

1. In `StudioX`, add the `Use Excel File` activity in the Designer panel after the `Enter Data` activity.

2. Update Display Name to Send Confirmation Emails.
3. In the Excel field, select the Notebook ➤ Indicate in Excel option. Select cell B8 from Scratchpad worksheet. As shown in Table 11-1, cell B8 contains the path of ConfirmationEmailTracker.xlsx file that contains details about all emails received.
4. Update the Reference field to Confirmation EmailTracker.
5. Uncheck the Save changes flag.
6. Uncheck the Create if not exists flag.
7. Check Template file and use the Folder icon to select C:\BookSamples\ProductDataEntryAutomation\Templates\ConfirmationEmailTracker.xlsx. This provides StudioX with the metadata of the file.
8. Next, we need to filter the Excel so that only the rows with Confirmation column value as Not Sent are visible. Add Filter activity in the body of Use Excel File (Send Confirmation Emails) activity.
9. Update Display Name to Filter Not Sent Emails.
10. In the Source field, hover over ConfirmationEmailTracker and select Emails [Sheet].
11. In the Column name field, hover over Range and select Confirmation. This is the column we are going to use for filtering.

12. Click `Configure Filter` and set the filter to `Is equal to Not Sent`. Figure 11-30 shows how the filter should look like, and Figure 11-31 shows the configuration of Filter activity.

13. Check the `Clear any existing filter` flag to ensure no other filters are applied that might return incorrect data.

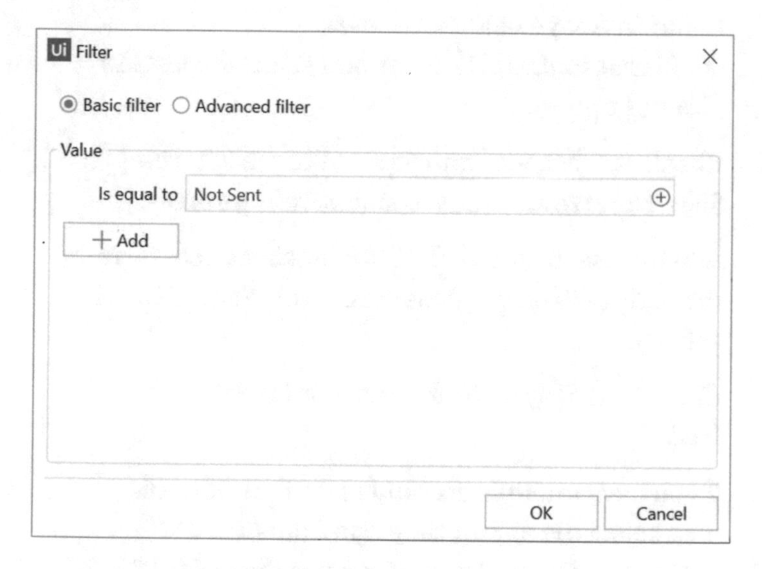

***Figure 11-30.** Excel filter on the Confirmation column*

***Figure 11-31.** Configuration of Filter Not Sent Emails*

14. Add For Each Excel Row activity in the body of Use Excel File (Send Confirmation Emails) activity after the Filter Not Sent Emails activity.

15. Update Display Name to Process Rows with Not Sent Status.

16. In the For each field, enter CurrentRow as the reference name of the current row.

17. In the In Range field, hover over ConfirmationEmailTracker and select the Emails [Sheet] option.

18. Check Has headers and Save after each row flags. Now we are ready to start sending emails.

19. Add the Use Desktop Outlook Account activity in the body of Process Rows with Not Sent Status activity.

20. Update Display Name to Send Confirmation Email.

21. If you have multiple accounts set up in Outlook, then select the appropriate one from Account dropdown. Otherwise, you can use the Default Email Account.

22. Leave the value in the Reference as field as Outlook.

23. Add Send Email activity in the body of Use Desktop Outlook Account activity.

24. Select Outlook in the Account field.

25. In the To field, hover over CurrentRow and select the From option. This will contain the email address of original sender.

26. In the Subject field, use Text Builder to type Inventory Confirmation.

27. In the Body field, use the Text option to create dynamic text. Figure 11-32 shows what the text should look like.

28. Uncheck the Save as draft flag.

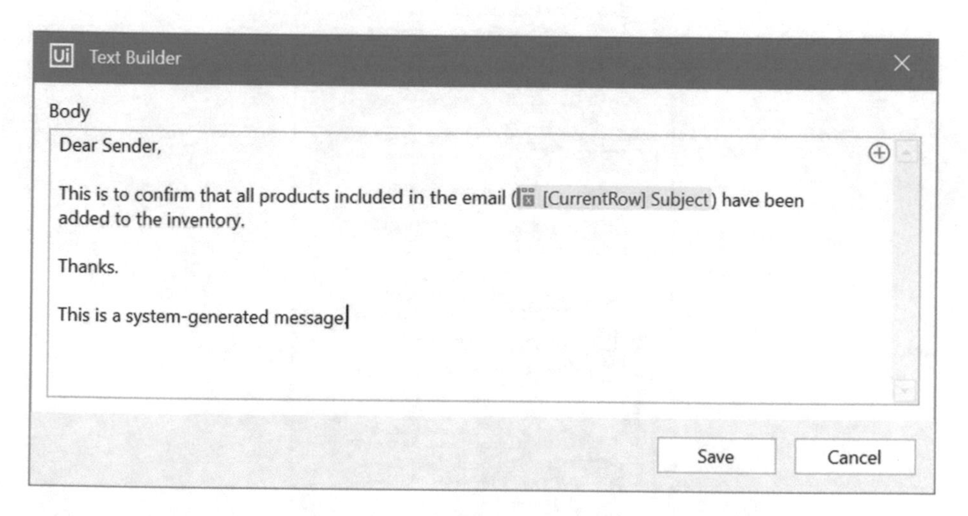

Figure 11-32. *Body of confirmation email*

29. Add a Write Cell activity in the body of Process Rows with Not Sent Status activity after the Send Confirmation Email activity.

30. Update Display Name to Update Confirmation Status to Sent.

31. In the What to write field, select Text option, and type Sent.

32. In Where to write, you need to provide a cell where data needs to be entered. Hover over CurrentRow and select the Confirmation option.

33. Check the Auto increment row flag.

Figure 11-33 shows the configuration of this step.

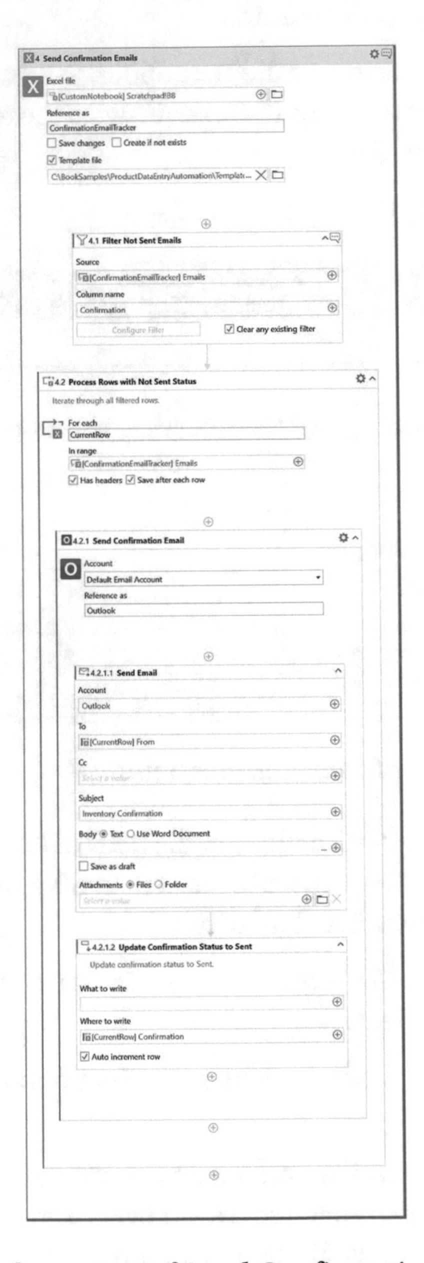

***Figure 11-33.** Configuration of Send Confirmation Emails step*

Test

Ideally, throughout the implementation, you should keep testing smaller portions. This section contains instructions on how to test the complete automation:

1. Open your Outlook and create a new email with the title `New Inventory MM/DD/YYYY`.
2. In the `To` field, type the email that you have configured for receiving emails in step 6 of the "Implementation" section.
3. Attach the `New Inventory.xlsx` to the email. This file is available at `C:\BookSamples\ProductDataEntryAutomation\Samples` location.
4. Send the email.
5. In `StudioX`, open your automation project, and click `Run` from the ribbon.

The automation will pick the newly sent email from Outlook. It will download the attachment to the newly created folder structure. The automation will launch the inventory management system in a new browser and start entering all products. Once all products have been entered, the automation will send an email.

CHAPTER 12

Invoice Generation Automation

In this chapter, we are going to automate the invoice generation task of an order management process.

Learning Objectives

This automation will help you understand the following topics:

- Creating folder structures, copying, and moving files
- Interacting with a desktop application
- Reading data from multiple tables
- Using Excel to temporarily store and manipulate data
- Entering singular and tabular data dynamically in a Word document
- Generating a PDF document from a Word document

A. Javed et al., *Robotic Process Automation using UiPath StudioX*,
https://doi.org/10.1007/978-1-4842-6794-3_12

Manual Process Overview

An Accounts Receivable Analyst (analyst) runs the invoice generation process every day. Figure 12-1 shows a high-level overview of the invoice generation task.

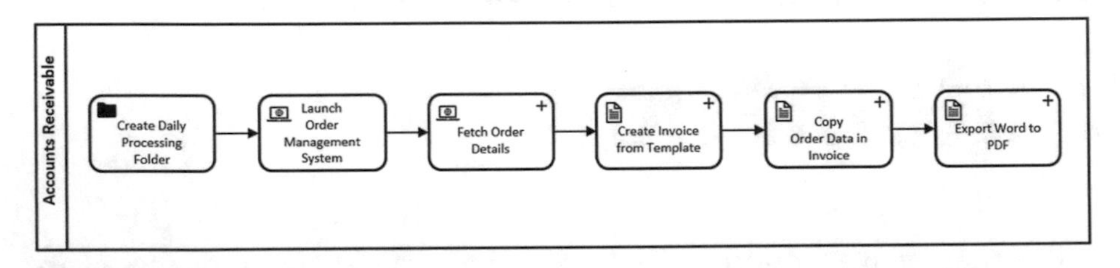

Figure 12-1. *High-level overview of the invoice generation task*

Since this task is performed every day, to keep the invoices organized by day, the analyst starts by creating a new folder for the current date in YYYYMMDD format under the DailyProcessing folder. Figure 12-2 shows the DailyProcessing folder structure.

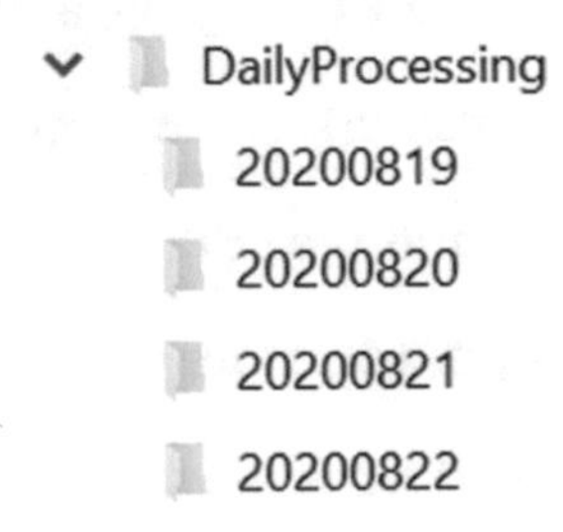

Figure 12-2. *Folder structure for managing daily invoice generation task*

Next, the analyst launches the Order Management System and filters for orders that have Pending invoices, as shown in Figure 12-3.

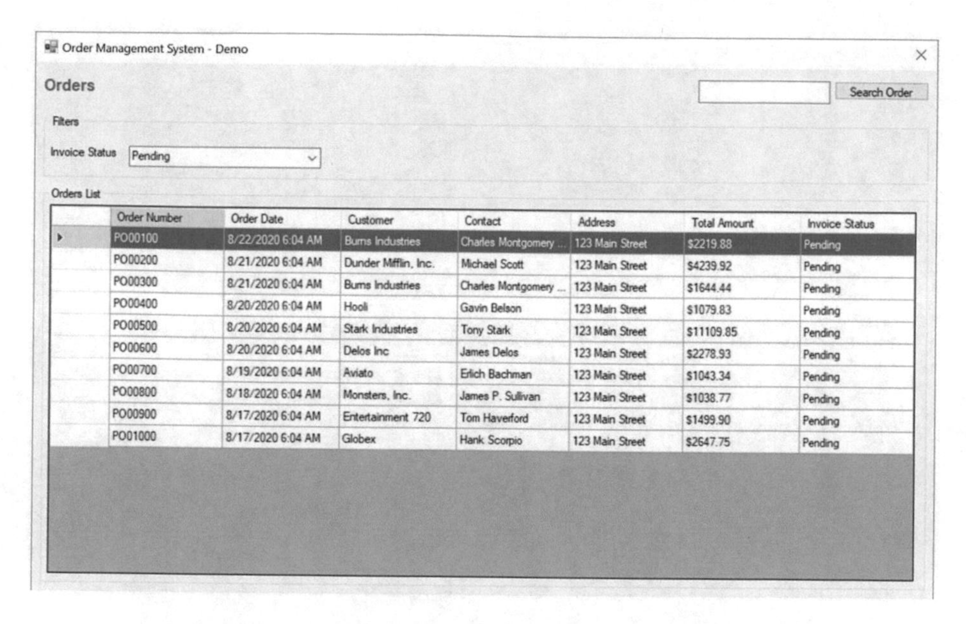

***Figure 12-3.** Orders filtered by Pending invoice status*

For each order that has Pending invoice status, the analyst enters the Order Number in the Search field and clicks the Search Order button. This opens the Order Details screen, shown in Figure 12-4.

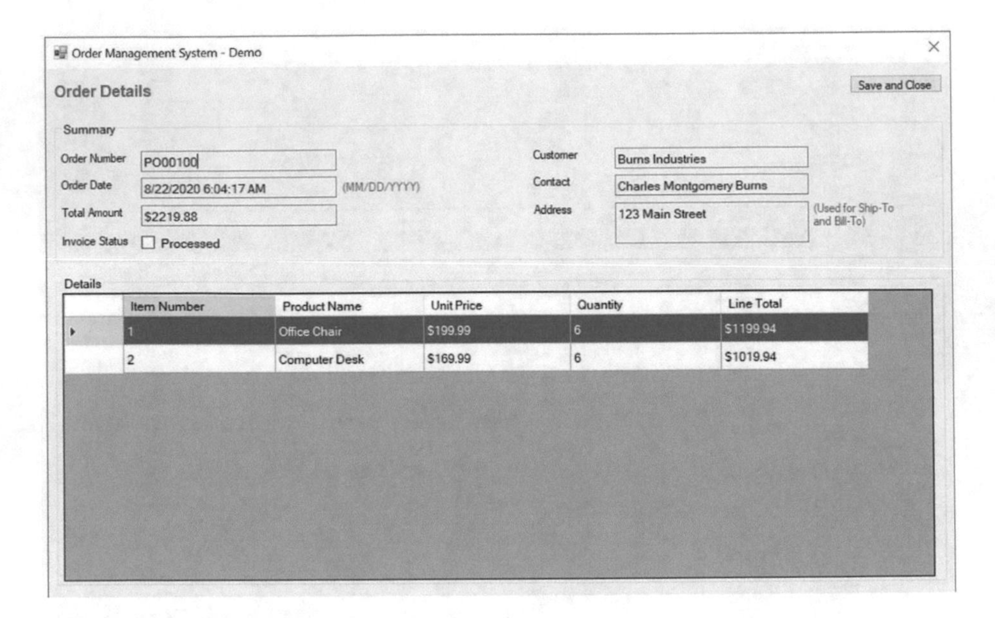

Figure 12-4. *Order Details screen*

In parallel, the analyst creates a new copy of the invoice from a Word template (DOCX). The invoice file is placed in the current date folder, and the file name reflects the invoice number.

The Invoice number is generated by replacing the PO prefix with INV in the Order Number, for example, PO00100 will become INV00100.

Next, the analyst copies data from the Order Details screen and pastes it in the invoice.

Once all the data has been copied to the invoice, the analyst exports it into PDF format. Figure 12-5 shows a sample of a generated invoice.

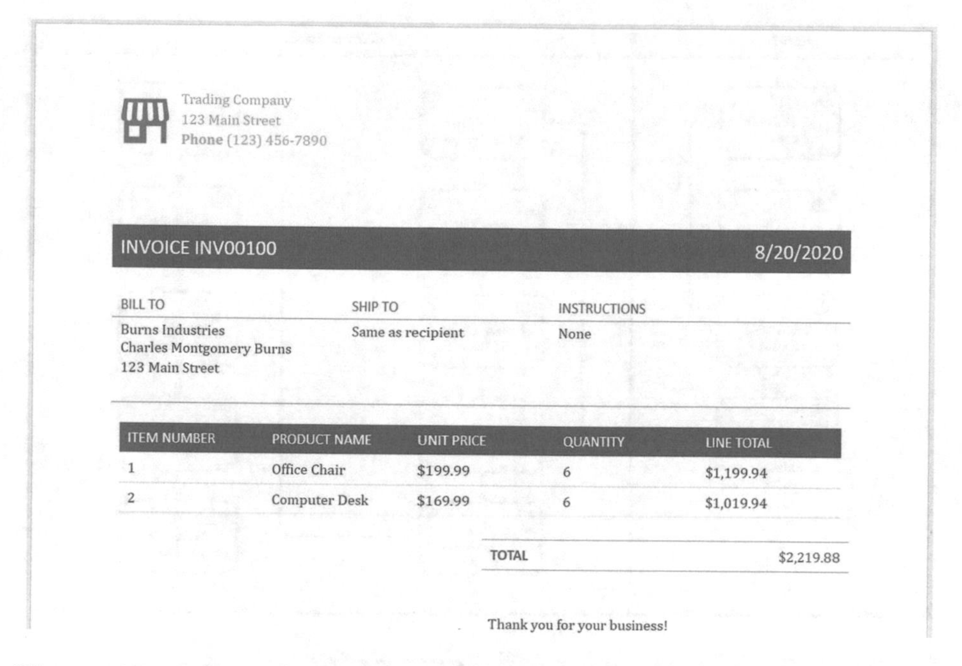

Trading Company
123 Main Street
Phone (123) 456-7890

INVOICE INV00100 8/20/2020

BILL TO	SHIP TO	INSTRUCTIONS
Burns Industries Charles Montgomery Burns 123 Main Street	Same as recipient	None

ITEM NUMBER	PRODUCT NAME	UNIT PRICE	QUANTITY	LINE TOTAL
1	Office Chair	$199.99	6	$1,199.94
2	Computer Desk	$169.99	6	$1,019.94

TOTAL $2,219.88

Thank you for your business!

Figure 12-5. *Sample of generated invoice*

Solution Design

This section provides an overview of how we are going to design the automation. Figure 12-6 shows the process flow diagram for automating the inventory management process.

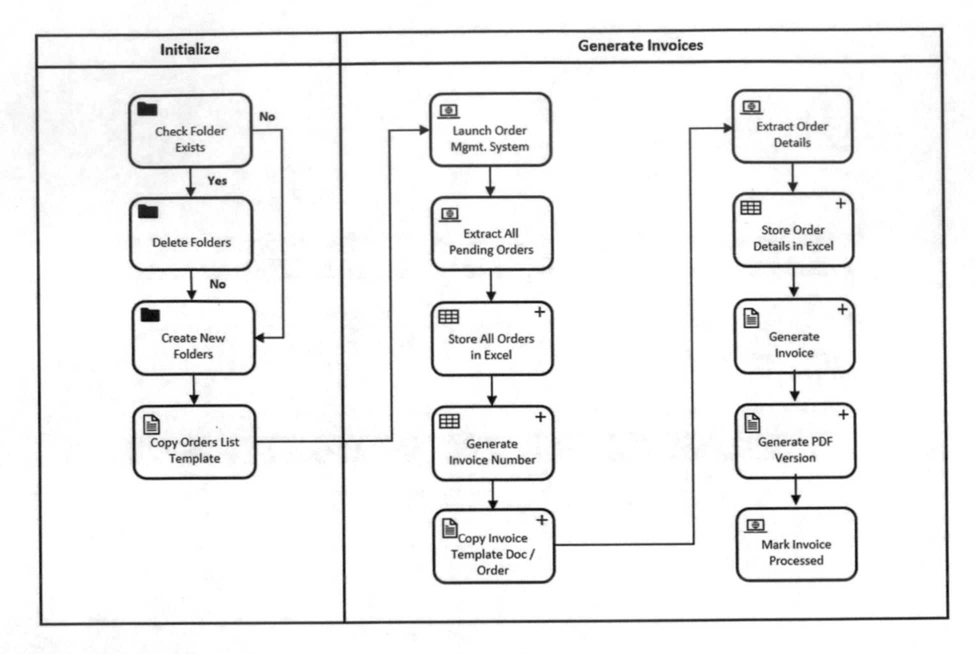

***Figure 12-6.** Solution design of the automation*

Dividing an automation into smaller and logical flows helps simplify the implementation and debugging process. As shown in Figure 12-6, we are going to divide the automation into two logical flows.

The automation will be run once a day. The only scenario where it will run multiple times in a single day is when it encounters errors and needs to be restarted. Each time the automation runs, we want to make sure that any data from the previous run is cleared.

Implementation

Now that we have a good understanding of how we are going to design the automation, we can get started with the implementation.

Step 1: Setup

Before we can start the implementation, download the project folder from the book's GitHub repository, which contains the demo application and associated templates.

Note This chapter assumes that the automation folder `InvoiceGenerationAutomation` is placed inside the `C:\BookSamples\` folder.

The `InvoiceGenerationAutomation` folder contains three subfolders as shown in Figure 12-7.

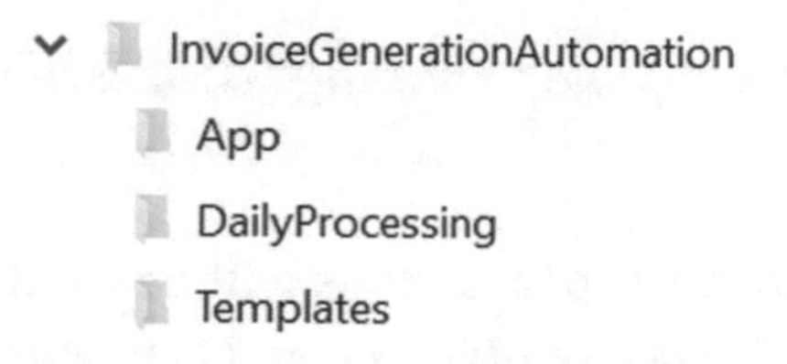

Figure 12-7. *Folder structure of invoice generation automation*

App

This subfolder contains the Order Management System. You can launch the desktop application by clicking `OrderManagement.exe`.

Templates

This subfolder contains three templates used by the automation.

The `OrdersList.xlsx` template will be copied by the automation to the `DailyProcessing\YYYYMMDD` folder. This Excel template will be used by the automation to store information for all orders pending invoice generation. Figure 12-8 shows the structure of this Excel template.

	A	B	C	D	E	F	G	H
1	Top Left Header Cell	Order Number	Order Date	Customer	Contact	Address	Total Amount	Invoice Status
2								

Figure 12-8. *Excel template to manage list of orders pending invoice generation*

A copy of the `OrderDetails.xlsx` template will be made in the `DailyProcessing\YYYYMMDD` folder for each order. This Excel will be used by the automation to store line item details of each order temporarily. Each file will be named based on the `Order Number`. Figure 12-9 shows the structure of this Excel template.

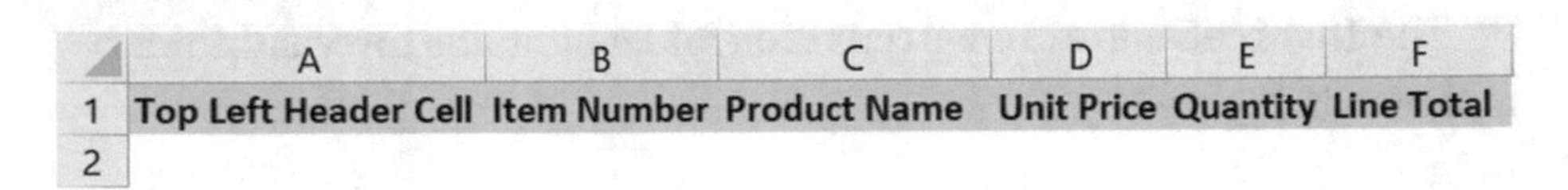

	A	B	C	D	E	F
1	Top Left Header Cell	Item Number	Product Name	Unit Price	Quantity	Line Total
2						

Figure 12-9. *Excel template to manage details of individual orders*

Finally, the `InvoiceTemplate.docx` will be used by the automation to generate invoice for each order. Figure 12-10 shows the Word template.

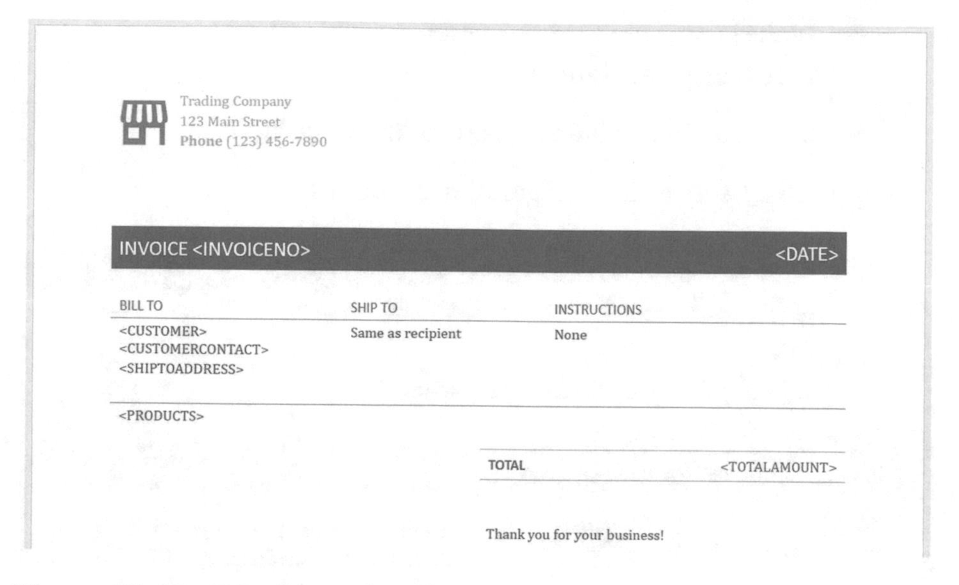

Figure 12-10. *Word template for invoice*

DailyProcessing

This subfolder is empty and will be used by the automation to create folders for the date when automation runs and storing generated invoices.

Step 2: Create Project

Next, we are going to create a new project in UiPath StudioX. Here are the steps to creating a new project. You do not have to use the exact same name, path, and description.

1. Launch `UiPath StudioX`.
2. From the `Backstage View`, click `Blank Task` to create a new project from scratch.
3. In the `Process name:` field, enter `Chapter_12_InvoiceGenerationAutomation`.

4. In the `Description:` field, enter `Automation for invoice generation task`.

5. Click the `Create` button to create the project.

Figure 12-11 shows the configuration of project.

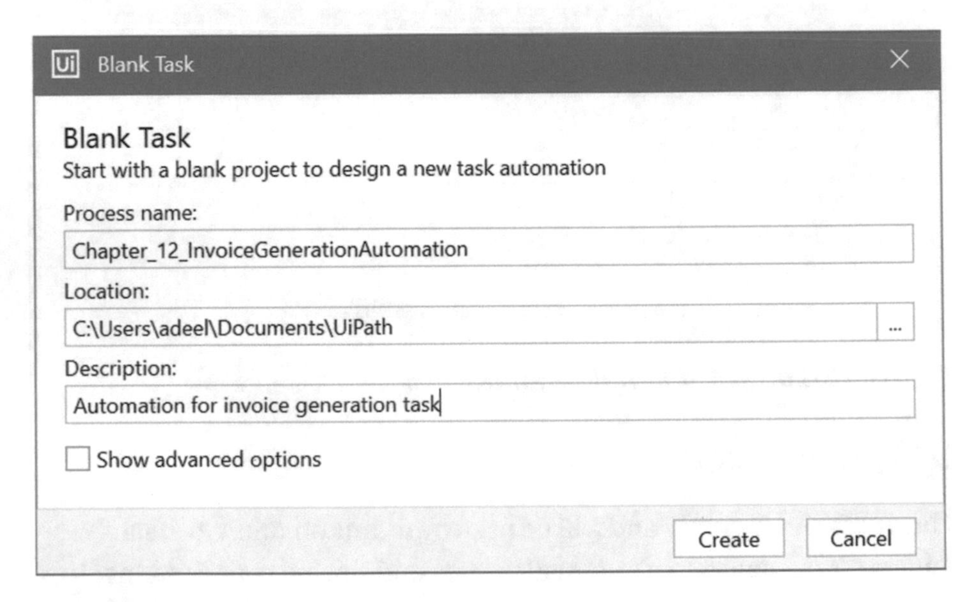

Figure 12-11. *Create new project*

Step 3: Set Up Project Notebook

This automation is going to use custom formulas for generating folders and file paths.

We are going to configure a custom notebook to manage all our custom formulas. Figure 12-12 shows the custom notebook with all formulas used by the automation.

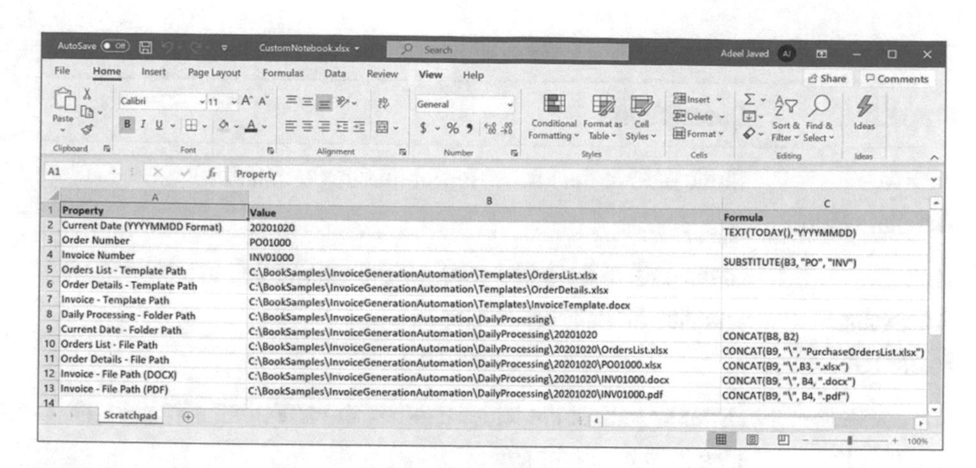

Figure 12-12. *Custom formulas in the Notebook*

Table 12-1 describes how each value shown in Figure 12-12 is used.

Table 12-1. *Description of each value in Scratchpad*

Property	Description
Current Date	The formula in cell B2 will generate the current date in YYYYMMDD format that will be used by all paths in subsequent rows.
Order Number	Cell B3 is used by other cells to generate invoice number and file paths. As the automation processes each order, this field will be populated with the Order Number currently being processed.
Invoice Number	The formula in cell B4 uses the numerical value from cell B3 and replaces the PO with INV to generate the invoice number.
Orders List – Template Path	Static path of the OrdersList.xlsx template.
Order Details – Template Path	Static path of the OrderDetails.xlsx template.

(*continued*)

Table 12-1. (*continued*)

Property	Description
Invoice – Template Path	Static path of the InvoiceTemplate.docx template.
Daily Processing – Folder Path	Static path of the folder used by the automation to store folders and files for daily processing.
Current Date – Folder Path	Dynamic path of the folder used by the automation to store all files for the current day.
Orders List – File Path	Dynamic path where the copy of OrdersList.xlsx will be placed.
Order Details – File Path	Dynamic path where the copy of OrderDetails.xlsx for current purchase order will be placed.
Invoice – File Path (DOCX)	Dynamic path where the MS Word copy of the current invoice will be created.
Invoice – File Path (PDF)	Dynamic path where the PDF version of the current invoice will be created.

Here are the step-by-step instructions to configure a custom notebook:

1. In `StudioX`, click the `Notebook` ➤ `Configure Notebook` menu from the ribbon on top.
2. In the `Notebook file` field, click the ... and select `C:\BookSamples\InvoiceGenerationAutomation\Templates\CustomNotebook.xlsx` file, as shown in Figure 12-13.
3. Click the `OK` button.

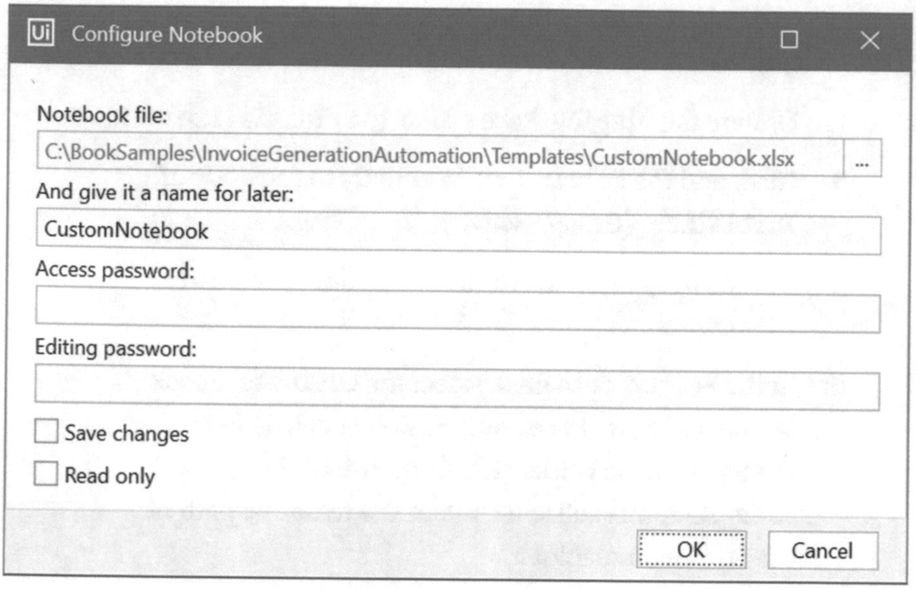

Figure 12-13. *Configure custom notebook*

Step 4: Initialize

The first activity that the automation needs to perform is to check if a folder for the current date (`YYYYMMDD` format) already exists under the `DailyProcessing` folder. The result of this operation will be stored in a `CurrentDateFolderExists` variable.

Next, based on the result, we are going to specify what actions the automation needs to perform. If `CurrentDateFolderExists` is true, then that means the folder structure already exists and needs to be deleted. On the other hand, if `CurrentDateFolderExists` is false, then the automation needs to create the necessary folder structure.

The final initialization step requires us to create a copy of `Templates\OrdersList.xlsx` in the `DailyProcessing\Current Date (YYYYMMDD)` folder. This file will be used by the automation to store a list of orders that are pending invoice generation.

1. In StudioX, add the Group activity to a blank process.
2. Update the Display Name property to Initialize.
3. Next, add the Folder Exists activity to the body of Initialize (Group) activity.
4. Update Display Name to Check Folder Structure Exists.
5. In the Folder path field, select the CustomNotebook ➤ Indicate in Excel option. Select cell B9 from the Scratchpad worksheet. As mentioned in Table 12-1, this cell will contain the dynamic path of the current date folder.
6. In the Save result field, use the Save for Later Use option and name your saved value as CurrentDateFolderExists.
7. Add an If activity after the Check Folder Structure Exists activity.
8. Update Display Name to Create Folder Structure.
9. Using the Condition Builder, create a new condition CurrentDateFolderExists is true.
10. Add a Delete Folder activity in the Then block of Create Folder Structure activity.
11. Update the Display Name to Delete Current Date Folder & Contents.
12. In the Folder name field, select the CustomNotebook ➤ Indicate in Excel option. Select cell B9 from the Scratchpad worksheet.

13. Add a `Create Folder` activity in the Then block of `Create Folder Structure` activity after the `Delete Current Date Folder & Contents` activity.
14. Update `Display Name` to `Create Current Date Folder`.
15. In the `Folder name` field, select the `CustomNotebook` ➤ `Indicate in Excel` option. Select cell `B9` from the `Scratchpad` worksheet.
16. Copy the `Create Current Date Folder` activity and paste it in the `Else` block of the `Create Folder Structure` activity.
17. Next, add the `Copy File` activity in the body of the `Initialize` activity after the `Create Folder Structure (If)` activity.
18. Update `Display Name` to `Create Orders List Excel Copy`.
19. In the `From` field, select the `CustomNotebook` ➤ `Indicate in Excel` option. Select cell `B5` from the `Scratchpad` worksheet. As mentioned in Table 12-1, this cell will contain the static path of the `OrdersList.xlsx` file.
20. In the `To` field, select the `CustomNotebook` ➤ `Indicate in Excel` option. Select cell `B10` from `Scratchpad` worksheet. As mentioned in Table 12-1, this cell will contain the dynamic path of the `OrdersList.xlsx` file.

Figure 12-14 shows the configuration of this step.

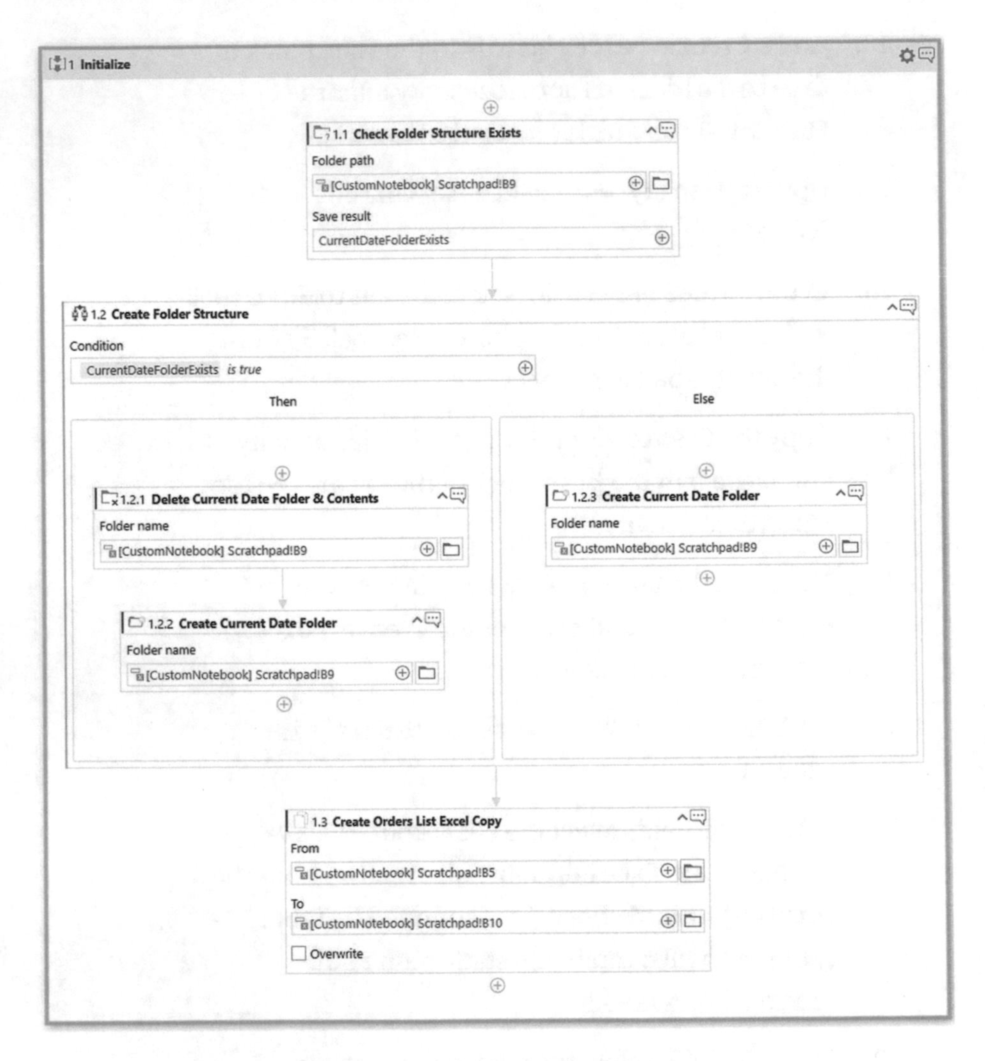

Figure 12-14. *Configuration of Initialize step*

Step 5: Generate Invoices

This is the parent activity that will contain all the remaining activities. This activity will handle interactions with the Order Management System and the generation of invoices.

Throughout the process run, the automation will be interacting with the Order Management System, so we will start by launching the application:

1. Launch the `Order Management System` application by clicking the executable file in the `C:\BookSamples\InvoiceGenerationAutomation\App` folder.
2. In `StudioX`, add the `Use Application/Browser` activity to the Designer panel after the `Initialize` activity.
3. Click `Indicate Application` link and point your mouse to the Order Management System application that you just launched. `StudioX` will automatically detect that you are trying to interact with a desktop application, and it will also populate the `Application path`.
4. From the `Properties` panel, select `Always` option for the `Options` ➤ `Open` property. This will ensure a new application instance is created each time the automation runs.
5. Update `Display Name` to `Generate Invoices`.

Figure 12-15 shows the configuration of this step.

Tip Based on your screen resolution, you might also need to add a Maximize Window activity to ensure that the application is fully visible.

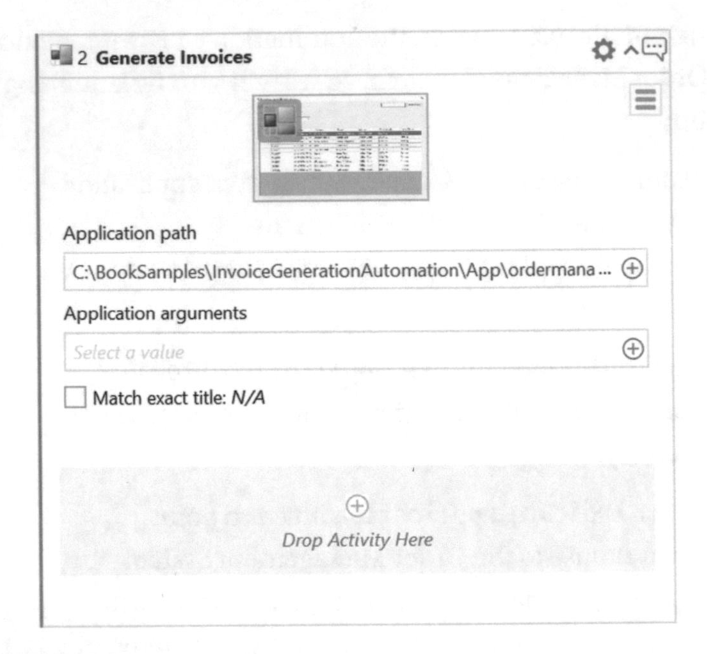

***Figure 12-15.** Configuration of Generate Invoices step*

Step 6: Extract & Process All Orders

Next, we are going to configure the automation to store the list of orders in the `OrdersList.xlsx` Excel file:

1. In `StudioX`, add the `Use Excel File` activity in the body of the `Generate Invoices` activity.
2. Update the `Display Name` to `Extract & Process All Orders`.
3. In the `Excel file` field, select the `CustomNotebook` ➤ `Indicate in Excel` option. Select cell `B10` from the `Scratchpad` worksheet.
4. Update the `Reference as` field to `OrdersList`.

5. Check the `Save Changes` flag to ensure the file gets auto saved after each change.
6. Uncheck `Create if not exists` flag.
7. Check `Template file` flag. Providing a template file will help `StudioX` understand the structure of the Excel.
8. Select `Browse for file` icon and select `C:\BookSamples\InvoiceGenerationAutomation\Templates\OrdersList.xlsx` file.

Figure 12-16 shows the configuration of this step.

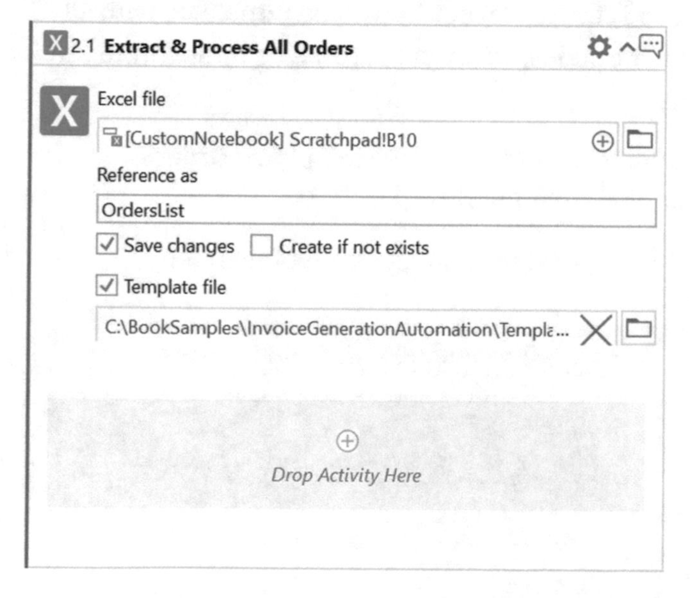

Figure 12-16. *Configuration of Extract & Process All Orders step*

Step 7: Filter Orders w/ Pending Invoices

The Order Management System might have some orders for which invoices have already been generated. So, to make sure that we only extract orders that require invoice generation, we are going to set the `Invoice Status` filter to `Pending`:

1. In `StudioX`, add the `Select Item` activity in the body of `Extract & Process All Orders` activity.
2. Click `Indicate target on screen (I)` link and select the `Invoice Status` dropdown in `Filters` section. Use label as an anchor.
3. The `Item to select` list will be populated with all filter options, so select the `Pending` option from the list.
4. Update the `Display Name` to `Filter Orders w/ Pending Invoices`.

Figure 12-17 shows the configuration of this step.

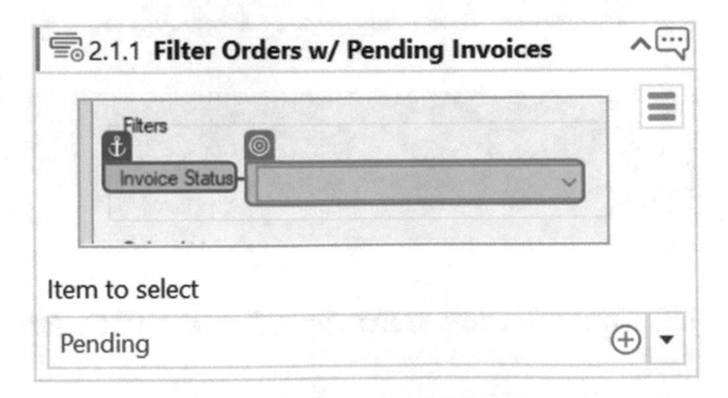

***Figure 12-17.** Configuration of Filter Orders w/ Pending Invoices step*

Step 8: Extract Orders Table Data

Next, we are going to configure the automation to extract all orders from the table and store them in `OrdersList.xlsx` discussed in the Initialize step:

1. In `StudioX`, add the `Extract Table Data` activity in the body of `Extract & Process All Orders` activity after the `Filter Orders w/ Pending Invoices` activity.
2. Click `Indicate target on screen (I)` link and this will start the `Extract Wizard`.
3. On `Select Element` step, click `Next`.
4. Click the header of `Order Number` column in `Orders` table on the Home screen of `Order Management System`.
5. The `Extract Wizard` will show the table of data in preview mode. Click `Finish`. This will end the table data extraction wizard.
6. From `Extract to` field, hover over `OrdersList`, and select the `Summary [Sheet]` option. This will store all table rows in the `Summary` worksheet of the `OrdersList.xlsx` file.
7. Update `Display Name` to `Extract Orders Table Data`.

Figure 12-18 shows the configuration of this step.

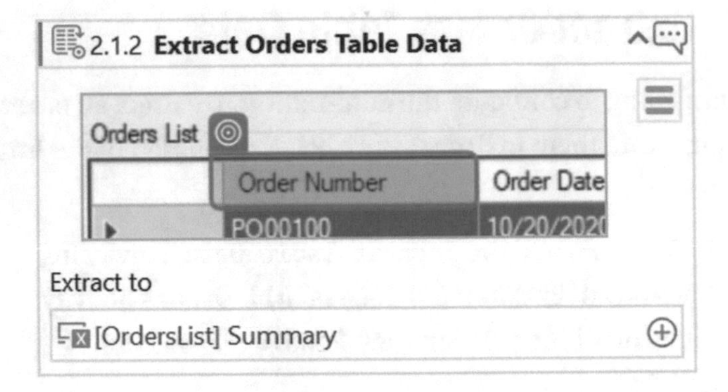

Figure 12-18. *Configuration of Extract Orders Table Data step*

Step 9: Process All Orders

At this point, we have a list of all orders for which we need to generate invoices. In this step, we are going to configure the automation to go through each order in the list. In subsequent steps, we will add what actions the automation needs to perform for each order.

1. In StudioX, add the `For Each Excel Row` activity in the body of `Extract & Process All Orders` activity after the `Extract Orders Table Data` activity.
2. Update `Display Name` to `Process All Purchase Orders`.
3. In the `For each` field, enter the `CurrentOrder` as the reference name of the current order or row.
4. In the `In Range` field, hover over `OrdersList` and select the `Summary [Sheet]` option.
5. Check `Has Headers` and `Save after each row` flags.

Figure 12-19 shows the configuration of this step.

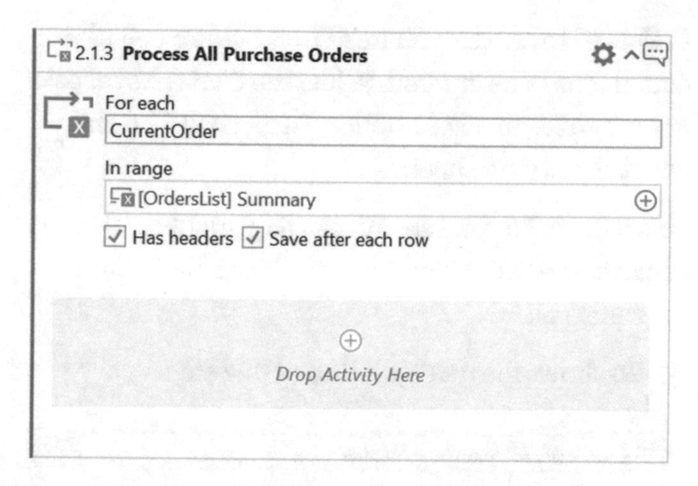

Figure 12-19. *Configuration of Process All Orders step*

Step 10: Generate Invoice Number

For each purchase order, we first need to create a corresponding invoice number. We are going to use a custom formula that we defined in the `CustomNotebook` ➤ `Scratchpad` worksheet in step 2. In this step, we are going to paste the current `Order Number` in cell `B3`, and the custom formula in cell `B4` will generate an `Invoice Number`:

1. In `StudioX`, add the `Write Cell` activity in the body of `Process All Orders` activity.
2. Update `Display Name` to `Generate Invoice Number`.
3. In the `What to write` field, hover over `CurrentOrder` and select `Order Number`; this will contain `Order Number` of the order currently being processed.

4. In `Where to write`, you need to provide a cell where data needs to be entered. Select the `CustomNotebook` ➤ `Indicate in Excel` option. Select cell `B3` from `Scratchpad` worksheet.

5. Leave the `Auto increment row` flag unchecked, because the Order Number will always be written to the same cell.

Figure 12-20 shows the configuration of this step.

2.1.3.1 **Generate Invoice Number**
What to write
[CurrentOrder] Order Number
Where to write
[CustomNotebook] Scratchpad!B3
☐ Auto increment row

Figure 12-20. *Configuration of Generate Invoice Number step*

Step 11: Create Order Details Excel Copy

For each order, we will need to store the details in a temporary Excel file. To do so, we will need to create a copy of `OrderDetails.xlsx` template in the `DailyProcessing\Current Date (YYYYMMDD)` folder. Each file will be named using the `Order Number`:

1. In `StudioX`, add the `Copy File` activity in the body of `Process All Orders` activity after the `Generate Invoice Number` activity.

2. Update `Display Name` to `Create Order Details Excel Copy`.

3. In the From field, select the CustomNotebook ➤ Indicate in Excel option. Select cell B6 from the Scratchpad worksheet. As mentioned in Table 12-1, this cell will contain the static path of the OrderDetails.xlsx file.

4. In the To field, select the CustomNotebook ➤ Indicate in Excel option. Select cell B11 from the Scratchpad worksheet. As mentioned in Table 12-1, this cell will contain the dynamic path of the Order Details Excel file.

Figure 12-21 shows the configuration of this step.

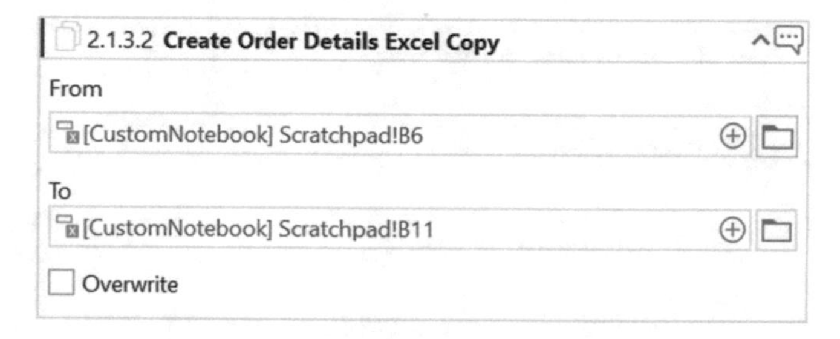

***Figure 12-21.** Configuration of Create Order Details Excel Copy step*

Step 12: Create Invoice Template Copy

For each order, we will also need to create a copy of the InvoiceTemplate.docx in the DailyProcessing\Current Date (YYYYMMDD) folder. Each file will be named using the Invoice Number generated by the custom formula:

1. In StudioX, add the Copy File activity in the body of Process All Orders activity after the Create Order Details Excel Copy activity.

2. Update Display Name to Create Invoice Template Copy.

3. In the From field, select the CustomNotebook ➤ Indicate in Excel option. Select cell B7 from Scratchpad worksheet. As mentioned in Table 12-1, this cell will contain the static path of the InvoiceTemplate.docx file.

4. In the To field, select the CustomNotebook ➤ Indicate in Excel option. Select cell B12 from Scratchpad worksheet. As mentioned in Table 12-1, this cell will contain the dynamic path of the invoice-specific Word document.

Figure 12-22 shows the configuration of this step.

Figure 12-22. *Configuration of Create Invoice Template Copy step*

Step 13: Extract Order Details & Generate Invoice

In step 11, we created a copy of OrderDetails.xlsx to store details of the current order. Next, we are going to enable the automation to use this Excel:

1. In StudioX, add the Use Excel File activity in the body of Process All Orders activity after the Create Invoice Template Copy activity.

2. Update Display Name to Extract Order Details & Generate Invoice.

3. In the `Excel file` field, select the `CustomNotebook` ➤ `Indicate in Excel` option. Select cell `B11` from the `Scratchpad` worksheet.

4. Update `Reference as` field to `OrderDetails`.

5. Check `Save changes` flag to ensure the file gets saved after each change.

6. Uncheck `Create if not exists` flag.

7. Check `Template file` flag. Providing the template will help StudioX understand the structure of the Excel.

8. Select `Browse for file` icon and select `C:\BookSamples\InvoiceGenerationAutomation\Templates\OrderDetails.xlsx` file.

Figure 12-23 shows the configuration of this step.

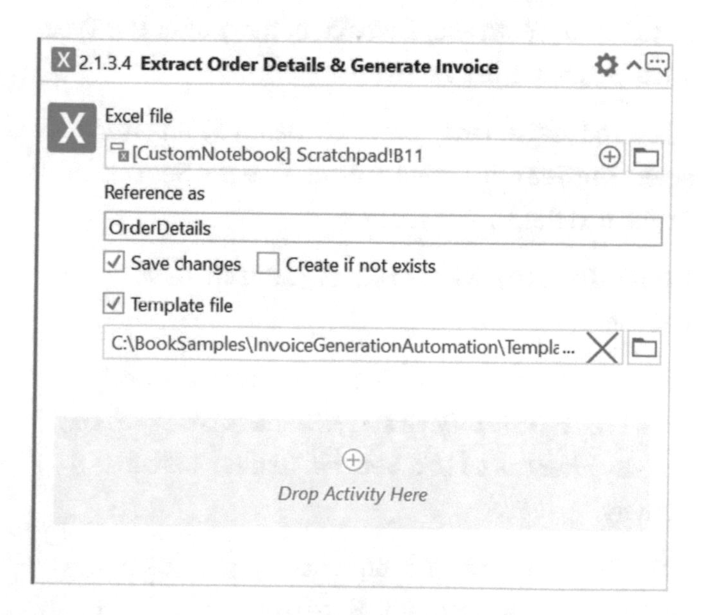

Figure 12-23. *Configuration of Extract Order Details & Generate Invoice step*

Step 14: Extract Order Details Table Data

Next, using the Order Number of the current order, we are going to fetch order details from the Order Management System:

1. In StudioX, add the Type Into activity in the body of Extract Order Details & Generate Invoice activity.
2. Click Indicate target on screen (I) link and select the Search text field on Home screen. Use the Search Order button as an anchor.
3. In Type this field, hover over CurrentOrder and select Order Number.
4. Update Display Name to Type Into Search Field.
5. Add a Click activity in the body of Extract Order Details & Generate Invoice activity after the Type Into Search Field activity.
6. Click Indicate target on screen (I) link and select the Search Order button. Use the Search Order text field as an anchor.
7. Update Display Name to Click Search Order Button.
8. Add an Extract Table Data activity in the body of Extract Order Details & Generate Invoice activity after the Click Search Order Button activity.
9. Click Indicate target on screen (I) link and this will start the Extract Wizard.

10. On the `Select a Value` step, click `Next`.
11. Point your mouse to and click the header of `Item Number` column in table on `Order Details` screen.
12. The `Extract Wizard` will show the data in preview mode. Click `Finish`. This will end the table data extraction wizard.
13. From `Extract to` field, hover over `OrderDetails` and select `Line Items [Sheet]` option.
14. Update `Display Name` to `Extract Order Details Table Data`.
15. The `Extract Order Details Table Data` activity adds an extra column `Top Left Header Cell` at the beginning of the Excel file. This column does not need to be copied to the invoice and serves no purpose, so we need to delete this before proceeding. Add a `Delete Column` activity in the body of `Extract Order Details & Generate Invoice` activity after the `Extract Order Details Table Data` activity.
16. Update `Display Name` to `Delete Extra Column`.
17. In the `Source` field, hover over `OrderDetails` and select `Line Items [Sheet]` option.
18. Check `Has headers` flag.
19. In the `Column name` field, use the `Text` option, and enter `Top Left Header Cell`.

Figure 12-24 shows the configuration of this step.

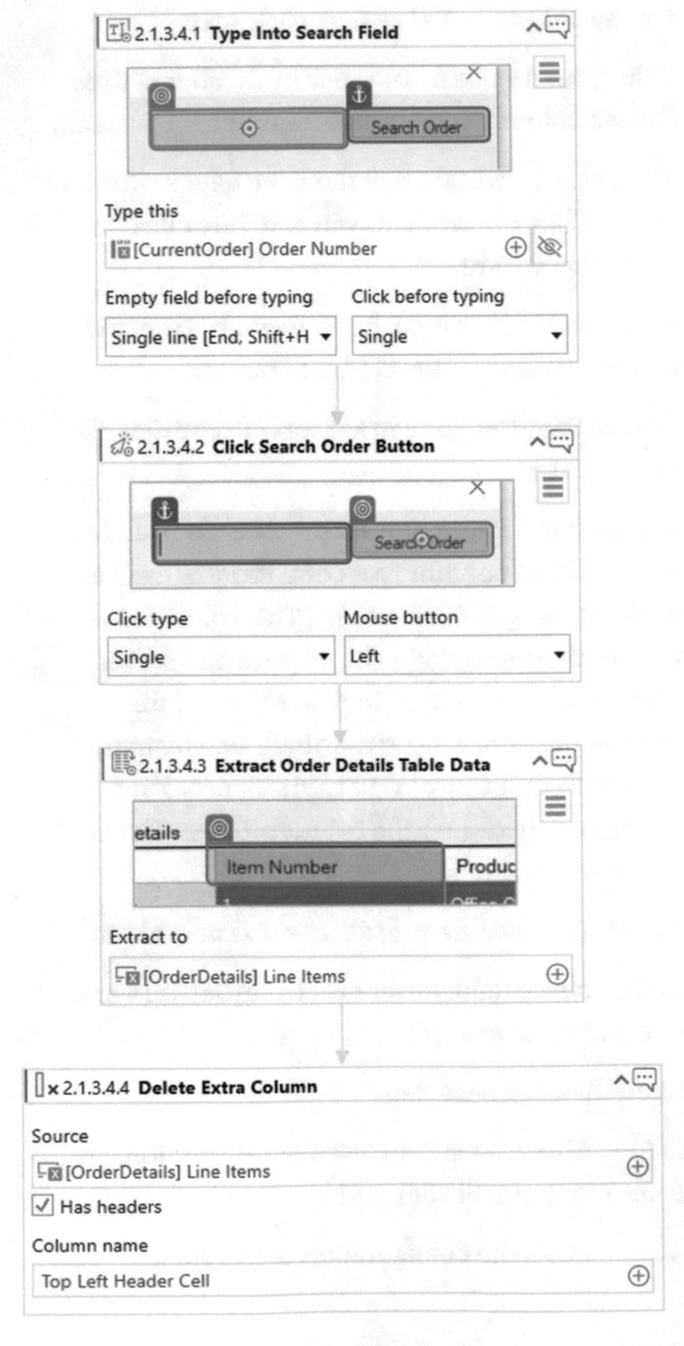

Figure 12-24. *Configuration of Extract Order Details Table Data step*

Step 15: Generate Invoice for Current Order

We have a copy of the invoice template and all the data that is needed for generating an invoice. Now, we can configure the automation to generate an invoice for the order currently being processed:

1. In StudioX, add the Use Word File activity in the body of Extract Order Details & Generate Invoice activity after the Delete Extra Column activity.
2. Update Display Name to Generate Invoice for Current Order.
3. In the Word file field, select the CustomNotebook ➤ Indicate in Excel option. Select cell B12 from the Scratchpad worksheet.
4. Uncheck Create if not exists flag.
5. Check Save changes flag to ensure file gets saved after each iteration.
6. Add a Replace Text in Document activity in the body of Generate Invoice for Current Order activity.
7. Update Display Name to Enter Invoice Number.
8. In the Search for field, select Text option, and enter <INVOICENO>. This is the placeholder in the invoice template that will be searched and replaced with value provided in the next configuration.
9. In the Replace with field, select the CustomNotebook ➤ Indicate in Excel option. Select cell B4 from Scratchpad worksheet. As mentioned in Table 12-1, this cell will contain the generated invoice number using a custom formula.

10. Add a Replace Text in Document activity in the body of Generate Invoice for Current Order activity after the Enter Invoice Number activity.

11. Update the Display Name to Enter Invoice Date.

12. In the Search for field, select Text option, and enter <DATE>.

13. In the Replace with field, select the CustomNotebook ➤ Indicate in Excel option. Select cell B2 from the Scratchpad worksheet. This will provide the value of the current date.

14. Add a Replace Text in Document activity in the body of Generate Invoice for Current Order activity after the Enter Invoice Date activity.

15. Update Display Name to Enter Customer.

16. In the Search for field, select Text option, and enter <CUSTOMER>.

17. In the Replace with field, hover over CurrentOrder and select the Customer option. This will provide the value of the customer for the order currently being processed.

18. Add a Replace Text in Document activity in the body of Generate Invoice for Current Order activity after the Enter Customer activity.

19. Update Display Name to Enter Customer Contact.

20. In the Search for field, select Text option, and enter <CUSTOMERCONTACT>.

21. In the `Replace with` field, hover over `CurrentOrder` and select the `Contact` option. This will provide the value of the customer contact for the order currently being processed.

22. Add a `Replace Text in Document` activity in the body of `Generate Invoice for Current Order` activity after the `Enter Customer Contact` activity.

23. Update `Display Name` to `Enter Ship To Address`.

24. In the `Search for` field, select `Text` option, and enter `<SHIPTOADDRESS>`.

25. In the `Replace with` field, hover over `CurrentOrder` and select the `Address` option. This will provide the value of the ship to address for the order currently being processed.

26. Add a `Replace Text in Document` activity in the body of `Generate Invoice for Current Order` activity after the `Enter Ship To Address` activity.

27. Update `Display Name` to `Enter Total Amount`.

28. In the `Search for` field, select `Text` option, and enter `<TOTALAMOUNT>`.

29. In the `Replace with` field, hover over `CurrentOrder` and select the `Total Amount` option. This will provide the value of the total amount for the order currently being processed.

30. So far, we have replaced all singular items; now we are going to insert a complete table from Excel in Word. Add an `Insert Data Table in Document` activity in the body of `Generate Invoice for Current Order` activity after the `Enter Total Amount` activity.

31. Update `Display Name` to `Insert Order Details Table in Invoice`.

32. In the `Table to insert` field, hover over `OrderDetails` and select the `Line Items [Sheet]` option. This contains all the data in Excel that we are going to insert in the Word document as a table.

33. In the `Insert relative to` field, select `Text` option from the dropdown.

34. In the `Text to search for` field, select `Text` option and enter `<PRODUCTS>`. This is the placeholder in invoice template where the table is going to be inserted.

35. In the `Text occurrence` field, select `First` from the dropdown.

36. In the `Position where to insert` field, select `Replace` option. We are going to replace `<PRODUCTS>` text with the actual table.

37. The invoice will be generated in Word format at this point. Next, we are going to configure the automation to export the invoice in PDF format. Add a `Save Document as PDF` activity in the body of `Generate Invoice for Current Order` activity after the `Insert Order Details Table in Invoice` activity.

38. In the `File path to save as` field, select the `CustomNotebook` ➤ `Indicate in Excel` option. Select cell `B13` from `Scratchpad` worksheet. As mentioned in Table 12-1, this cell will contain the dynamic path of the PDF file.

Figure 12-25 shows the configuration of this step.

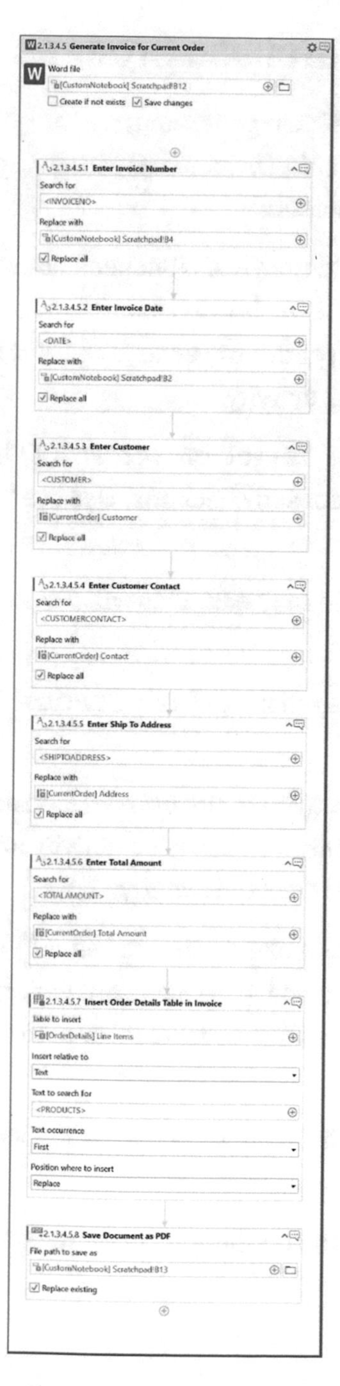

Figure 12-25. *Configuration of Generate Invoice for Current Order step*

Step 16: Mark Purchase Order as Processed

Now that the invoice has been generated, we are going to mark the Invoice Status of the current order in Order Management System to Processed and close the Order Details screen:

1. In StudioX, add the Check/Uncheck activity in the body of Extract Order Details & Generate Invoice activity after the Generate Invoice for Current Order activity.
2. Click Indicate target on screen (I) link and select the Processed checkbox. Use the label as an anchor.
3. Set Action field to Check. This will mark the invoice as Processed.
4. Update Display Name to Mark Invoice As Processed.
5. Add a Click activity in the body of Extract Order Details & Generate Invoice activity after the Mark Invoice As Processed activity.
6. Click Indicate target on screen (I) link and select the Save and Close button. For anchor, use the header area where the Save and Close button is located.
7. Update Display Name to Click Save and Close Button.

Figure 12-26 shows the configuration of this step.

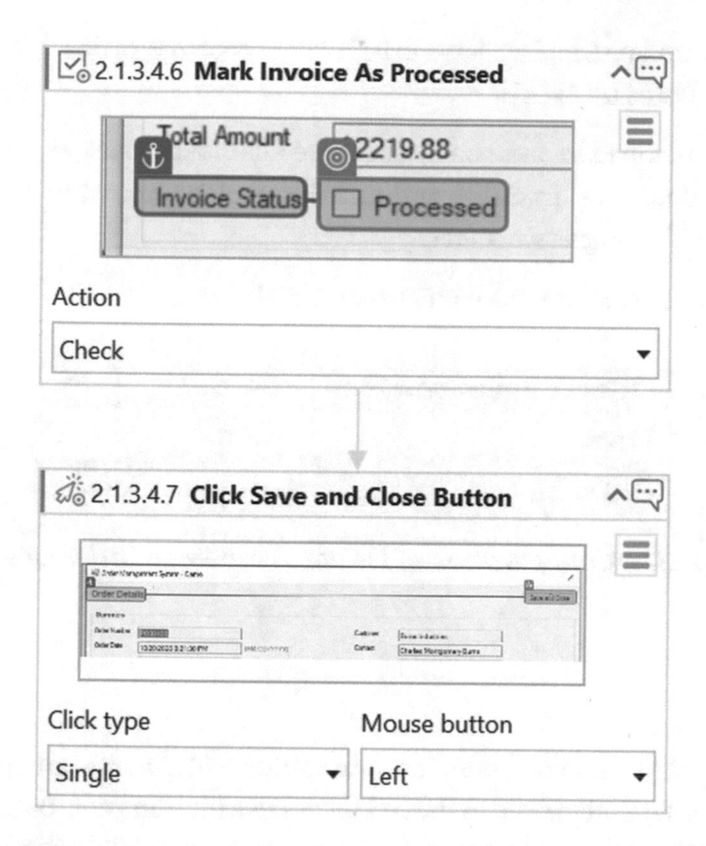

***Figure 12-26.** Configuration of Mark Purchase Order as Processed step*

Step 17: Delete Temporary Order Details Excel

This is the final step of the automation. Now that the invoice has been generated for the order currently being processed, we can safely delete the `OrderDetails.xlsx`. This file was temporarily created just to store order details for the current order.

1. In `StudioX`, add the `Delete File` activity in the body of `Process All Purchase Orders` activity after the `Extract Order Details & Generate Invoice` activity.

2. Update `Display Name` to `Delete Temporary Order Details Excel`.

3. In the `File name` field, select the `CustomNotebook` ➤ `Indicate in Excel` option. Select cell `B11` from the `Scratchpad` worksheet.

Figure 12-27 shows the configuration of this step.

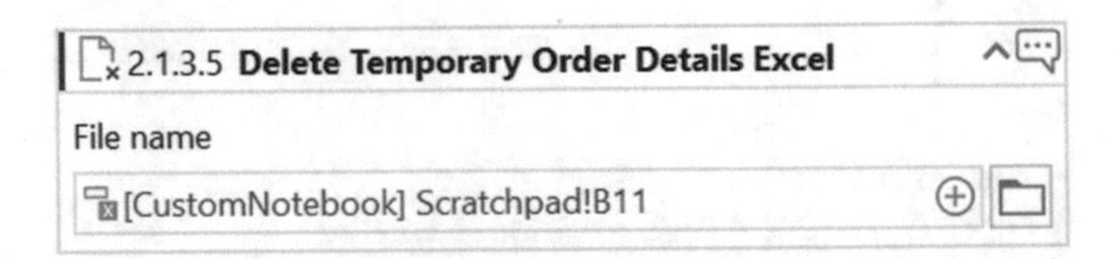

***Figure 12-27.** Configuration of Delete Temporary Order Details Excel step*

Test

Once you click the `Run` button, the automation will generate an invoice for each order and place it in the `DailyProcessing\Current Date (YYYYMMDD)` folder. Figure 12-28 shows the final output of the automation.

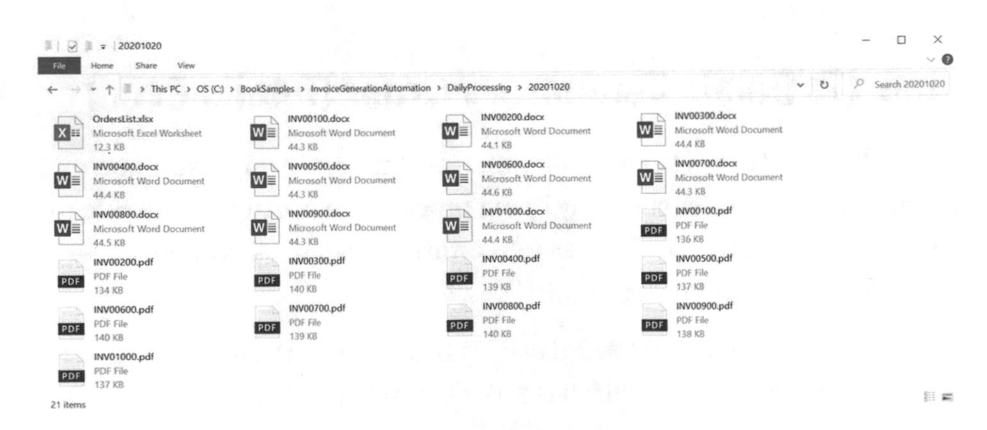

***Figure 12-28.** Output of the automation run*

Index

A, B

C

D

A. Javed et al., *Robotic Process Automation using UiPath StudioX*,
https://doi.org/10.1007/978-1-4842-6794-3

E

F

G

H

I, J, K, L

M

N, O

P, Q

R

S

T

U, V

W, X, Y, Z

Printed in the United States
by Baker & Taylor Publisher Services